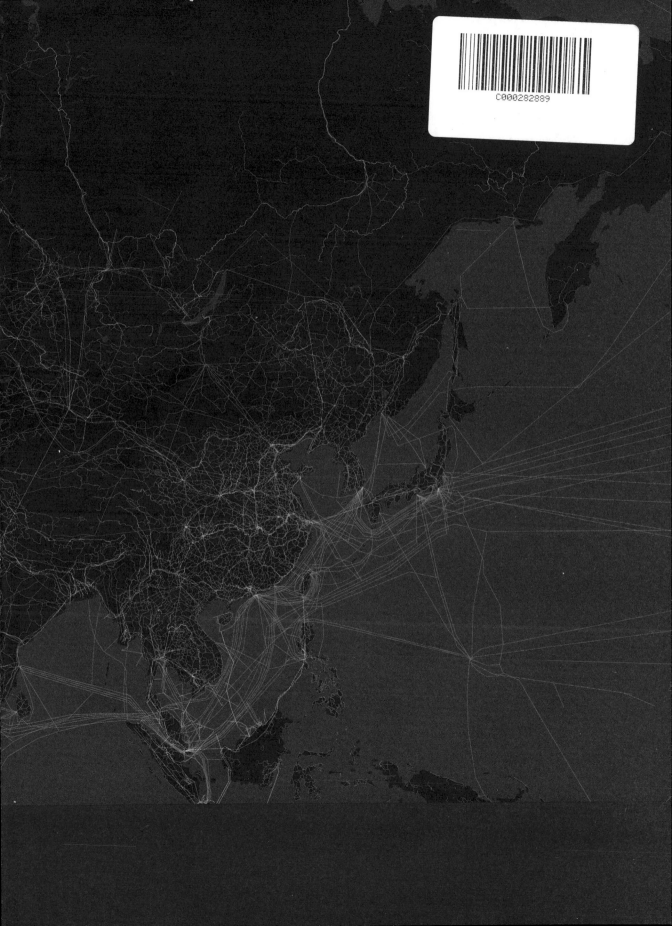

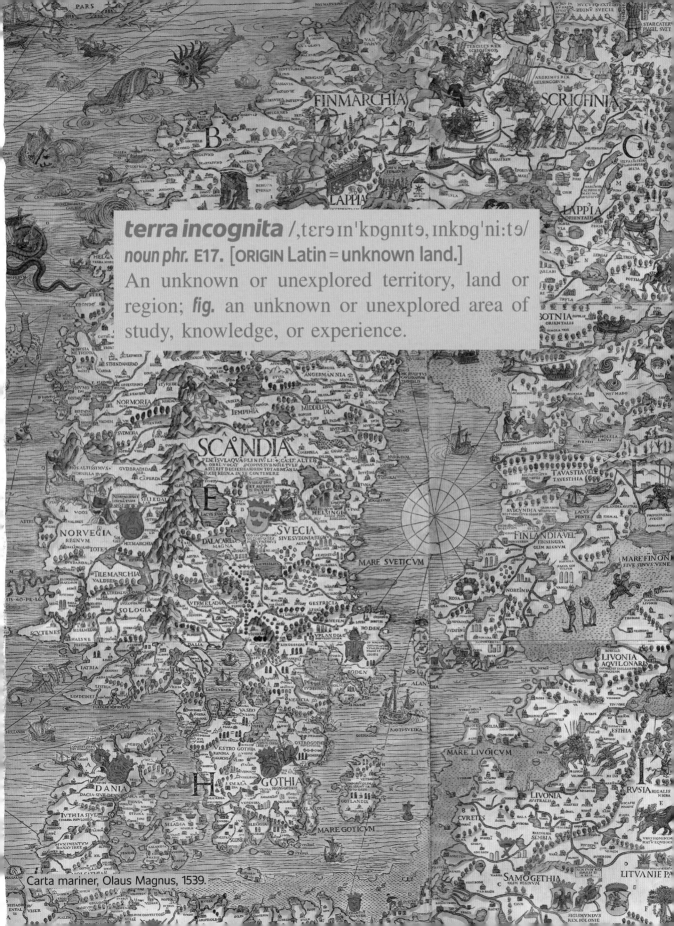

terra incognita /ˌtɛrə ɪnˈkɒgnɪtə, ɪnkɒgˈniːtə/ *noun phr.* E17. [ORIGIN Latin = unknown land.] An unknown or unexplored territory, land or region; *fig.* an unknown or unexplored area of study, knowledge, or experience.

Carta mariner, Olaus Magnus, 1539.

Refugee flows, 2015.
Each dot represents
17 separate refugees.
They highlight the high
concentration of flows
in Africa, the Middle East
and Europe.
UNHCR, 2016

Terra Incognita

100 Maps to Survive the Next 100 Years

IAN GOLDIN & ROBERT MUGGAH

EarthTime images by CREATE Lab:
Paul Dille, Ryan Hoffman and Gabriel O'Donnell

C

CENTURY

1 3 5 7 9 10 8 6 4 2

Century
20 Vauxhall Bridge Road
London SW1V 2SA

Century is part of the Penguin Random House group of companies
whose addresses can be found at global.penguinrandomhouse.com.

Copyright © Ian Goldin, Robert Muggah, Paul Dille, Ryan Hoffman and Gabriel O'Donnell 2020

Ian Goldin, Robert Muggah, Paul Dille, Ryan Hoffman and Gabriel O'Donnell
have asserted their right under the Copyright, Designs and Patents Act, 1988,
to be identified as the authors of this work.

First published by Century in 2020

www.penguin.co.uk

A CIP catalogue record for this book is available from the British Library

Typesetting and design by Roger Walker
Features EarthTime images by CREATE Lab
Features additional images by Roger Walker

ISBN 9781529124194

Printed and bound in Italy by L.E.G.O. S.p.A.

Penguin Random House is committed to a sustainable future for our business, our readers
and our planet. This book is made from Forest Stewardship Council® certified paper.

With thanks to the following for granting permission to use their data and/or imagery in this book:

World Bank; Public Religion Research Institute; Goren, Erkan; Catalogue of Endangered Languages; Worldmapper.com; Tatem, Andy; Review on Antimicrobial Resistance; Open Streets; Overpass API; European Space Agency; Zucman, Gabriel; UN DESA Population Division; UN Comtrade; Institute for Health Metrics Evaluation; UN FAO Statistics and Information Branch; Oxford Martin School, Technological and Economic Change; Our World in Data; Pew Research ;Google; United Nations High Commissioner for Refugees; Kudelski Group; Parag Khanna and Jeff Blossom, Harvard World Map; DHL; Organisation for Economic Co-operation and Development; United States Geological Survey; NASA; NASA Earth Observatory; Berkeley Earth; Joint Research Centre; Netherlands Environmental Assessment Agency (PBL); Electronic Data Gathering, Analysis, and Retrieval system; VIIRS satellite; Sentinel-2; Intergovernmental Panel on Climate Change; Climate Central; S&P Global Inc.; National Oceanic and Atmospheric Administration; US Census; ACS 2016; National Historical Geographic Information System; Igarape Institute; International Renewable Energy Agency; National Renewable Energy Laboratory; IEA ; Ookla (TM); International Federation of Robotics; North Atlantic Treaty Organization, Wikipedia; Freedom House; Gallup; Pew Research Center; Center for Systemic Peace; The Armed Conflict Location & Event Data Project; Stockholm International Peace Research Institute; Global Terrorism Database, University of Maryland; Southern Poverty Law Center; Australian Strategic Policy Institute; Campaign to Stop Killer Robots; Global Covenant of Mayors for Climate and Energy; Reba, M., Reitsma, F. & Seto, K.; Damian Evans & Cambodian Archaeological Lidar Initiative, Inc; Lebreton, L., van der Zwet, J., Damsteeg, J. et al; Klasing, Mariko J. and Milionis, P. (Nickol, Michaela and Kindrachuk, Jason; Federico, Giovanni and Antonio Tena-Junguito; Pesaresi, M., Melchiorri, M., Siragusa, A., & Kemper, T.; UN HABITAT MDG Database; Horace Dediu; Comin and Hobijn; ITU World Telecommunication/ICT Development Report and database; Internet Census 2012, Carnbot Project; Alvaredo, Facundo, Lucas Chancel, Thomas Piketty, Emmanuel Saez, and Gabriel Zucman; Kummu, M., Taka, M. & Guillaume, J. Gridded; Thomas Piketty, Emmanuel Saez, Gabriel Zucman; Refinitiv; Krueger, A.; The World Economic Forum; World Wealth and Income Database, Joshua S. Goldstein; Federation of American Scientists; Marshall M., et al. POLITY IV PROJECT; Uppsala Conflict Data Program; Peace Research Institute Oslo; Transboundary Freshwater Dispute Database; UN Office for Disarmament Affairs; UN Office on Drugs and Crime; UN Peacekeeping; Ngs/National Geographic Creative; Citi Research and Abramitzky and Boustan; Philip's Atlas of World History, Patrick Karl O'Brien; CIA World Factbook; Food and Agricultural Organization of the UN; World health Organization; National Institutes of Health; Center for Disease Control.

Also by Ian Goldin

Age of Discovery: Navigating the Storms of Our Second Renaissance

Development: A Very Short Introduction

*The Pursuit of Development: Economic Growth,
Social Change and Ideas*

Is the Planet Full?

*The Butterfly Defect: How Globalization Creates Systemic Risks,
and What to Do about It*

*Divided Nations: Why Global Governance is Failing,
and What We Can Do about It*

*Exceptional People: How Migration Shaped our World
and Will Define Our Future*

Globalization for Development: Meeting New Challenges

The Case for Aid

Also by Robert Muggah

Stabilisation Operations, Security and Development

Open Empowerment: From Digital Protest to Cyber War

Global Burden of Armed Violence

*Security and Post-Conflict Reconstruction:
Dealing with Fighters in the Aftermath of War*

*Relocation Failures in Sri Lanka:
A Short History of Internal Displacement and Resettlement*

No Refuge: The Crisis of Refugee Militarization in Africa

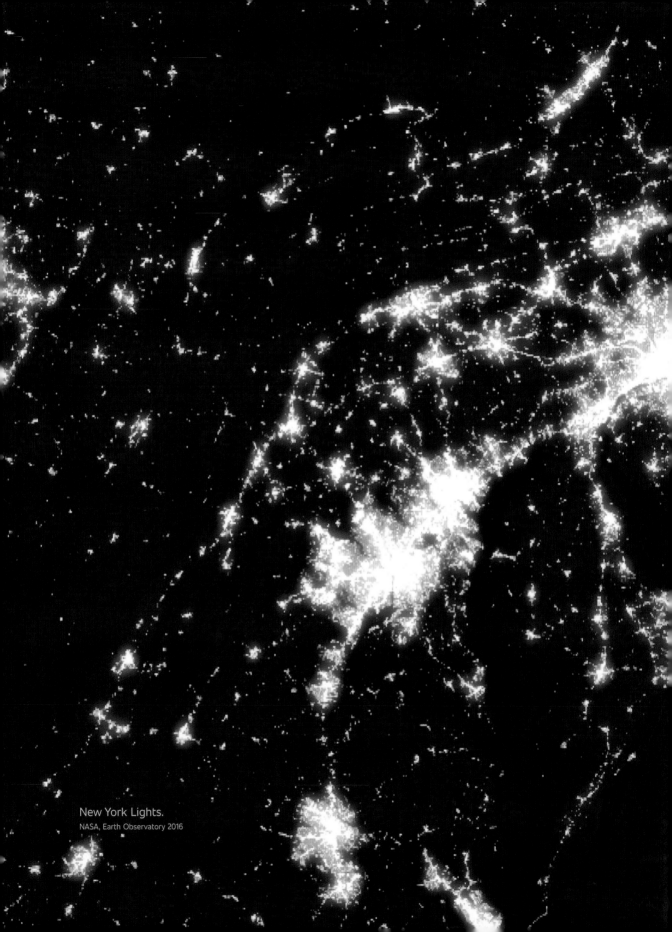

New York Lights.
NASA, Earth Observatory 2016

Contents

100 Maps

Palm oil mainly comes from two countries, Indonesia and Malaysia, where palm plantations have replaced tropical forests.
UN Comtrade, OEC, 2017

Tess, Olivia and Alex
May you thrive along your journeys

IAN GOLDIN

Henry and Betty
Who fiercely embrace parts unknown

ROBERT MUGGAH

Acknowledgements

We are most grateful to Ben Brusey, our patient and highly effective editor at Penguin, who nurtured this book from its birth to publication, with support from Kasim Mohammed and Jess Ballance. Illah Nourbaksh, at CREATE Lab, together with his colleagues Paul Dille, Ryan Hoffman and Gabriel O'Donnell, provided many of the maps that define this book. Roger Walker expertly designed the images and laid-out the manuscript. Sincere thanks are also due to Adam Ferris for his research assistance.

This volume visits a vast array of subjects, several of which reside at the outer edges of our expertise and experience. To help us navigate and course correct when we were off track, we drew on a vast array of friends and colleagues whose enthusiastic and careful inputs were essential. While we alone take responsibility for the views expressed in this book, we are deeply grateful to the vast number of people who have so generously helped us, and especially Max Roser and his inspired team at Our World in Data.

Ian thanks the Oxford Martin School that has continued providing a wonderfully nurturing and supportive environment for intellectual exploration. He could not wish for a better place to explore ideas and learn than Oxford University and Balliol College, which facilitated a sabbatical at the MIT Poverty and Media Labs. Ian wrote-up parts of the book while at the Rockefeller Centre at Bellagio and the Stellenbosch Institute for Advanced Studies.

Robert wishes to thank his many colleagues at the Igarapé Institute, SecDev Group and World Economic Forum for their unrelenting support. He is grateful to the University of British Columbia's Liu Centre where he started drafting chapters while on sabbatical serving as the Lind Fellow and New York University's Center for International Cooperation. The Chicago Council on Global Affairs and the Canadian Global Affairs Institute have also supported him over the years.

As always, all books come at a cost, including time that might otherwise have been spent with friends and family. Ian and Robert are particularly grateful to their partners and children for so selflessly giving up nights and weekends in favour of this book. With juggling calls across the Atlantic and the pressures of time, there were too many times when Ian was not available for Tess, Olivia and Alex, whose constant love, support and tolerance sustains him. Robert's debts to his family and friends are many, and he is thankful for their inputs to early drafts. His wife, Ilona, and daughter, Yasmin-Zoe, offered unrelenting support and heroic patience.

Ian Goldin, Oxford
Robert Muggah, Rio de Janeiro
June 2020

Arctic
Ocean

NORTH
AMERICA

EUROPE

ASIA

North
Atlantic
Ocean

North
Pacific
Ocean

AFRICA

SOUTH
AMERICA

Indian
Ocean

South
Pacificc
Ocean

South
Atlantic
Ocean

Coronavirus pandemic:
the outbreak from
31 November 2019
to June 2020.

Southern
Ocean

ANTARCTICA

Arctic
Ocean

North
Pacific
Ocean

USTRALIA

Preface

In the first five months of 2020, COVID-19 spread to 188 countries killing over 400,000 of the more than 8 million individuals who tested positive, a small fraction of those infected. Maps of the coronavirus, like others featured throughout this book, highlight the ways we are connected and face common threats, as well as opportunities, in the twenty-first century.

In early 2020 when more than two-thirds of the world's population were in lockdown, we were transfixed by a map. The bright red and black display, produced by researchers at Johns Hopkins University and ESRI, tracks the evolution of the most devastating virus of the past 100 years.[1] Chilling to look at, the map provides a real-time count of the number of people who are infected and killed by COVID-19. While the speed of disease outbreak caught the world by surprise, the virus itself was not unexpected. For years, Bill Gates and many others, including the two of us, had warned whoever would listen about the threat of a massive pandemic, even anticipating where it might start and how it would mushroom around the world.[2] While offering a terrifying hint of the scale of the 2019–2020 coronavirus, the data also provides a glimmer of hope. The number of recoveries, hinting at the heroic efforts of doctors, nurses, ambulance drivers, paramedics and other first responders who risk their own lives to save people who are struggling to breathe.

We submitted this book to Penguin Random House as the COVID-19 pandemic was making its deadly rampage around the world. After the virus was first detected in Wuhan in December 2019, Robert joined a team of epidemiologists and statisticians to help model its trajectory around the world, with a view to identifying ways to contain it. When stock-markets started to crater, wiping out over \$9 trillion in a single week, Ian took to the airwaves to remind listeners about past crises and how to end them. Among our biggest enemies then as now were conspiracy-peddling and science-denying politicians and pundits. While we cannot predict how this crisis will end, we know it will have lasting effects on globalisation and many aspects of our lives. The COVID-19 pandemic reveals the systemic risks that accompany accelerated connectivity. It is a reminder of how our fates are inextricably bound together and that cooperation, while not inevitable, is essential. Denialism and inaction on the many known existential threats we face – be they pandemics, climate change or weapons of mass destruction – are not just dangerous, they are downright criminal.

The COVID-19 pandemic accelerated and disrupted our lives in unfathomable ways. Many of us cancelled plans, took time off work, cared for sick relatives, and lost loved ones before their time. The impacts were often sudden and substantial. In early March 2020, Robert was on a sabbatical in New York City. After crunching the global numbers on COVID-19, he decided to examine more closely the situation in his own backyard. Unsettled by his own statistical projections, he took a quick stroll around future 'hot spots' only to find the streets jam-packed with people completely oblivious to social distancing. Robert packed-up his apartment that same day, hired a truck, and moved his family across the border to his native Canada where he worked with Ian, who was locked-down in Oxford in the UK, to finalise the book. A few days after Robert left New York, Canada and the US shut their border and flights to Brazil, where he normally lives, were cancelled. Within a few weeks, New York became the global epicentre of COVID-19, registering tens of thousands of fatalities and a quarter of all infections in the US.

While throwing all our lives upside down, the COVID-19 pandemic starkly highlights the purpose of this book which is to identify and visualise the world's most pressing global challenges and solutions. COVID-19 will not kill globalisation, but it will certainly reshape it.[3] The outbreak also exposes the vices and virtues of globalisation. In a hyper-connected world, contagious viruses move more rapidly and widely than ever before. Their ability to infect not just people, but also politics and economics, is instantaneous. Owing to highly integrated global supply chains, the disruption of just one producer can have cascading effects with unpredictable consequences. Yet despite calls to de-globalise and shorten supply chains, the world is bound together by the internet and communities committed to sharing information about how to fight damaging outbreaks. The sheer diversity and abundance of producers, sellers and consumers can bolster resilience: even when several are disrupted, the multiplicity of networks and nodes keeps the system chugging along. We experienced these impacts first-hand as this book went to press: Penguin Random House had to change plans and move printers to Italy from China, where printing presses were largely out of action for a few months due to the pandemic.

Despite being more exposed to dangerous viruses than ever before, we have never been better prepared. Governments, businesses and societies are in good shape to weather global crises. The ostrich-like leaders of Brazil, Belarus, Nicaragua, Turkmenistan and even China and the US delayed efforts to fight the COVID-19 pandemic, but, for the most part, science informed government responses, including health moves to slow its impact on vulnerable populations.[4] This is hardly the first time humanity has had to face down the first horseman. Hundreds of millions of people have been killed by plagues and pestilence over the course of history. Of the 15 major pandemic outbreaks over the past 500 years, the 1918–1919 influenza pandemic is considered one of the worst. It infected at least one in three people on the planet and killed up to 50 million of them, though no one knows exactly how many died.[5] The first victim, believed to be a US soldier, was traced back to an army base in Kansas in March 1918. Within months the virus had spread from the US by ships taking troops to fight in the trenches of France and Germany, killing millions. By the end of the year, a second vastly more destructive wave had spread to Australia, Japan, India and eventually China. As the old saying goes, history does not repeat, but it certainly rhymes.

A century ago, the world was totally ill-prepared for a pandemic outbreak and there were no molecular biologists or virologists who could determine the genetic sequence of influenza. Indeed, most doctors did not even know that the disease was caused by a virus at all. Nor were there any antiviral drugs or vaccinations available to protect people from contracting the disease – these would only emerge decades later. There were no global institutions like the World Health Organization (WHO), or even competent national ones like the Centers for Disease Control and Prevention (CDC) and its counterparts in the

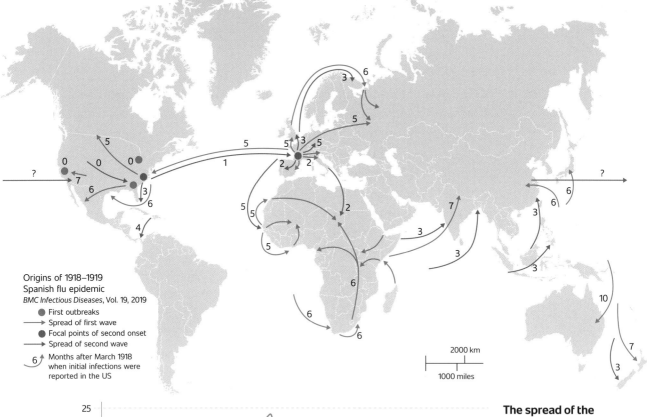

Origins of 1918–1919
Spanish flu epidemic
BMC Infectious Diseases, Vol. 19, 2019

- ● First outbreaks
- → Spread of first wave
- ● Focal points of second onset
- → Spread of second wave
- ↶ 6 Months after March 1918 when initial infections were reported in the US

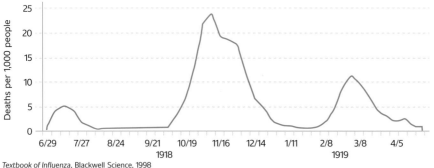

Textbook of Influenza, Blackwell Science, 1998

The spread of the 1918–1919 influenza pandemic ('Spanish Flu') [6]

The 1918–1919 influenza pandemic, or Spanish Flu, claimed up to 50 million lives. At least 675,000 US residents died from the virus, three times more than from the entire First World War. It spread in three waves around the world, first from North America to Europe and then to Africa and Asia.

Americas, Europe, Africa and Asia, to survey and track the spread of new illnesses. Instead, information about the pandemic was almost non-existent, not least because of war-related censorship.

Today, despite incredible progress across most metrics of well-being, the international community failed to deter the spread of COVID-19. Many of the measures adopted by countries like China, Italy, France, Spain, the UK, and the US to contain COVID-19 during the first few months of 2020 bore a striking resemblance to the ones introduced between 1918 and 1919, and even during the Black Death of the fourteenth century. Then, as now, physical distancing was encouraged, quarantines were imposed, affected communities were isolated, handwashing was promoted, and strategies were introduced to prevent large crowds assembling. While they did not know exactly what the cause of the mass death was – many people believed it was 'bad air', or miasma – local authorities understood that healthy people needed to be separated from sick ones. In the absence of a vaccine and given the sorry state of global leadership, this is probably the best that today's most advanced societies could muster.

The impacts of pandemics always extend well beyond population health. The Spanish Flu had a sharp effect on real economic activity and slowed growth in many large countries during the 1920s.[7] As bad as the Spanish Flu was, there are several reasons why COVID-19 will be even more disruptive to global health, politics and economics than past pandemics. The most obvious is that the world is now dramatically more interdependent than it was in the early twentieth century. At that time, the commercial aviation industry had yet to get off the ground, so most people travelled using ocean liners or by rail. Tourism, as the global industry we know it as today, was tiny.[8] International trade, which had been growing quickly before the First World War, had collapsed.[9] Many populations, exhausted by war and sick from pollution, were already susceptible to influenza. The Spanish Flu led to labour shortages and wage increases, but it also may have hastened the end of the war and inspired the creation of social security systems.

The COVID-19 outbreak is more dangerous economically because the world's economies are more intertwined than ever before. In 2003, when the SARS epidemic struck, China contributed just 4 per cent to the global economy, whereas now its share is over 16 per cent. China is at the centre of global supply chains, and its tourists spend over $260 billion annually abroad.[10] Surging trade and travel within countries and across national borders has lifted billions of people out of poverty. But all this interdependence has a dark side. Within months of breaking out, COVID-19 had fragmented supply chains for more than 75 per cent of US companies and disrupted manufacturing globally.[11] Calls for moving from 'just in time' to 'just in case' management grew louder, as did efforts to decouple from China. The recovery will be slow and painful, especially for the most vulnerable. All the while, COVID-19 will exacerbate inequalities within and between countries.[12]

The COVID-19 pandemic exposed stark inequalities in the nature of governance and the tattered state of social contracts in many countries around the world. Countries with functional democracies and high levels of trust such as New Zealand did not just bend the curve, they crushed it. Meanwhile, nations overseen by populist leaders such as Brazil experienced soaring rates of infection and death. At the time this book went to press, the US was the world's worst affected country. Like everywhere else, it was the poorest members of society – especially black, Asian and minority ethnic households – who bore the brunt of the virus. The world was reminded of the multiple burdens facing minorities after George Floyd, a black man in the US, was filmed being suffocated to death by a police officer. For years the Black Lives Matter movement has channelled outrage and indignation felt by many citizens about police brutality, but to limited effect. Coming as it did in the middle of a fully-fledged pandemic, the murder of Mr. Floyd ignited a raging bonfire of anger and despair in over 350 cities in the US. Within days the protests had spread worldwide to dozens of countries, sparking desperately needed conversations about racism, police violence, justice and inequality.

The pandemic is also going to have far-reaching geopolitical ramifications. Part of the reason for this is that the economic impacts will prolong political instability, deepen social unrest linked to austerity and provoke protectionist backlashes.[13] While COVID-19 should teach us the value of cooperation, so far the response from powerful countries has been disjointed and frequently aggressive. Ever since the Second World War the US has played a leadership role during global crises, but in the wake of COVID-19 its leadership actively undermined the WHO and adversaries and allies alike. The chances of global cooperation eroded further when the US escalated its trade war with China and accused its rival of concealing information about the pandemic. The pandemic not only intensified pre-existing frictions between the US and China, it also weakened the cohesion of the European Union, including its core principle of freedom of movement.

While there are many downsides associated with COVID-19, the rapid response of the scientific community to the pandemic underlined some of the virtues of globalisation. Owing to the availability of digital connectivity, high-tech laboratory testing, and new genomic technologies, scientists sequenced the virus's genome in a matter of weeks. International networks of epidemiologists, microbiologists and mathematicians worked around the clock to share projections, research and best practice. The crisis also triggered acts of solidarity between and within some countries, including the sharing of expertise and equipment such as masks and respirators. For a moment, it looked as if the global crisis could reinvigorate collective action, much as the 2008 financial crisis spurred the formation of the G20. But the crisis has revealed deep deficits in global leadership, and fault-lines in the social contracts of many countries. It has also exposed the fragile state of multilateralism. The United Nations Security Council did not even agree to meet to discuss the global crisis until 100 days had passed and over 40,000 people had died.

We are entering an extraordinarily complex and uncertain new era. No one knows what comes next. The International Monetary Fund (IMF) predicts that a great depression is looming and that it will affect all economies, especially emerging ones. But there are also reasons to believe that some of its worst effects can be avoided if we act decisively and avoid the mistakes of the past. Much depends on when (and if) the world discovers a vaccination and effective antiviral drugs. Without them, the coronavirus will be around for years, possibly forever. Even in its short existence, COVID-19 has rewired the future of travel, work, education, health and the way we interact, though it is still not clear how dramatic these changes will be. The new normal could well be governed by very different rules. This is not all bad news. Surveys conducted after the pandemic was announced in the UK revealed that only 9 per cent of Britons want to go back to the way things were after lockdown ends.[14] Also, carbon and nitrogen dioxide emissions registered unprecedented declines in many parts of the world, bringing clean air to polluted cities.[15] Despite facing enormous difficulties, mayors and city councils around the world have started re-imagining city life that could be greener and more sustainable.[16]

One of our biggest concerns in the emerging COVID-19 era is the widening gap between the accelerating threats presented by systemic risk and the slow pace of global, national, state and city-level preparedness and response. At a moment when global threats are rapidly escalating, many governments are turning their back on cooperation and abandoning the international system that has prevailed since the middle of the twentieth century. The United Nations has withered and suffers from decades of neglect. Not surprisingly, it has struggled to address the devastating scale of the health and cascading economic, food and humanitarian demands that COVID-19 has spawned. The World Bank and IMF have pumped billions of dollars to countries in need of financial help, but this has proved inadequate as over 2 trillion dollars is required. The multilateral system today is governed by divided nations and nothing short of radical reform and renewal can fix it.[17]

COVID-19 could not have come at a worse time. It hit at a moment when many countries were drained by austerity and privatisation. In order to mitigate the 2008 financial crisis and subsequent Euro crisis, central banks had turned on the printing presses and reduced interest rates to close to zero, but having fired off their monetary ammunition their scope for action is now greatly reduced. Many governments and firms were mired in excessive debt burdens even before the coronavirus emerged in December 2019. The collapse of economic activity that it precipitated will compound the underlying economic challenges countries face and undermine their ability to address immediate, let alone future needs. And well before the arrival of COVID-19, popular support for democracy was flagging, reaching some of the lowest levels since records began.[18] Societies around the world were already fragmented by polarisation and deepening mistrust of their elected (and unelected) leaders.[19]

Just when global cooperation is essential it is desperately wanting, with a vacuum of global leadership. As we show in this book, the two biggest powers, the US and China, seem to be on a collision course and careering towards a Cold War 2.0. The European Union is also diversifying its supply chains away from China and abandoning a strategy designed to expand economic relations with the Asian giant. China, for its part, is adopting a new 'slow living' paradigm as the economy decelerates to levels not seen in four decades. Meanwhile, the United Nations is a paralysed bystander as are the G7 and G20 which struggle to do more than provide photo opportunities for world leaders. With no clear endpoint in sight, many countries remain inwardly preoccupied and on their own path. Global institutions require urgent reform to be made fit for twenty-first century challenges. But with governments starving them of the leadership, legitimacy and resources they require, they are falling further and further behind.

The inadequacies of the world's global institutions are eroding global resilience and exposing all of us to risks, especially infectious disease outbreaks and climate change. This is occurring despite the fact that these twin threats are the most likely source of future crises. Making matters even worse, international and national regulatory policies have largely ignored the risks associated with the concentration of finance in a small number of locations. Rather than being geographically dispersed, the headquarters of global competitors are in neighbouring buildings. A similar concentration of risks exist when it comes to the production of life-saving medicine. When a pandemic isolates the US from its suppliers, the risks to population health are amplified. Similarly, when a hurricane closes Wall Street, or vital supplies or services ares disrupted in certain parts of the world, the risks to the global economy and all our collective wellbeing are amplified by our tight connections.[20]

Pandemics are unlike many other global risks in that they can originate anywhere in the world. In the case of threats such as climate change, cyber-attacks, and antibiotic resistance, a small set of actors, including companies, cities and communities, can create coalitions to greatly reduce risks.[21] Not so with pandemics, where the capacity to monitor and intervene to isolate outbreaks at source is vital, as is the need for national action to quarantine and treat affected individuals. For countries that do not have this capacity, vigilant surveillance and rapid international response is required to contain the outbreak at source. The sharing of information and resources globally is vital for this, not least to support the poorest countries. The WHO has this mandate, but it has been crippled by a lack of reforms and inadequate support from its shareholders, which are the governments of the world.

COVID-19 starkly reveals how the world's integrated and complex systems are only as strong as their weakest link. Ultimately, higher fences and thick bunker walls will not stop pandemics, or any global threat for that matter. What is being tested is our collective will to cooperate in an unequal world. The stakes could not be higher. We are entering uncharted waters. In writing this book, it is our belief that maps can help forge better co-operation. Maps such as the red and black visualisation at the start of this preface can communicate complex ideas in powerful yet simple ways. Our hope is that these maps and images, 100 of which we list at the front of this book, can educate and guide us to new places of insight and understanding. Ultimately, they can offer hope. So let us now turn to the new maps of the twenty-first century. Let us be guided by them. Let us begin.

Ian Goldin
Robert Muggah
June 2020

Carta mariner, Olaus Magnus, 1539.

Introduction

'You can't use an old map
to explore a new world'
– Albert Einstein

For most of human history we had literally no idea where we were. What could not be directly observed was unknown. That started to change when our ancestors began painting their surroundings (including the nearest wild game or water sources) on their cave walls. About 3,000 years ago, humans started committing their earliest maps of the world to papyrus. The areas that were unexplored were populated with fantastic and terrifying beasts – dragons, serpents and lions. Until recently, most of the planet was *terra incognita*. With the help of navigational aids and the explosion of scientific discovery, all that started to change. While we still have much more to learn, our complex world – its climate, biodiversity and complicated human systems – is rapidly being revealed. We now have reached a state of *terra cognita*. We are in a known world.

In this book we use maps to explain some of our gravest existential challenges and a few of the most inspiring solutions. We are living through a period of disorientating uncertainty and boundless opportunity. Before the arrival of COVID-19, the climate crisis had risen to the top of the agenda, fuelling a boom in concern about changes to our atmosphere, glaciers, oceans, ecosystems and life on Earth. And while global warming poses an unprecedented threat within our lifetime, this is not the only emergency we face. In a globalised era, we are confronted with a vast range of threats, from extreme inequality and infectious disease outbreaks to political extremism, privacy-killing surveillance and explosive violence. Each of these risks is interconnected, which is why engaging with all of them simultaneously, while hard to do, is more essential than ever.

What if we could clarify, contextualise and visualise the world's biggest dangers so that they could illuminate and inform, instead of sow fear and deepen divides between and within communities? Imagine if we could mobilise information in a way that helps reduce uncertainty, improve understanding and transparency, and aide future decision-making? What if we could build a set of accessible new tools to navigate our future more confidently, rather than hide from it and risk moving backwards? One way to distil clarity from complexity is to map out our biggest challenges. Remarkable improvements in geographic imaging and data sciences have given us the means to plot our gravest threats, creating visualisations to reveal the past, present and future. And it's not just physical terrain that is opening up to exploration – but environmental, social, economic and political dimensions as well.

Map-making is an ancient impulse. From the moment *Homo sapiens* learned to communicate they have used maps to make sense of their world. Well before the invention of writing, we used them to explain our relationships to one another and to our environment, planet and cosmos. Whether etched into stone or used to navigate distant solar systems, maps have been instrumental in pointing us in the right direction. Yet maps then as now come with a health warning: they cannot be taken at face value. This is because maps can often conceal as much as they reveal.

Maps that changed the world

Several millennia before the invention of satellite imaging and remote sensing, humans navigated the high seas by estimating the distances between land masses. One of the most influential maps of all time – Claudius Ptolemy's *Geographia*, created in the second century – was the first to show longitude and latitude. Although *Geographia* assembled all known geographical knowledge at the time, it was misleading. Stretching from the Canary Islands to the Gulf of

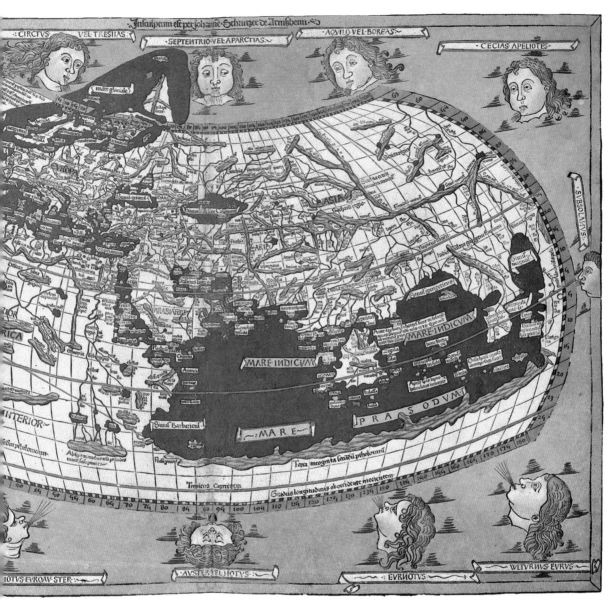

Ptolemy's *Geographia*
(AD 150) by Johannes
Schnitzer – 1482.

Thailand, it severely misjudged the size, location and shape of most countries and waterways. It also located the world's meridian – the longitudinal centre – somewhere in West Africa. Even so, *Geographia* was reproduced by cartographers for centuries, including by Johannes Schnitzer in 1482, offering a fascinating, if profoundly distorted, understanding of the planet.

Geographia profoundly shaped how we conceived the world. The earliest European explorers relied on it during the fifteenth and sixteenth centuries to power a new age of discovery. It helped Portuguese explorers, for example, to chart new coastlines of Africa and ultimately sail around the continent to enter the Indian Ocean. It was also used by Christopher Columbus, who extended the length of the degrees of Ptolemy's map, pushing the latitudes further eastward. Despite these design improvements, Columbus's map was still misleading, with global consequences. For one, it led Columbus and his crew to believe that Asia was closer to Europe than was actually the case. When Columbus finally reached the Caribbean in 1492, he had travelled as far as he thought Asia was from Europe. This explains why he was so dismayed when he learned that Hispaniola was not actually Japan. The rest, as they say, is history.

Orbis Terrae Compendiosa Descriptio by Gerardus Mercator – 1569.

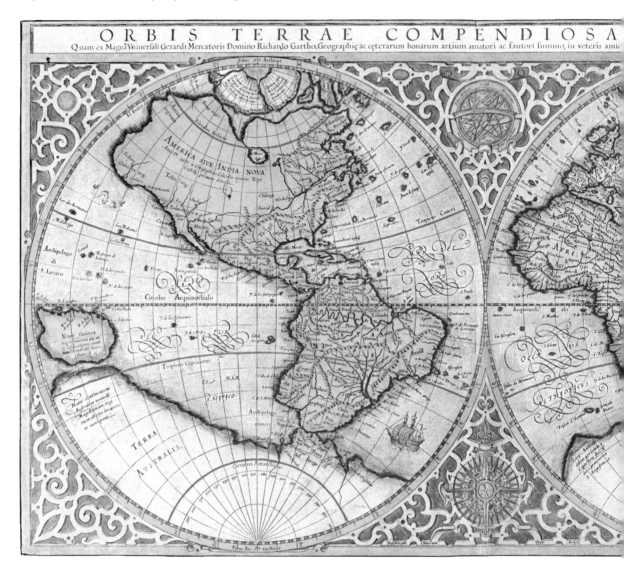

Another transformational map emerged in the sixteenth century and was produced by the great cartographer, geographer and cosmographer, Gerardus Mercator. His innovation was to use a cylindrical projection and keep latitude and longitude lines at consistent 90-degree angles on a constant course (known as rhumb lines), allowing mariners to navigate the world's oceans more efficiently. Map-makers had struggled for years to stitch two-dimensional charts on to three-dimensional spheres. While revolutionary, Mercator's maps were frustrating to use: sailors had to constantly recalculate their routes to compensate for the built-in deficiencies of their charts.

Despite offering a more accurate navigational aid than those of

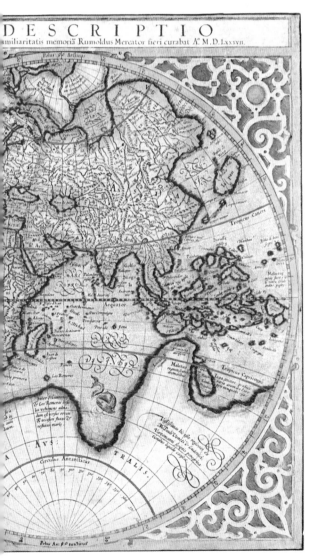

his predecessors, Mercator's maps come with a catch – they dramatically distort the size of objects as the latitude increases from the equator to the poles, where the scale becomes infinite. The constancy of latitude and longitude in Mercator's maps explains why Greenland and Antarctica, for example, appear much larger than they really are relative to land-masses closer to the equator. Take Alaska, which seems to be the same size as Brazil, when in fact it is just one-fifth of the size. As the map on page 7 shows, Canada and Russia alone seem to take up about 25 per cent of the world's surface when they occupy just 5 per cent. Notwithstanding these limitations, which were once highlighted in the American TV series *The West Wing*, Mercator's 'atlas' caught on and it is the basis for most maps in use today.

One of the most popular maps ever made was published by a New Jersey-based company, Geographia, in the early 1940s. It is a more elaborate version of the Mercator maps and still adorns the walls of university campuses and elementary schools the world over. A common criticism of Geographia, and other Mercator maps, is that it distorts the size of land-masses and reinforces a Euro-centric view of the world – one in which the northern hemisphere dominates the southern. By contrast, the Gall-Peters projection, a format that gained a following among cartographers in the 1970s, reveals some of the dramatic limitations of the Mercator maps. Specifically, every country in the Gall-Peters map has the correct size relative to each other. As a result, it shows how Africa and Brazil, among other countries, are much larger than other maps would have us believe. Western Europe is also not remotely as large as South America – it is just half the size.

Maps are undergoing yet another revolutionary change in the twenty-first century: they are increasingly digital, multidimensional and based on remote sensing technologies. Google Maps, an app launched in 2005, now with over one billion users a year, warps its

Right: **Mercator and Gall-Peters maps** (projected vs actual sizes).

Below: **The World by Geographia – 1942** Standard map of the World. Geographia Map Company.

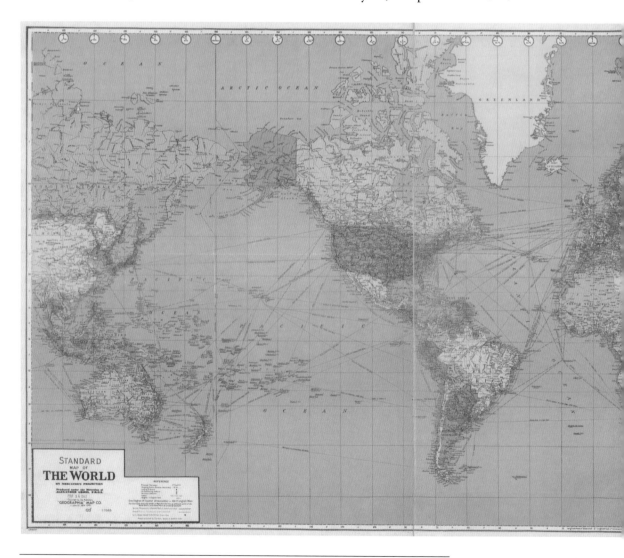

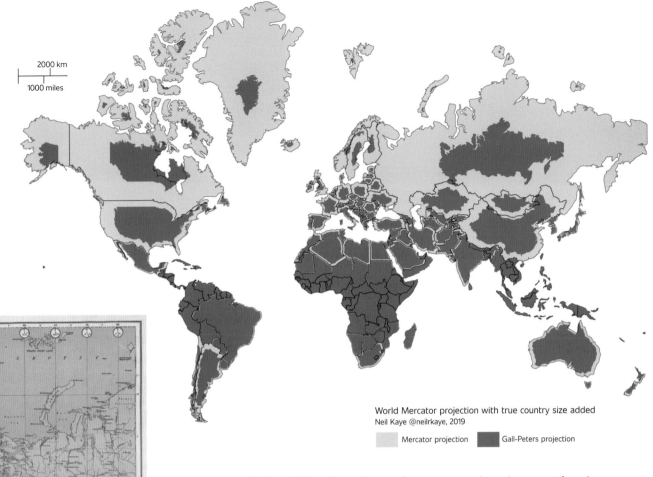

World Mercator projection with true country size added
Neil Kaye @neilrkaye, 2019

Mercator projection   Gall-Peters projection

images around the globe showing a closer approximation to what it truly is – a bumpy spheroid. As we can see on the next page, it also offers observers options to view the world divided into administrative units or from the perspective of satellites. The flat two-dimensional rendering was also upgraded to a three-dimensional version in 2018.

If maps can reveal aspects of our past, can they help us navigate the future? The short answer is that the maps featured in this book are designed to guide readers to greater understanding, not to a physical place. The longer answer, which we tackle at length in these pages, is that maps reveal many profound inequalities that, if left unaddressed, could overwhelm us all. It is by understanding how problems are connected and concentrated in time and space that we can begin to develop solutions. We are living in anguished and perplexing times as old certainties are unravelling. The far-reaching consequence of the COVID-19 pandemic and the rapid pace of change means that many of our mental and physical maps are no longer fit for purpose. While we should not abandon the fundamental principles of map-reading laid down by Ptolemy, Mercator, Gall and

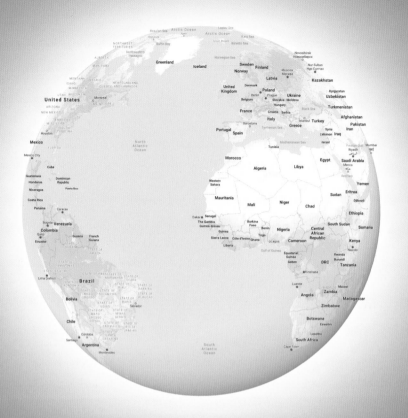

Peters, Google, and so many others, new maps are urgently needed to help orientate ourselves and put us on a surer path to a better destiny.

Even amid all the bad news surrounding the COVID-19 outbreak, there are good reasons to be an informed optimist. It is an extraordinary moment to be alive, still probably the best time in human history. We are experiencing multiple and simultaneous technological, social, economic and political revolutions. Notwithstanding the horrendous impacts of coronavirus, most people are still living longer and healthier lives than at any time since the first *Homo sapiens* started walking the earth over 200,000 years ago. As the chapter on health shows, average life expectancy across the globe has increased by a staggering twenty years over the past five decades. It took from the Stone Age to the middle of the twentieth century to achieve similar improvements. Although the devastating economic consequences of COVID-19 will be felt for years to come, there is real potential to close the remaining gaps and provide a better and more meaningful life for all our planet's population. The chapter on inequality, however, reveals that despite unmistakable progress and promise, many people, especially in wealthier countries, are feeling a foreboding sense of doom and gloom. This sense of powerlessness has increased dramatically in the wake of COVID-19.

Although humans have always lived with uncertainty, today is different. At no time in our history have the decisions made by a single generation been so consequential for the survival of succeeding ones. The most pressing questions are the stuff of everyday conversation: how can we build a more inclusive and resilient world in the wake of COVID-19? Can we reverse greenhouse gas emissions and mitigate the dangers of sea-level rises to our coastal cities? Will artificial intelligence (AI) be controlled or will it control us? Can we avoid catastrophic nuclear or biological warfare and terrorism by nation states or even lone extremists? How do we extract the advantages of globalisation without giving rise to destructive populism and deepening polarisation? Will we reduce refugee flows that ensures their dignity and rights, and promote sensible approaches to migration? Can veganism save the planet?

We need to take a deep breath. The magnitude of these questions can feel overwhelming, even debilitating. When confronted with the news on infectious disease outbreaks, climate change, automation and cyber-warfare, many of us feel disempowered and afraid. It is

Google's three-dimensional maps – 2018.
Google Maps / Google Earth.

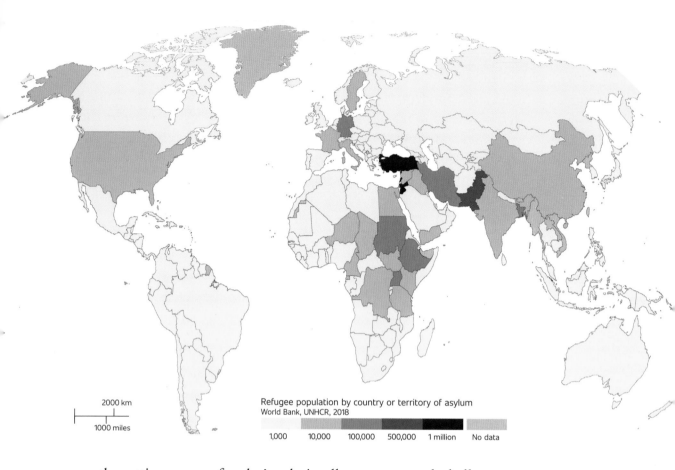

Refugee population by country or territory of asylum
World Bank, UNHCR, 2018

1,000 10,000 100,000 500,000 1 million No data

worth putting some of today's admittedly monumental challenges in perspective. Just consider the issue of refugees, a topic that has gripped Western Europe and North America like few others. There are roughly 26 million refugees in the world today, more than at any other time since the Second World War. Owing to misleading newspaper headlines, Westerners believe that most of them are seeking sanctuary in their cities. Partly as a result, attitudes towards people on the move are hardening, contributing to the spread of reactionary nationalism and its unhappy symptoms – restrictive migration policies, detention camps and high walls.

Yet even a cursory glance at our maps shows that if anything, refugees are much more widely dispersed than is commonly believed. In fact, roughly 90 per cent of all refugees and asylum seekers currently reside in Africa, the Middle East, Asia and Latin America, usually close to their country of origin. Less than 10 per cent of refugees ever make it to wealthier countries. In 2020, the US with its 325 million residents intends to accept no more than 18,000 refugees. This is less than

Refugee population by country or territory of asylum – 2018.

0.006 per cent of the global total, and its lowest intake in recorded history. Meanwhile, Jordan, which has a population of 9.7 million, is home to more than 650,000 Syrian refugees who crossed the border to escape unimagin-able suffering. Jordan's tally is the equivalent of the UK receiving more than 4 million refugees or Canada accepting 2.5 million virtually overnight. Rather than treating refugees as a burden, Jordan is treating their arrival as an opportunity to diversify and jump-start its economy.

Even though refugees do not pose an existential threat to national identity or liberal democracy, populists and pundits are working over-time to convince us otherwise. With refugee flows slowing to a trickle in the US, the White House's call for travel bans and the mobilis-ation of soldiers to 'defend' the southern border from Mexican and Central American asylum-seekers smacks of political pyrotechnics. Meanwhile, rejections of asylum applications outnumber approvals by a factor of two to one in European countries and the debate over migration policy is centre-stage. The tone set by world leaders matters, as the stigmatisation of refugees and migrants is polarising even the most diverse and tolerant communities. These calls will grow even louder when the economic pain of the COVID-19 pandemic sets in. As we shall see in the chapters on migration, geopolitics and culture, a sudden influx of people can be disruptive if poorly managed. Such fears have helped fuel reactionary movements from Brexit in the UK to the rise of ultra-right nationalists in France, Germany, Hungary and Poland. Discrimination against foreigners is also a common trope of leaders from Australia and Brazil to India and the Philippines.

Interactive maps to make sense of our world

The early twentieth-century advertising executive, Fred Barnard, is credited with coining the phrase 'a picture is worth a thousand words'. He was right. When people explore maps they are able to make connections that might have been missed in written form. Our approach to mapping is to blend satellite images with a vast array of data collected from national research institutes, leading universities, private companies and non-governmental organisations. The visualis-ation of the outcomes of human behaviour mapped on to the contours of our planet can be at once dazzling, disconcerting and counterintuitive. Once you start to see what is really going on, things can start making a lot more sense. Take the case of the 2019–20 bushfires that ravaged Australia, burning tens of millions of hectares of land and killing over a billion animals. Maps such as those produced by the Japanese weather satellite Himawari-8, help reveal the sheer scale and dimensions of the fires. You begin to move from a state of alarm and anxiety, to feeling enlightened, moved and potentially empowered enough to take action.

500 km
250 miles
NASA-FIRMS

250 km
100 miles
Himawari 8. JMA

Australia's bushfires from space – 2020.

Most of the maps in this book were produced by EarthTime, an online data visualisation platform developed by the CREATE Lab at Carnegie Mellon University. The primary data source includes more than five million satellite images acquired over the past three decades by different Landsat satellites as well as the Sentinel-2A satellite. These satellites circle Earth fourteen times a day and transmit huge volumes of data. The EarthTime maps combine high-definition satellite imagery with more than 2,000 data layers from partners such as the United Nations, European Union, NASA, the US Census Bureau, universities, research institutes and private partners such as Google, to show the patterns of natural change and human impact from global to local levels.

This book is comprehensive, but far from exhaustive. Mostly written before the COVID-19 pandemic first appeared, it features fourteen chapters focusing on some of the most important pre-occupations of our time. Each one highlights several key challenges we face and suggests solutions grounded in the latest scientific evidence. The world has changed remarkably over the past few decades and dramatically in 2020, but many of the most significant changes are yet to come. To flourish on our wounded and interconnected planet, we all must get better at interpreting data and becoming more acquainted with new ways of separating the signal from the noise. This requires each and every one of us to improve our digital and cartographical literacy. Data platforms, including new maps, cannot tell us precisely what to do. But they can inform our judgement and provide much needed perspectives to help us navigate these uncertain times.

Even as governments, businesses and societies learn to adjust to a world of COVID-19, we collectively have the potential to make this the best century in human history. But if we fail to manage cascading risk, in the words of Martin Rees, Astronomer Royal, this may well be our final century. The future depends on our ability to harvest the unprecedented opportunities ahead and understand and overcome potentially catastrophic threats. The world has gone through tumultuous transformations before. The fall of the Berlin Wall and the invention of the World Wide Web three decades ago fundamentally altered the global order and how all of us interact, mostly for the better. There is a danger that our past successes may lead to complacency regarding the future. The assumption that the arc of history bends in favour of the just and virtuous is mistaken. We need to celebrate progress achieved, but also identify the multiplication and convergence of risks in a clear-headed way. We are at a tipping point: whether we step forward, or fall dangerously backwards, is entirely up to us.

Three themes run through every chapter of this book. The first relates to the impact of globalisation. The accelerated movement of people, products and ideas has profoundly affected our cultures and commercial relations, and not always for the better. The second theme is growing inequality. Although living standards virtually everywhere have risen, inequality is growing in many countries as is the gulf between the richest and poorest people. The figures are staggering. Some 2,150 billionaires are wealthier than 60 per cent of the world's population. Just 42 of them own more wealth than 3.7 billion of the poorest people on Earth. We are also seeing deepening inequalities in relation to lifespans, access to food and exposure to crime. Our postcode predicts our lifespans and life opportunities, a fact made painfully obvious by the COVID-19 pandemic. The final theme is that of the implications of rapid shifts in new technology. The stunning evolution of AI, robotics, genomics and biotechnology are rewiring everything – from our politics to our health and education – in exciting and disconcerting ways. The scientific community's response to COVID-19 is testament to this: within three months of its detection there were more than 30,000 peer-review and unpublished studies being shared around the world. Can maps help us learn the lessons of the past to make a better future?

Navigating maps to a new understanding

The travel writer Bill Bryson once said: 'What is it about maps? I could look at them all day.' As all navigators and orienteers (including Mr Bryson) know, there are a few dos and don'ts when it comes to reading maps. Without them, we can swiftly find ourselves making simple mistakes and getting profoundly lost. At the outset, it is important to select the right type of map: a nautical chart is useless to a mountaineer, just as a topographical map serves little purpose for someone driving between cities.

You should always check the legend – what is often called the 'key' – before trying to interpret the map. Legends typically consist of symbols or colour codes for cities, mountain ranges, forested areas, or to denote fresh and salt-water. Throughout this book, legends feature colour codes to help readers understand changes, for example, in forest cover over time, in the extent of industrial and natural fires (seen from space), in the pathways of organised violence in war zones, and in the variation of child mortality in countries around the world.

It is also important to always consider what is not included in a map. All maps are imperfect projections and every one of them has some distortion – the larger the area covered, the greater the distortion. It may be tempting to draw conclusions too quickly from what is featured, confusing correlation and causation. Although some countries may register high life expectancies and excellent school test scores, this does not mean that all people are benefiting equally. We must always be sceptical, interrogating what is included and constantly questioning what is missing and why. Our maps are far from perfect, and we hope you can find ways to improve them.

Map-readers should know who created their map. Maps are not neutral or impartial; no matter how sophisticated, they are symbolic representations of selected characteristics of a place. Map-makers create maps for different purposes and their perspective will be reflected in the content of each one. In the past, there were just a few well-known cartographers. They were courted by royalty, and published their findings in expensive leather-bound volumes – Mercator's 'Atlas' in the fifteenth century or the 'Atlas Nouveau' in the eighteenth. Today, maps come in all shapes and sizes, are developed by all manner of experts and lay people, and are digitally reproduced for billions of people. While the widening access to maps is cause for celebration, understanding their provenance is more important than ever.

Maps are one tool among many to help us move towards greater insight and understanding. For centuries, navigators have relied on a combination of methods – assessing the location of constellations and planets, observing the direction of the wind, interpreting their compasses and sextants, and drawing on local wisdom – to find their way to a destination. Maps can help give us safe passage, but we must not depend on them alone. For the reasons given in this section and elsewhere, we rely on a variety of techniques to tell the story of our world – to make it *terra cognita*.

The world is more connected than ever before, with underwater
fibre-optic cables, rail networks and pipelines.
Kudelski Group, Parag Khanna and Jeff Blossom, Harvard World Map

Globalisation

Cross-border flows define globalisation

Globalisation lifted billions out of poverty

Globalisation spreads opportunities and risks

Globalisation is not dead, but it is under threat

Our connected world
The white dots are clusters of phones, computers and other connected devices, while the purple lines are submarine communication cables connecting continents. The extent of connectivity within densely populated places is immediately apparent, as is the absence of connectivity in deserts and rural areas, as well as those where poverty is prevalent, as is the case in Africa.

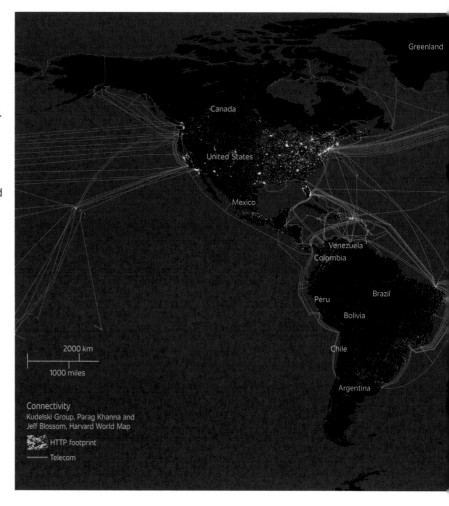

Introduction

Globalisation is widely debated but seldom defined. At the most general level, globalisation translates into the sum of human activities crossing national borders.[1] These flows can be economic, social, cultural, political, technological or even biological, as is the case of the spread of pandemics such as coronavirus or HIV/AIDS. The international movement of ideas is the least visible but among the most powerful dimensions of globalisation. The collision and mingling of ideas shapes how societies are organised, the way economies and companies are managed, and what people understand and think.

The map of internet connectivity depicts a key element in globalisation: how information flows around the world. The visible white dots show clusters of connected devices, including computers, servers

and phones. Underwater telecommunication cables are shown as purple lines criss-crossing the oceans. The sheer intensity of these networks sheds light on just how interconnected we are. We can see luminescence in North America, Europe and South and East Asia, as well as the eastern seaboard of Australia and Latin America. Over 90 per cent of North Americans and 88 per cent of Europeans have access to the internet.[2]

The map also reminds us how some parts of the world are less connected than others. For the most part, dark areas are sparsely populated, such as the Arctic, Amazon and Sahara. The UK telecoms regulator Ofcom has ruled that a broadband speed of 10Mbps is the minimum requirement for consumers to participate in a digital society.[3] Yet just 4 of the 40 African countries surveyed registered broadband speeds of over 5Mbps. Only one African country, Madagascar, appears above 10Mbps in the speed league, and it has under 2 per cent of the continent's population.[4] It is no accident that the fastest growing parts of the developing world happen to be the most connected, notably South and East Asia, the coastal regions of Africa and cities in Latin America. Places left in the dark are falling further behind.

The tearing down of the Berlin Wall and collapse of the Soviet empire over three decades ago was expected to globalise liberal democratic principles and values. Francis Fukuyama's famous book in 1992 heralded 'the end of history' and the triumph of open markets.[5] There was a widely held expectation that geopolitical convergence and the newly invented World Wide Web would shrink the world, underpinning a notion of 'the death of distance'.[6] All this was supposed to spread opportunities, creating what Thomas Friedman described as a 'flat' world.[7] Unfortunately, these scenarios did not pan out as planned.[8]

The euphoria of thirty years ago has given way to a bitter realisation. Globalisation has not delivered what many pundits in the US and Western Europe promised. COVID-19 has spread throughout the arteries of globalisation, creating an unprecedented health and economic emergency everywhere. Meanwhile, although the world's top 1 per cent have benefited enormously, many citizens feel worse off as a result of rising income inequality.[9] These concerns have become even more widespread since the financial crisis of 2008. More than a decade later, average wages in the US, UK and southern Europe were still below their pre-crisis levels.[10] People have good reason to be anxious. Globally, as has been the case for generations, higher incomes don't just indicate financial advantages, they also correlate with higher life expectancies, better job prospects, better schools and other entrenched advantages.

For a while it seemed that globalisation was leading to a confluence of views not only between Europe and North America, but also with emerging economies, as China, India and much of Asia, Africa and Latin America embraced more open markets. The 2008 financial crisis reflected the high point of co-ordinated multilateral action, with the G20 rapidly agreeing a global rescue package that brought the world back from the financial precipice. But this was the last hurrah, as the crisis precipitated a decade of stagnating middle-class wages and sharply rising in-country inequality. Rising anger with the incompetence of bankers, politicians and the panoply of experts managing the system fuelled a populist backlash against globalisation.

Far from leading to a meeting of minds within and across our societies, globalisation has caused ideological and political divides. This is undermining the potential for collective decision-making and identifying solutions. Globalisation, as the financial crisis demonstrated, absolutely demands more co-ordination to manage systemic risks. It also requires collective action to contain the negative consequences of the spread of economic growth and opportunity. Examples of the kinds of externalities that need to be managed include infectious disease outbreaks, rising antibiotic resistance and climate change. At a time when our increasingly fragile planetary ecosystems are crying out for greater co-operation, the political fallout of globalisation means we have less.

Over the past decade globalisation has become a term of abuse and is being blamed for a wide variety of ills. As heretical as this may sound, the problem is not too much globalisation, but too little. Places where people do not have access to adequate finance, to open trade, to low-cost internet provision, to quality education and health care, and to the latest ideas, have a higher exposure to infectious disease, and lower incomes and opportunities than those who do.[11] We also need better globalised governance precisely to manage the problems that know no borders. In this chapter, we first examine what is meant by globalisation – exploring its evolution over time. We then consider different dimensions of globalisation, how it has reshaped finance and international aid, and ways we might better manage it in the future.

Waves of globalisation

Globalisation is not a twenty-first century invention. It has been around for centuries. Economic historians typically use different measures to explain distinct episodes of globalisation involving the flows of products, services, finances, technologies, people and most of all ideas across national borders.[12] Although there have been previous waves of globalisation, what we are currently witnessing is a tsunami by comparison. Over the past three decades the cross-border trade in physical products as well as virtual flows of finance and communication have soared to levels that would have been unimaginable in the past.

The image below is a reproduction of a map produced by Francis Galton in 1881. It shows how long it used to take to transport people and goods from London to the rest of the world.[13] For example, one could travel to Europe or the North African coast within 10 days, whereas Sydney, Shanghai or much of South America would require a minimum of 40 days. Today, any major city globally can be reached within 36 hours by plane and is connected on high-speed internet. As we shall see throughout this book, the physical and digital connection of people and places has spurred on revolutions in health, education, culture, technology and so much more.

What makes the latest phase of globalisation so distinct from past iterations? Scale certainly has something to do with it. But as suggested already, part of the problem is that there is still disagreement about

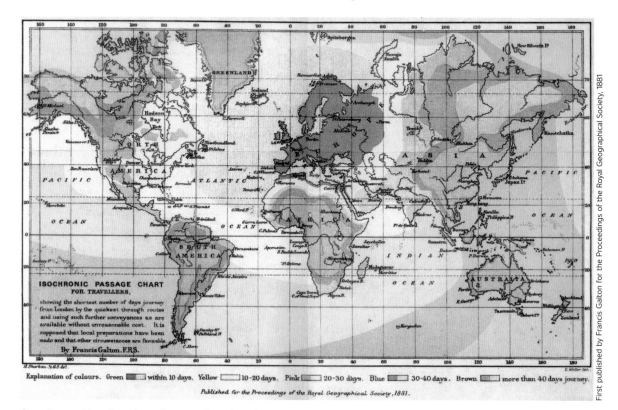

Slow boats: How long it took to get from London to elsewhere in 1881
Francis Galton's first known isochronic map was published by the Royal Geographic Society in 1881 to show how many days it took to travel from London to different parts of the world. It assumes favourable travel conditions, that travel arrangements have been made in advance, and that travellers had the necessary finance. From the map one can see that one could reach anywhere in Europe within 10 days, but that it took over 40 days to reach Sydney, Tokyo or Beijing.

how globalisation is defined, much less when it began.[14] Some scholars contend it first got underway when *Homo sapiens* started their long journey out of Africa hundreds of thousands of years ago. Others claim it started only after China and Eurasia ramped up their trade in luxury products during the first century BC, what is commonly referred to as the Silk Road. The spread of religion and trade (especially in spices) from the seventh to the fifteenth centuries is another possible start period. So is the 'discovery' of the Americas and explosion of scientific innovation between the fifteenth and nineteenth centuries. To be sure, globalisation is poorly understood and we don't solve all these questions in this book.

What we can say is that globalisation has not been a smooth progression. Technological innovations, such as the marine compass, the lightbulb, the combustion engine, the internet and modern fibre-optics, have propelled particular surges. So too have changes in policies, institutions and local preferences that have alternately accelerated and even reversed globalisation. In the fifteenth century, for example, the Chinese emperor Hung-Hsi banned maritime expeditions.[15] Chinese leadership of maritime commerce soon gave way to European, with the Renaissance and the so-called Age of Discovery. This brought potatoes, tomatoes, coffee and chocolate to Europe, while also devastating the lives of pre-Colombian civilisations and subjecting millions of Africans to the horrific barbarism of the slave trade.[16]

The first wave of modern globalisation got underway in the late nineteenth century.[17] Before then, less than 10 per cent of global economic output was due to trade.[18] The renowned economist John Maynard Keynes referred to the period from the second half of the nineteenth century to the First World War as the birth of the modern world economy.[19] During this time the integration of financial markets was accompanied by advances in railways and shipping, with the adoption of steam engines, and the development of telegraph communication. European colonial systems moved into a higher gear.[20] Unlike in our current phase of globalisation, migration was at an all-time high relative to the size of the global population. With up to a third of the Irish, Italian and Scandinavian populations migrating, and more than a million people crossing the Atlantic each year, this period came to be known as the Age of Mass Migration.[21]

Previous waves of globalisation were short-lived, with sharp reversals leading to disaster. The graph on the next page depicts the 'trade openness index' of global trade to GDP between 1500 and the present. It shows a level of about 30 per cent in the second half of the 1800s and then the collapse resulting from the First World War. The war and then the Great Depression brought this wave of globalisation crashing down. The rejection of globalisation during that period, with rising nationalism in Europe, sowed the seeds for the Second World War. In the twenty-first century reactionary nationalism and protectionism are once again an increasing threat to globalisation. And it is not just existential threats such as pandemics, but also cascading financial crises and climate change that could stop globalisation in its tracks.

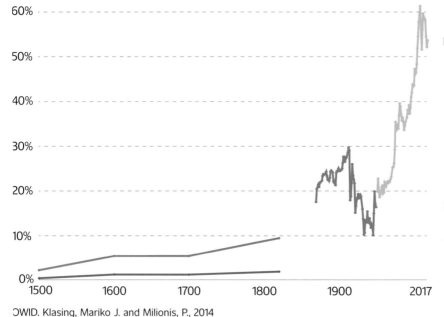

Penn World Tables (9.1)

Klasing and Milionis (2014)

Estevadeordal, Frantz, and Taylor (2003) (upper bound)

Estevadeordal, Frantz, and Taylor (2003) (lower bound)

OWID. Klasing, Mariko J. and Milionis, P., 2014

Five centuries of globalisation and reversals
The figure shows one way of thinking about globalisation in terms of how open countries have been to trade at any time over the past 500 years. The 'trade openness index' depicted here is defined as the sum of world exports and imports, divided by world GDP. Each differently-coloured line represents a distinct source of the information.

The second great wave of globalisation started after the Second World War. It continued gaining momentum until the financial crisis of 2008, when trade accounted for 62 per cent of global GDP. In 2019 trade as a share of global GDP was 59 per cent which, although off the peak reached before the financial crisis, was still high by historical standards. The COVID-19 crisis led to a sharp reversal in 2020, with trade falling by at least a third. Whether this sharp reduction marks a temporary setback or a longer-term resetting of trade on to a slower trajectory and the beginning of the end of the current wave is the $80 trillion dollar question. The answer depends whether COVID-19 can be overcome and the extent to which a vaccine and control measures can be introduced, allowing for the resumption of travel and trade. It also depends on whether the expansion of pro-tectionist policies is sustained in advanced countries leading to further trade tensions, miscalculations and crises.

Technological developments are also critical in determining the forward march of globalisation. Declining energy and transportation costs, and more uniform systems, have allowed manufacturers to outsource production to multiple locations in different countries. Even before COVID-19, the fragmentation of supply chains may well have reached its limits and 3D printing and robotics are facilitating the reshoring of manufacturing and services that previously had benefited from specialised production facilities in lower-cost loca-tions. At the same time, consumer preferences were moving towards individualised, locally produced items that could be rapidly delivered, as well as personal services that could not be easily traded. These trends are likely to be accelerated in a COVID-19 world.

Ultimately, the scale and pace of twenty-first century globalisation hinges on several factors. The latest phase benefited tremendously from significantly reduced costs of transportation and communication. As the graph on the next page shows, sea freight rates plummeted with the development of containers sharply reducing the costs of ocean transport by allowing enormous quantities of products to be shipped directly from a factory to a wholesale or bulk purchaser. Likewise, dramatic reductions in the cost of communication and air travel allowed business and tourism travel to grow to over 1.5 billion trips in 2019 compared to 430 million in 1990.[22] These sectors plummeted catastrophically with the outbreak of COVID-19.

The rapid expansion of trade during the latest phase of globalisation is depicted in the figure showing the global value of exports. It reveals how exports climbed steeply from the 1980s, interrupted only by the briefest of blips following the 11 September 2001 terrorist attacks in the US. But after peaking immediately prior to the 2008 financial crisis they have remained in limbo, and in 2020 collapsed due to COVID-19. Until comparatively recently, the international rules agreed by the

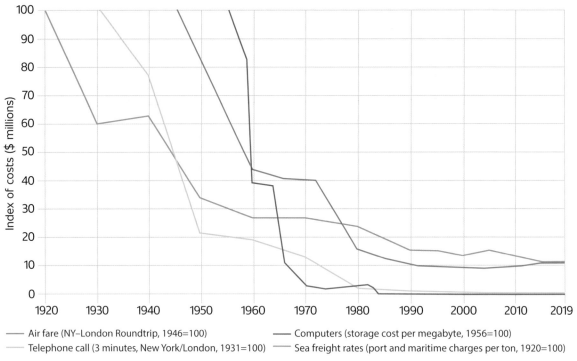

The Geography of Transport Systems, 4th edition

Cheaper communication and transport
The extraordinary decline in the cost of communication and transport over the past 100 years is shown above, with the vertical axis showing an index of costs, from 0 to 100.

allied nations at Bretton Woods in July 1944, to promote financial and trading arrangements after the Second World War, worked reasonably well. As we will discuss in the chapter on geopolitics, the World Trade Organization, together with the International Monetary Fund (IMF) and the World Bank, played a pivotal role in keeping trade humming. But as the commitment to multilateralism among many powerful countries has waned and the institutions have failed to adjust to the rise of new powers, the effectiveness of existing institutional frameworks is increasingly in doubt.

The latest phase of globalisation stands out for the speed of economic integration. In the maps on page 29, each dot represents $10 million worth of traded physical or tangible products, rather than non-tangible services such as financial and information flows. A comparison of 1990 and 2018 reveals the remarkable increase in trade within and between Europe, North America and Asia. In 1970, about

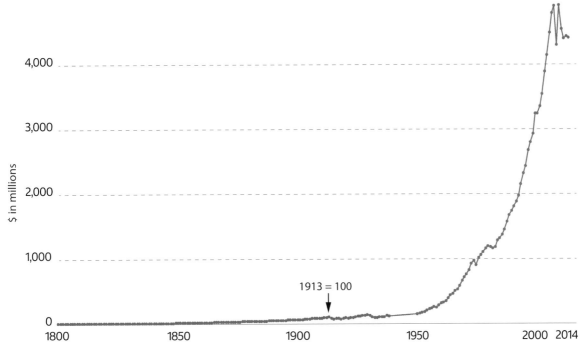

4,000

3,000

$ in millions

2,000

1,000

1913 = 100

0

1800   1850   1900   1950   2000   2014

OWID. Federico, Giovanni and Antonio Tena-Junguito, 2016

Rising and stalling trade

The extraordinary growth and then stalling of trade is apparent from the figure.
This depicts the value of world exports, at constant prices, after inflation, with
the value in 1913 set as $100 million. In the early years of our millennium exports
peaked at 508 times the level they were a hundred years before, but since then
have stalled.

a quarter of global economic activity or GDP was traded, down from
the estimated 30 per cent share a century before. Today the proportion
is roughly double what it was a century ago.[23] The composition of
imports and exports has also changed significantly over the past few
decades, with services comprising a growing share. The rise of Asia's
trading influence reflects the fact that it has been one of the principal
beneficiaries of globalisation. China accounted for barely 0.6 per cent
of global trade in 1970 and today has the largest share, at around 13 per
cent.[24] China is also a far more open economy than in the past, with
trade accounting for 38 per cent of its economic activity compared to
26 per cent in the US.[25]

The pace of globalisation sped up from the late 1980s onwards.
Key developments fuelling this acceleration included the fall of the
Berlin Wall and the collapse of the Soviet Union. Also critical was
the creation, in 1993, of the single market, in which twelve European

Union countries in 1993 committed to a process which would create free movement of goods, capital, services and labour, known as the four freedoms of the EU. In 1994, the North American Free Trade Agreement (NAFTA) came into force, creating the world's largest free trade area, between Mexico, Canada and the US. This fundamentally reshaped North American economic relations by facilitating unprecedented integration between the developed economies of Canada and the US and Mexico's far less developed one. Under NAFTA (now renamed as the US-Mexico-Canada Agreement or USMICA), US trade with its neighbours has more than tripled, growing far more rapidly than its trade with the rest of the world, adding several billion dollars of growth to the US economy each year.[26]

These and other monumental structural shifts and trade deals occurred at a time of technological leaps associated with the invention of the World Wide Web in 1989. The development of the HTML, URL and HTTP system allowed computers to communicate with one another and opened-up the internet to everyone, not just the scientific community.[27] This, along with exponential improvements in computing power and declining costs, facilitated the more rapid spread of ideas, including freedom and democracy. As the chapter on geopolitics explains, in 1989 there were more autocracies than democracies. Two decades later the balance had been reversed and democracies outnumbered autocracies by a ratio of two to one.[28] This does not mean, however, that all is well with democracies around the world. Surveys suggest that dissatisfaction is reaching historic high levels.[29]

These major political and technological trans-formations have touched every aspect of our lives. Books like the one you are currently reading would not have been possible without the internet. It represents a collaboration between Robert in

Trade 1990 and 2017
The reduction in trade barriers and opening of China and other Asian countries to trade in the 1990s is shown in the extent of the lines joining countries. These show the value of goods traded, with each dot representing $10 million of manufactured goods. Minerals, oil and services traded (for example financial flows) are not included.

1990

2017

Miscellaneous manufactured articles
UN Comtrade, 2018

● $10,000,000 per dot

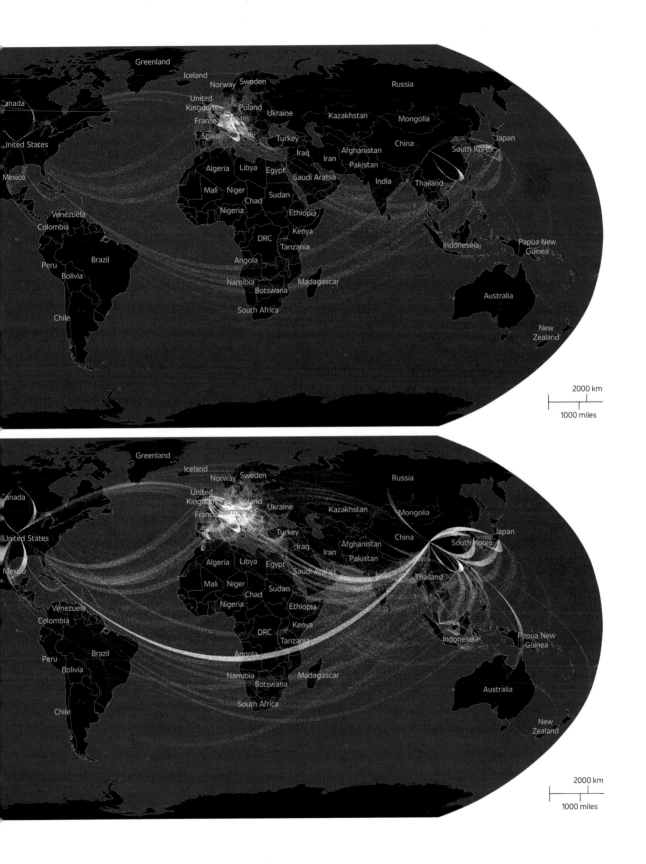

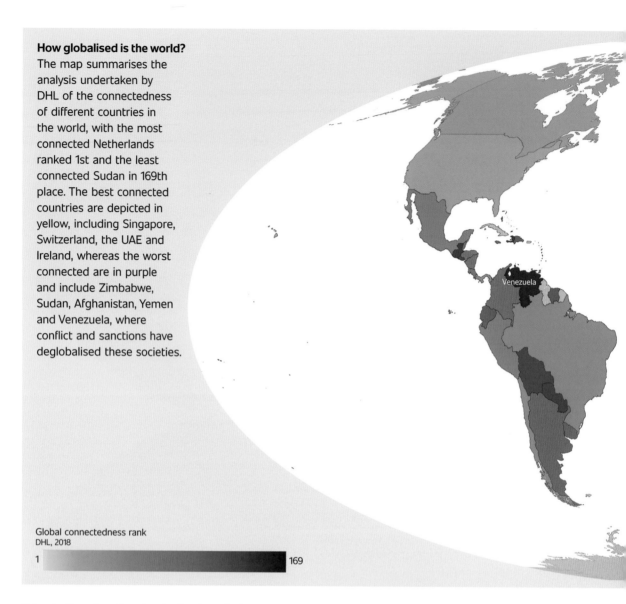

How globalised is the world?
The map summarises the analysis undertaken by DHL of the connectedness of different countries in the world, with the most connected Netherlands ranked 1st and the least connected Sudan in 169th place. The best connected countries are depicted in yellow, including Singapore, Switzerland, the UAE and Ireland, whereas the worst connected are in purple and include Zimbabwe, Sudan, Afghanistan, Yemen and Venezuela, where conflict and sanctions have deglobalised these societies.

Global connectedness rank
DHL, 2018

1   169

Rio de Janeiro, Ian in Oxford and the Create Lab team in Pittsburgh. It draws on information shared through inter-national institutions like the United Nations and World Bank and captured using sophisticated survey methods and satellite imagery collected by teams from the US to Japan. Our personal experiences – travelling to far off places over the past few decades – were facilitated by low-cost air travel and extensive transportation infrastructure.

The way our world is connected can be inferred from how we exchange ideas, products and services with one another. This map was adapted from the logistics company DHL to reflect the extent to

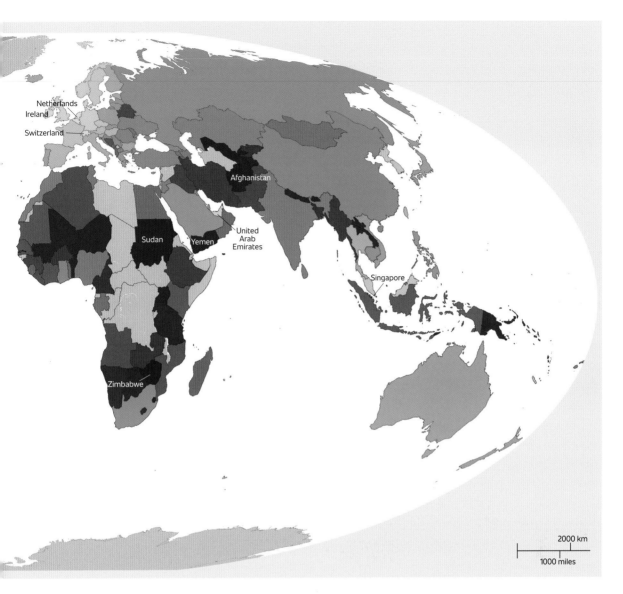

which companies in 169 countries engaged in activities that involved cross-border movements of products, capital, information and people.[30] The map captures the 'depth' of connectedness, which summarises the extent to which a country's population, trade, finance and information flows are international rather than domestic. This is combined with an analysis of the depth of the linkages, with a higher score given to links with more countries, which implies that countries are globally, not only regionally, integrated.

The top ten countries with the deepest levels of connectedness are: the Netherlands, Singapore, Switzerland, Belgium, United Arab

Emirates, Ireland, Luxembourg, Denmark, the UK and Germany. The top five slots are held by relatively small but open economies that have thrived by developing a broad range of products and services targeting a large variety of global consumers. Europe appears to be the most globally connected region, with eight out of the ten most connected countries, while the US is comparatively more insular.[31] The greater integration of many European countries with each other and the world reflects the extent to which the single market, together with the trade, capital market and other reforms associated with globalisation, has transformed their economies.

The globalisation of finance

Historically, the flow of finance across national borders mainly reflected the money paid to import goods or services and the revenues received for exports. Meanwhile, investment flows were principally for the financing of mining, oil and other forms of production, and were repaid from exports. While domestic investment (nationals investing in their own countries) continues to account for more than 75 per cent of investment flows, foreign investment, loans and aid are increasingly significant for many economies.[32] However, relying on foreign finance also has drawbacks. The experience of developing countries during the 1997 financial crisis, which led to an economic collapse in Indonesia, South Korea and Thailand, and the 2008 financial crisis, suggests that the countries that were more dependent on foreign flows suffered higher costs and that these were disproportionately borne by poorer people.[33] This is because financial flows tend to be more vulnerable than trade to speculation and sentiment swings and therefore are more volatile.[34]

The sharp increase in financial flows across national borders over the past twenty years coincided with the liberalisation of capital controls and deregulation of new financial instruments such as credit derivatives, which serve both risk management and speculative purposes. The failure to manage the combination of newly harnessed computing power and the ability to trade seamlessly across national borders was a major contributor to the financial crisis of 2008.[35] Whereas in the past, financial flows grew broadly in tandem with economic activity, in recent decades this relationship has been severed.

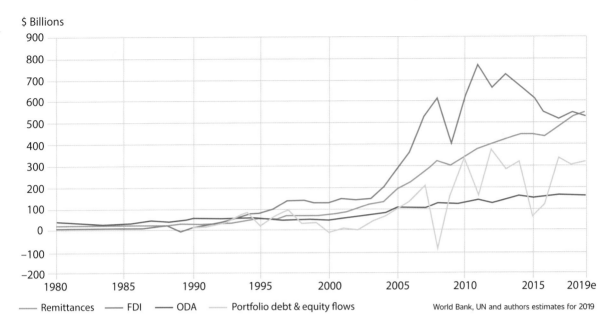

$ Billions

Remittances — FDI — ODA — Portfolio debt & equity flows

World Bank, UN and authors estimates for 2019

Financial flows to developing countries, 1980–2019 (current US$)
Increased financial flows are a vital dimension of globalisation, with these rising significantly after the end of the Cold War. Foreign Direct Investment (FDI) is the most significant but benefits a relatively small group of countries and sectors, notably mining and oil. Remittances which migrant workers send home have recently exceeded foreign investment. The volatility of debt and equity flows and the stagnation of Official Development Assistance (ODA) flows, which are the least significant of the financial flows, is also evident.

Capital flows have been increasing at a much faster clip. The rapid rise in flows from developed to developing countries is reflected in this figure. It shows that flows were virtually flat prior to the late 1980s and have risen dramatically since then, though are prone to much greater fluctuation.

The financial crisis of 2008 resulted in the collapse of stock markets, as investors dumped their holdings. In the US, fifteen major banks failed, including Lehman Brothers, which had been founded in 1850. Others were put on life-support by the federal government, and the head of the IMF warned that the global financial system was teetering on the 'brink of a financial meltdown'.[36] At such times, investors tend to panic, believing things can only get worse. The result was that an all-time record of shares were sold. Over a few days the Dow Jones stock average fell an unprecedented 18 per cent and the S&P stock market index by more than 20 per cent, with similar record-breaking collapses internationally.[37]

While most financial flows collapsed, remittances sent home by migrants stayed firm. Remittances have continued to rise, reaching a record high of $707 billion in 2019 with those sent to low and middle-income countries accounting for $551 billion of these flows.[38] The top remittance recipients were India, with $82 billion, China ($70 billion), Mexico ($39 billion), the Philippines ($35 billion) and Egypt ($26 billion).[39] Remittances provide vital lifelines during periods of adversity. In the Pacific island country of Tonga, for example, in 2019 remittances accounted for 39 per cent of national income. In Haiti they represented 34 per cent of GDP, in Nepal 30 per cent, and in El Salvador and a number of other Central American countries around 21 per cent.[40]

In fact, in 2018 remittances overtook foreign investment for the first time as the largest financial flow to developing countries. While remittances are primarily directed to private individuals, foreign investment mainly goes to private firms, with both being important for economies. Foreign investment flows follow long-term cycles, with the crisis having a delayed impact, as growth in demand slowed, reducing the attractiveness of investment. As a result, foreign investment peaked in 2015, seven years after the crisis, with total global investment flows reaching $2 trillion before declining to $1.2 trillion in 2018. Those going to developing economies peaked in 2015 at around $700 billion but by 2018 had fallen back $673 billion, with $478 billion of this going to Asia and only $38 billion to Africa.[41]

The transformation of finance in recent decades shifted the relative weight of different financial markets, both in relation to each other and to their underlying economies. For example, just before the COVID-19 pandemic, the flow of euros was over six times the size of the European economy, whereas twenty years ago, shortly after the euro was adopted, it was not much bigger than the combined economies of its members.[42] Sterling, which for more than a century dominated foreign transactions, was recently overtaken by the euro, and Japanese yen transactions have been outpaced by Chinese renminbi. This trend is expected to continue, with Brexit accelerating the UK's decline.[43] Owing to the sheer scale of its debt, the US continues to dominate global flows.[44]

To the extent that foreigners buy debt, or bonds, countries can live beyond their means. The US is the world's most indebted country, with the government owing more than $23 trillion as of 2019. This is about 80 per cent of the size of the US economy, and as public

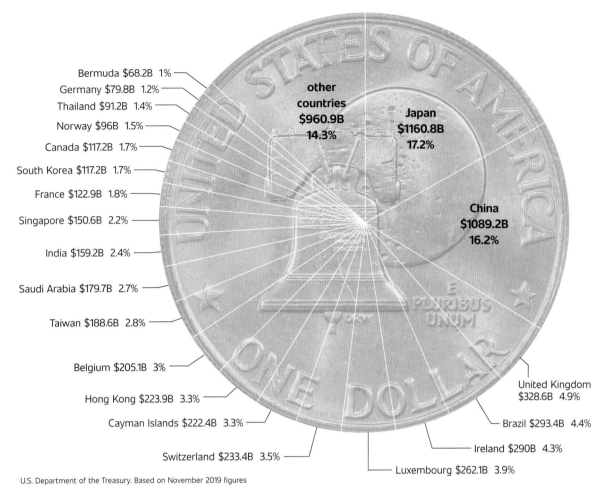

Bermuda $68.2B 1%
Germany $79.8B 1.2%
Thailand $91.2B 1.4%
Norway $96B 1.5%
Canada $117.2B 1.7%
South Korea $117.2B 1.7%
France $122.9B 1.8%
Singapore $150.6B 2.2%
India $159.2B 2.4%
Saudi Arabia $179.7B 2.7%
Taiwan $188.6B 2.8%
Belgium $205.1B 3%
Hong Kong $223.9B 3.3%
Cayman Islands $222.4B 3.3%
Switzerland $233.4B 3.5%

other countries $960.9B 14.3%
Japan $1160.8B 17.2%
China $1089.2B 16.2%

United Kingdom $328.6B 4.9%
Brazil $293.4B 4.4%
Ireland $290B 4.3%
Luxembourg $262.1B 3.9%

U.S. Department of the Treasury. Based on November 2019 figures

Who owes who? Foreign holders of US Debt, November 2019
The US has borrowed over $23 trillion from the rest of the world, and as the pie above shows, its biggest creditor is Japan, which it owes over $1.1 trillion, followed by China. With over half of China's foreign reserves in US dollars, it has little incentive to try to destabilise the US economy or push the value of the dollar down.

spending rises and tax cuts reduce revenues, the deficit is expected to balloon further and even exceed the size of the economy in a few years.[45] The growing dependence of the US on foreign lenders means that the dollar con-tinues to dominate foreign exchange. Just before the outbreak of of the COVID-19 pandemic, about 30 per cent of US debt was held by foreigners.[46] As the image shows, before COVID-19, China held almost $1.1 trillion, which is over 16 per cent of US debt. Despite the latest trade war, this underlines the extent of the interdependence between the two countries.[47] Japan in 2019 had $1.1 trillion of US bonds, making it the biggest creditor.[48]

Giving

By contrast, foreign aid includes multiple financial flows, ranging from grants and concessional loans from governments and international agencies to assistance from non-governmental organisations and philanthropists. Gifts in kind, such as medicines or water pumps, and technical assistance also qualify as in-kind support measured as part of 'official development assistance', or ODA.[49] The volume of aid, its purpose and its sources have changed markedly in recent decades. Until the 1990s it was mainly dominated by geopolitical, military and colonial relationships. Then in the early 1990s, with the end of the Cold War, the pendulum moved away from tied aid and towards a more targeted framework supporting programmes owned by the recipient countries and communities.[50] Until recently, most aid was scattered and fragmented. Efforts to co-ordinate aid flows picked up during the 1990s. By 2000 this was articulated in the Millennium Development Goals, which were followed in 2015 by the Sustainable Development Goals.[51]

The graphs below depict the good and the bad news regarding ODA flows tracked by the Organisation for Economic Co-operation and Development (OECD), a club of the world's wealthiest countries. A United Nations resolution in October 1970 committed the rich countries to a target of allocating at least 0.7 per cent of their national

Stingy rich world: Aid flows to developing countries

Although on average rich countries are richer than they ever have been, and developing countries are better managed, the share of their income that people in rich countries give to developing countries has not increased for fifty years, as is depicted in the ratio of ODA to Gross National Income (GNI) of the 30 countries convened by the OECD Development Assistance Committee (DAC). The total amount of ODA in billions of dollars has climbed slowly and unsteadily, as countries have got richer, and this too has recently stalled.

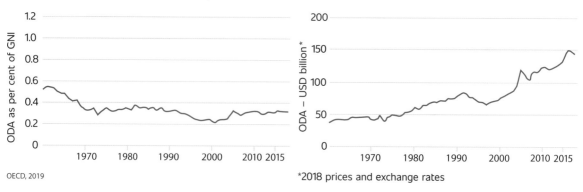

OECD, 2019

*2018 prices and exchange rates

income to foreign aid by 1975, which only the Netherlands and Sweden achieved.[52] Fifty years after it was first agreed, only four countries have consistently met the target: Denmark, Luxembourg, Norway and the United Kingdom. In fact, the share of rich countries' incomes devoted to ODA has actually declined from an average of 0.5 per cent in 1970 to around 0.3 per cent in 2019.[53]

Because the total size of the traditional donor economies has grown in absolute terms, overall assistance has risen from $36 billion in 1960 to $143 billion in 2018.[54] The European Union gave the most, but the US was the biggest single national contributor, providing $33 billion in 2018, equivalent to just 0.16 per cent of its national income. By comparison, Sweden gave $5.8 billion, which was more than 1 per cent of its national income. The upper map on the next page illustrates ODA flows with the lighter ends of the transfers showing the origin countries and the green the destination. As is evident, the majority of flows are from Western Europe, North America and Japan and head towards African countries. There are also more limited flows from Europe and the US to the Middle East, and from the US to Central America and the Caribbean. The map shows the significant but still small Chinese aid programme, with its transfers concentrated on Sub-Saharan Africa, and the significant flow of aid from Turkey and Saudi Arabia to nearby countries, as well as from Australia to Papua New Guinea.

The rise in what the OECD describes as 'non-traditional' donors reflects their growing economic and political clout. As described in the geopolitics chapter, the Chinese-led Belt and Road Initiative is the most significant expression of this emergence of new aid relationships, with the associated investments mainly concentrated in road, rail and maritime infrastructure across Asia, Africa, Europe and even Latin America.[55] Other new donors, including India, Qatar, South Korea, Brazil and Turkey, have been increasingly active. In 2015, for example, Jordan, Lebanon, and especially Turkey's support for Syrian refugees, much of it provided through humanitarian aid, was highly significant. Many of the concerns that were previously identified as having undermined the effectiveness of aid from the OECD countries have resurfaced with non-OECD donors. These include the dependencies that aid can engender, the unequal power relations between donors and recipients, as well as the appropriateness of technical advice and foreign standards, or lack thereof.[56]

In order to get a sense of its sheer scale, it is useful to graphically visualise the distribution of aid flows. This map of ODA in 2018, the latest year for which data are available, reveals how war-torn Syria was by far the biggest recipient, with more than $10 billion, equivalent to $567 per Syrian. The non-traditional donor, Turkey, provided more than $6.7 billion of this, followed by Germany, which contributed $769 million, although it should be noted that this includes support for Syrian refugees resettled in German cities.[57] The Syrian civil war is a reminder not just of the way conflicts consume aid dollars – especially in the form of humanitarian assistance – but also how violence can throw development into reverse. Before open warfare broke out in 2011, Syria was one of the most advanced countries in the Middle East and barely received any aid at all.

Good, bad and ugly globalisation

The impacts of the current wave of globalisation are increasingly uneven. To be sure, no force in history has generated more benefits, more quickly, to more people. The tide of change initiated after the Second World War brought with it unprecedented human progress in nutrition, literacy, longevity and other key indicators of well-being. And since 1990, more than 2 billion people have been lifted out of dire poverty, most of them in East and South Asia. As we shall see in the health chapter, average life expectancies across the world have increased by more than fifteen years in a single generation, since 1990. During that same period, in excess of 3 billion more people have learned to read and write.[58] These and other

Aid flows from rich to poor countries

This map illustrates the flow of aid from donors, where each dot represents a million dollars. It displays the direction of aid from donors depicted with white dots, to recipients, represented in green. The map shows the extent to which the donors shower assistance on a small set of countries, and how the sources are concentrated in Europe, the US, Japan and the Gulf states, while the destinations are mainly in the Middle East and North Africa, and to a lesser extent, for the US, in the Caribbean, and for Australia in Papua New Guinea.

ODA, 2018
OECD, 2019
● Each dot equals $1,000,000

Where Aid goes

The red dots identify the destinations of the aid in 2018, with the concentration in Africa and the Middle East being apparent. Conflict is associated with rising humanitarian aid, including for refugees who flee to neighbouring countries. The largest circle on the map is Syria which received $10.7 billion, while other notable recipients of aid include Yemen $7.6 billion, Iraq $2.2 billion and Turkey $1.1 billion. The smallest circle is Equatorial Guinea, which received $5.27 million.

Haiti

Destinations of Aid, 2018
OECD, 2019
● Size of dot corresponds to amount of aid received

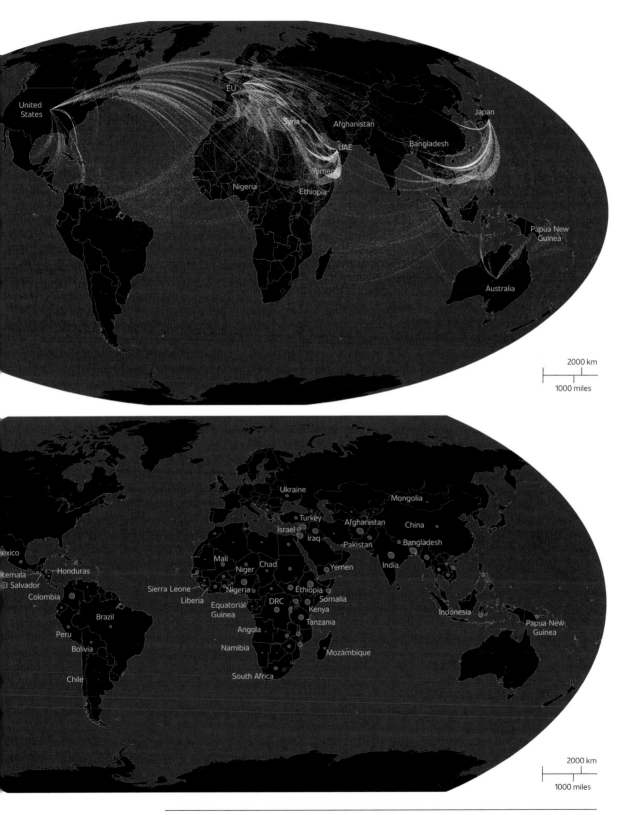

United States

EU

Syria

Afghanistan

Japan

UAE

Bangladesh

Yemen

Nigeria

Ethiopia

Papua New
Guinea

Australia

2000 km

1000 miles

Ukraine

Mongolia

Turkey

Afghanistan

China

Israel

Iraq

Bangladesh

Pakistan

exico

Mali

Niger

Chad

Yemen

India

temala

Honduras

El Salvador

Colombia

Sierra Leone

Nigeria

Ethiopia

Somalia

Brazil

Liberia

DRC

Kenya

Indonesia

Papua New
Guinea

Equatorial
Guinea

Peru

Angola

Tanzania

Bolivia

Namibia

Mozambique

Chile

South Africa

2000 km

1000 miles

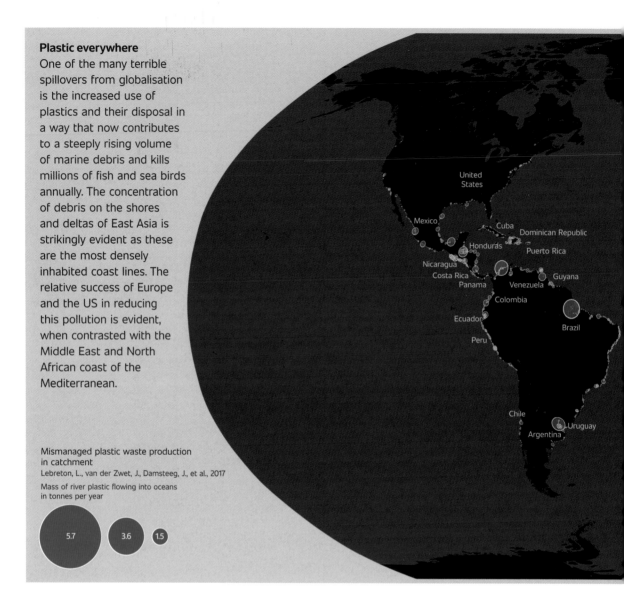

Plastic everywhere

One of the many terrible spillovers from globalisation is the increased use of plastics and their disposal in a way that now contributes to a steeply rising volume of marine debris and kills millions of fish and sea birds annually. The concentration of debris on the shores and deltas of East Asia is strikingly evident as these are the most densely inhabited coast lines. The relative success of Europe and the US in reducing this pollution is evident, when contrasted with the Middle East and North African coast of the Mediterranean.

Mismanaged plastic waste production in catchment

Lebreton, L., van der Zwet, J., Damsteeg, J., et al., 2017

Mass of river plastic flowing into oceans in tonnes per year

5.7   3.6   1.5

positive findings have been reviewed extensively in seminal volumes by the late physician Hans Rosling and psychologist Steven Pinker, among others.

The juxtaposition with the recent past is hard to miss. During the 1970s and 1980s, China, India, Brazil and many other developing countries displayed great suspicion about globalisation. Yet now it is the 'left-behind' in wealthy countries who are most forcefully rejecting globalisation, calling for a 'return' to better times. Those most disillusioned with the current phase of globalisation are largely confined to rich countries in North America and Western Europe.

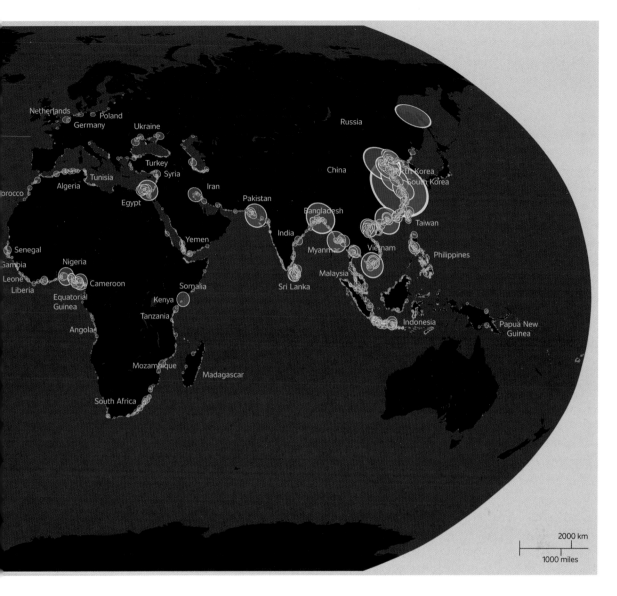

This is not so surprising. Whereas incomes in rich countries have barely doubled in the past thirty years, average incomes have increased sixfold in the rest of the world, and in China have increased almost thirty times (from $310 to $8,827).[59]

The point is that the latest episode of turbo-charged globalisation has also exacerbated risks. As legal cross-border flows intensify, so too do illegal flows. As we shall see in the chapter on violence, money laundering, tax evasion and other illicit transfers have boomed in the wake of globalisation.[60] Likewise, the internet has not only spread creative ideas and given rise to innovation, it has become the conduit for fake news, identify theft, ransomware, radicalisation and extremism. The sprawling dark web – likely several times larger than the surface web – also fosters all manner of illicit activities, from drug trafficking and cyber-crime to paedophilia. The trade in small arms and ammunition that fuels conflicts and crime, human smuggling, and the trade in contraband and endangered wildlife are estimated to account for at least 10 per cent of all trade.[61]

Another unexpected outcome of contemporary globalisation is that individual activities can have genuinely global consequences. Decisions taken on Wall Street can devastate the lives of workers and pensioners all over the planet. Plastics washing up on seashores and floating in the oceans is a massive global problem that will span generations. The extent of plastic debris on China's coastline is evident in the map on the previous page. This is one of the reasons why China, once a dumping ground for the world's waste, has banned the production and sale of single use plastics and the import of used plastic. Although more and more countries are putting restrictions on plastics, the extent to which such detritus is building up off the West African, Mediterranean and other coasts is striking.

Responsible and inclusive globalisation comes with obligations. It demands greater awareness and care about how our own habits and actions impact communities from the local to the global scale. What may satisfy our wants as individuals – consuming more fossil fuel energy, taking more antibiotics, eating more meat and tuna or using disposable plastics – is fundamentally irrational from the perspective of ensuring the health and stability of our communities and the wider world. Globalisation has brought us closer together. But as we shall see, it is also driving us further apart. To survive, all of us will need to find ways to co-operate. And this means thinking about the consequences of our actions not just on ourselves, but on each other, our communities, and our planet.

In the end, our future well-being is not compatible with rampant individualism or reactionary nationalism. If we are to thrive, we need to grapple with the complexities unleashed by globalisation. We need to demand greater understanding and tolerance from our leaders. The 2008 financial crisis represented, among other things, a catastrophic failure to manage globalisation. It led to a severe deterioration of trust in our leaders to manage rapid change. What drives support for populists and authoritarians on the left and right is a belief that the traditional authorities (and the elite who support them) have let them down and are increasingly disconnected from day-to-day realities. The goal of reactionaries is to upend politics as usual. We have sympathy with frustrated protesters around the world, even if we disagree with some of their tactics and policy prescriptions. The old guard *have* let them down. Inequality *is* growing. The rise of AI and rapid technological change *does* threaten jobs. Globalisation *is* badly managed and it increases the risks of climate change, pandemics and other dire threats. Meanwhile, the race to the bottom in tax, regulation and safety nets is leaving governments with fewer resources and is leading to a retrenchment of vital welfare provisions.

As the world emerges from the COVID-19 crisis, there is an opportunity to reset globalisation. The fight is not between capitalism and socialism. Our priority must be to fundamentally reshape capitalism to prioritise the welfare of people – especially the most vulnerable – and our planet, not shareholder profits. Addressing the biggest challenges facing the world requires more co-operation, not less. The world requires more openness to ideas, not less. To keep improving the human condition, we need more finance, vaccines and other goods and services to benefit poor countries and reach poor people, not less. To succeed, the forces of globalisation must benefit those who have hitherto been excluded. Whether we sink or swim depends on both our individual and collective actions. A new globalisation needs to be nurtured. If it does not thrive, neither will we.

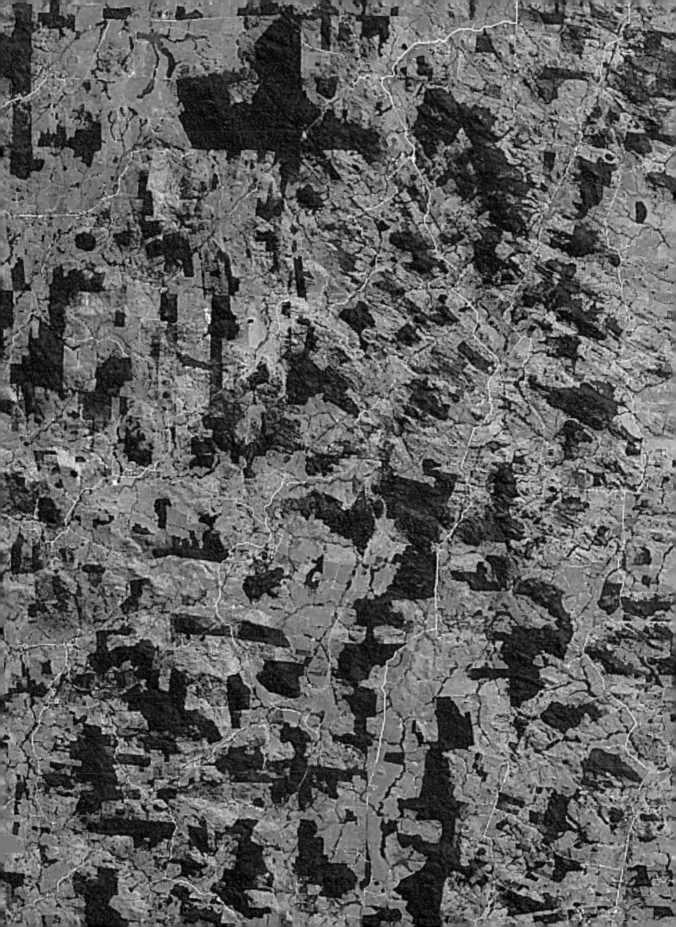

Climate

Melting glaciers and sea ice

Temperature anomalies

Forest fires and gas flares

Deforestation and degradation

Rising seas and sinking cities

Deforestation in the Brazilian Amazon – 2019.
Google / Google Earth Engine, USGS, NASA, ESA

1 km

1 mile

The Blue Marble – **1972**.

NASA / Apollo 17 crew

Introduction

The delicate ecosystems sustaining life on earth are starting to collapse. In less than two centuries, *Homo sapiens* catastrophically altered the climate of our 4.3 billion-year-old planet.[1] We are well past the debate about *whether* humans are to blame. The burning question now is *how fast* we can change course? The sheer dimensions of what comes next are mind-bogglingly complex. Many scientists believe we may already have passed irreversible tipping points. Apocalyptic media headlines confirm our worst fears: monster icebergs and stranded polar bears; mass extinctions of animals and insects; over-heating and polluted cities; raging tempests and firestorms; and acidic oceans and dying coral reefs. The news is so dreadful that it is tempting to declare defeat and shutter ourselves from the news. But that would be precisely the wrong thing to do. We are facing serious challenges, yes, but we can potentially lessen the damage of what's to come. It is more important than ever to connect the dots and invest in solutions that can quickly scale up. This is where maps come in. They can help us see the big picture, revealing patterns and relationships in new and unexpected ways.

For centuries, maps have helped civilisations interpret and navigate the unknown. They have shown us where we are and where we want to go. Some of them have fundamentally changed the way we understand the world. Many of the maps featured in this chapter can be traced back to a single photograph taken at 5.39 a.m. EST on 7 December 1972. The image, credited to a member of the Apollo 17 mission, was eventually dubbed the 'Blue Marble'. It captures the fragile beauty of Earth – the only habitable planet in our solar system – from 18,000 miles away. For many people around the world, it was the first time they saw a colour image of the world. The photograph became a symbol of the modern environmental movement and one of the most widely shared images of all time. It is a reminder of the ways in which photos and maps can open people's minds to dazzling insights and future possibilities. They can stop us in our tracks, but also impel us to action.

In this chapter, we take a whirlwind tour of some of the earth's most vital ecosystems, the fates of which will determine our common future. Our first stop is the Himalayas, the earth's 'third pole' and key to the survival of billions of people across Asia. From there we travel to Greenland, the Arctic and the Antarctic, to show how monumental ice sheets are melting faster than at any time in recorded history. These changes are accelerated by rampant greenhouse gas emissions that are themselves compounded by unprecedented forest fires and industrial pollutants visible from space. We then examine how global warming is also impacting the oceans that cover more than 70 per cent of the planet and contain 97 per cent of all water on Earth. As sea levels rise, our maps reveal how many of our most connected cities – home to 4 billion people and centres of global capital and finance – may soon be submerged. That is unless we take steps to prevent this from happening right now.

The COVID-19 crisis distracted the world at a pivotal moment in the fight against climate change. It delayed vital international meetings between governments and allowed opportunists to speed-up deforestation and other forms of resource exploitation. But the pandemic has also generated some climate dividends. By dramatically slowing global travel and trade, it has also reduced carbon dioxide and nitrogen dioxide emissions. For the first time in a generation, residents of some Indian cities were able to see the peaks of the Himalayas, and Beijing residents breathed clean air. It is still too early to know precisely how the response to the pandemic has affected climate change, but there are early signs that it may have had some net positive effects.

Melting Himalayan glaciers – 1984 and 2019.

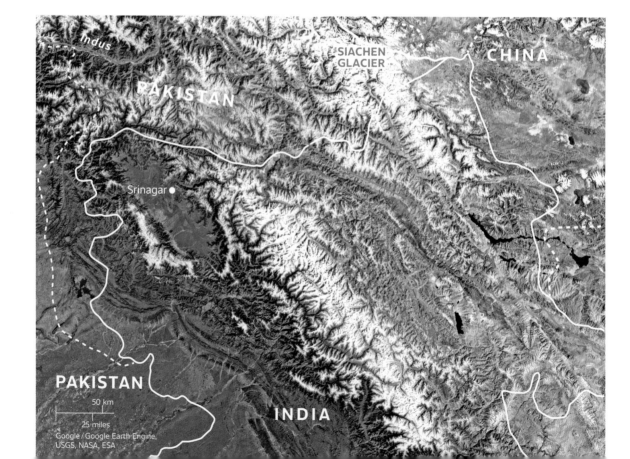

The Himalayas – Asia's shrinking water tower

We decided to climb the Himalayas to better understand the gravity of the climate emergency. Our immediate priority was, naturally, not to die trying. Since the early 1950s, more than 25,000 people have tried summiting the 'eight-thousanders', the fourteen peaks rising 8,000 metres (26,000 feet) above sea level.[2] Thousands of mountaineers have failed and at least 900 of them have died trying to make the ascent, mostly from avalanches, falls and prolonged exposure to cold. Aware of these grim statistics, we decided to hike *around* (rather than over) the towering Annapurna and Manaslu summits. Even at around 4,000 metres the nights were bitterly cold and the days were disconcertingly warm. We avoided accidents in the jagged peaks above but a terrifying new reality revealed itself in the valleys below.

The Himalayan glaciers are melting. As these images from NASA satellites taken in 1984 and 2019 show, the majestic white glaciers have dramatically contracted. As we traversed rocky moraines once blanketed in snow, our Sherpa guides pointed out the hollow imprints where vast mountains of ice had recently receded. They told us that the retreat had sped up during their lifetimes. Climate scientists

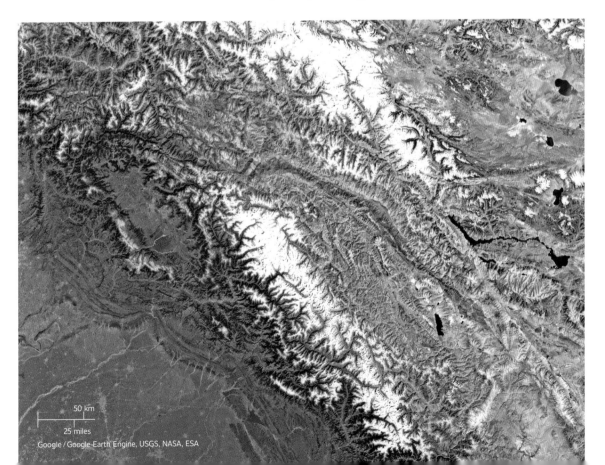

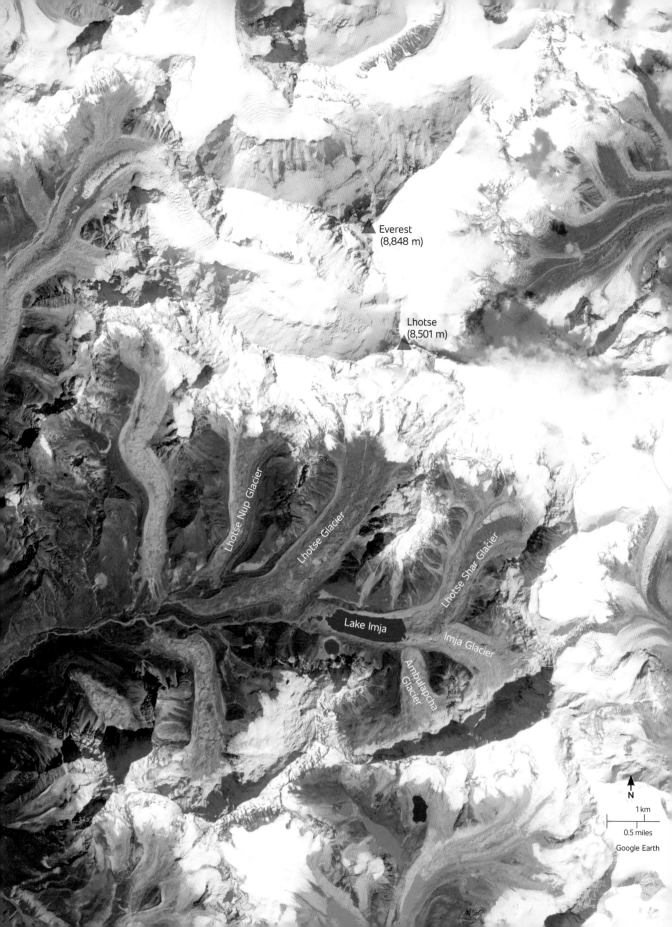

Everest
(8,848 m)

Lhotse
(8,501 m)

Lhotse Nup Glacier

Lhotse Glacier

Lhotse Shar Glacier

Lake Imja

Imja Glacier

Ambulapcha Glacier

N

1 km

0.5 miles

Google Earth

back up the Sherpas' personal anecdotes with hard empirical data.[3] After comparing detailed field studies stretching back to 1975, their research shows that the world's third largest deposit of ice and snow is disappearing.[4]

What does all this glacial change have to do with those of us living thousands of miles away? The answer turns out to be a lot. As the maps on pages 48–9 show, the Himalayas are vast. The 1,500-mile range spans five countries – China, India, Pakistan, Nepal and Bhutan – which together account for over 40 per cent of the world's population and at least 20 per cent of the global economy. The ice locked-up in the Himalayas is the single largest surface deposit of fresh water after the Arctic and Antarctic, feeding several of the world's great rivers, including the Indus, Yellow, Ganges, Brahmaputra, Irrawaddy, Salween, Mekong, Yangtze and dozens of tributaries. In short, the Himalayas provide physical, and spiritual, sustenance to billions of people stretching from the Arabian Sea to the Bay of Bengal.

The Himalayas are called the 'third pole', since they are the world's biggest ice deposit after the Arctic and Antarctic. The sprinkling of white frosting that you see on the satellite map represents at least 15,000 glaciers that hold the equivalent of 4,600 square miles of fresh water. Some of these magnificent ice fields – Gangotri,[5] Khumbu,[6] Lhotse,[7] Yamunotri[8] and Zemu[9] – are in full retreat. Almost a fifth of the entire Himalayan ice surface has contracted since the 1970s. In its place are exposed rock faces, gravel plains, and huge glacial lakes. A good example is the Lhotse glacier right next to Mount Everest, and Lake Imja that is expanding with unnerving speed below.

The disappearing Lhotse glacier and the expanding Lake Imja Satellite image taken October 4, 2010.

The Lhotse glacier is a beast, even by Himalayan standards. As you can see on the map, it is located on Everest's south face and adjoins the Imja and Ambulapcha glaciers to the east and south-east. It extends from the peak of the mountain roughly 13,000 feet down and is pockmarked with hundreds of sinkholes, ponds, wet sediment and alpine shrubs.[10] It has also receded 100 feet a year over the past three decades. When glaciers melt, they become unsteady. In 2016, the Lhotse glacier went rogue and released more than 70 million cubic feet of water in less than an hour.[11] Melting glaciers create vast pools that can swell to bursting point, sending torrents of water, mud and debris hurtling down to densely populated villages below.

More than a thousand new high-altitude lakes have been formed by melting glacial ice in the Nepal region alone. They expanded, on

100 km

50 miles

Water change 1984–2018
Google / Google Earth Engine, USGS, NASA, ESA, JRC, 2019

Water decrease | Water increase

average, 70 per cent in the past decade.[12] As you can see on the map, areas in green indicate where water levels have accumulated over the past three decades on the Himalaya Plateau. Much of this is due to glacial melt. Take the case of Lake Imja in Nepal, which has tripled in size since the 1980s and is now almost 500 feet deep and holds more than 2.6 billion cubic feet of water, the equivalent of 30,000 Olympic-sized swimming pools. In a part of the world renowned for its earthquakes, even a minor tremor could trigger a humanitarian disaster. The cash-strapped Nepalese government started draining the basin after it received a multimillion-dollar grant from the United Nations and the World Bank's Global Environment Facility.[13] The military engineers quickly discovered this was much easier said than done owing to the altitude, rocky terrain and unpredictable seismic activity in the area.

Tibetan plateau – high-altitude lakes formed by melting glacial ice.

So why are the Himalayan glaciers melting so fast? Greenhouse gases are a big part of the problem, especially those containing nitrate, sulphate and carbon particles. 'Black carbon',[14] or soot, is right behind carbon dioxide and methane as a leading cause of climate change.[15] We could see evidence of black carbon just about everywhere in the Himalayas, including in the air we breathed, where it joins other industrial pollutants to form foul brown clouds that are visible from space.[16] Most of the soot there is belched from coal-fired power plants, vehicles, burning fields and cooking stoves across China and India. Black carbon doesn't only warm the atmosphere and trap heat, it also settles on white surfaces contributing to what is known as the 'albedo effect'.[17] Put simply, sooty snow absorbs more sun than white snow and heats up the earth. These feedback loops are hazardous to planetary health.

The future for the Himalayas – and those who depend on them – is bleak.[18] Despite the unlikely scenario that global emissions are reduced and warming is limited to 1.5 degrees Celsius in the next two decades, scientists predict that more than a third of all its glaciers will disappear by the end of the century.[19] If emissions continue unabated and temperatures increase above 2 degrees Celsius, at least two-thirds of the glaciers are doomed. This spells disaster for more than 240 million people who permanently reside in the mountain plateaus stretching across the Himalayas and the Hindu Kush. It is also going to imperil almost 2 billion people who directly and indirectly rely on the ten big glacier-fed rivers in the valleys below, and the 3 billion others who depend on food grown in those very same river basins.[20]

The impacts of melting glaciers won't all be felt at once; they will come in waves. At first, the accumulation of water will swell rivers and trigger floods. Soon after, river flows will sharply decline, shrinking food production and starving hydroelectric power plants. Although billions of people across Asia are at risk, the great melt will hit the poorest and most vulnerable of them hardest, especially in Bangladesh and Nepal, where populations are growing rapidly. And that's not all. The drying out of rivers will severely reduce subsistence farming and agricultural productivity and very likely trigger mass migration and social unrest,[21] a subject discussed at length elsewhere in this book. Simmering tensions between geopolitical rivals such as China, India and Pakistan could boil over into open conflict as water availability evaporates.

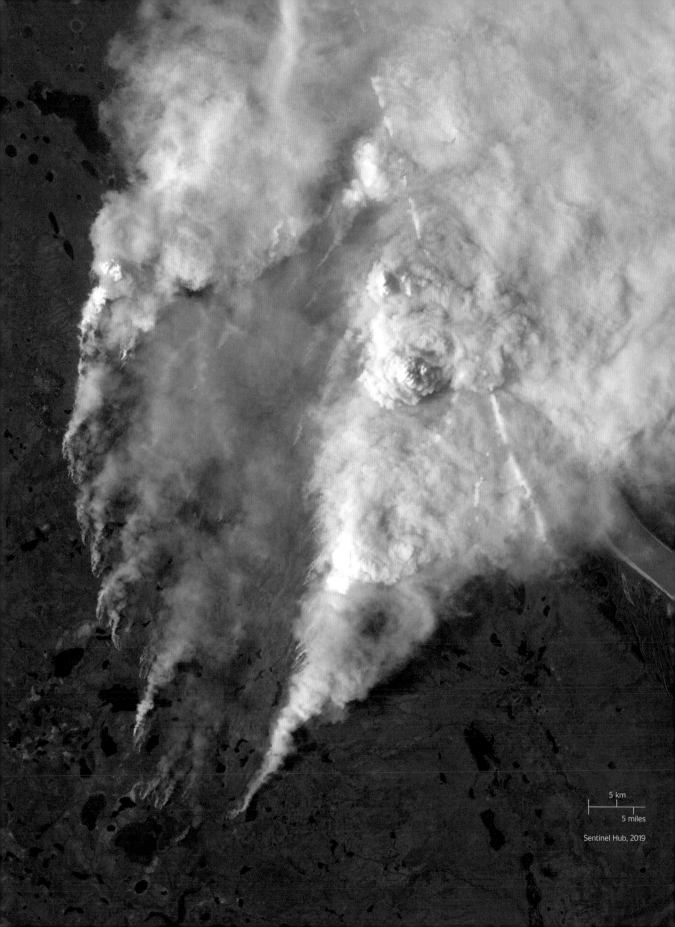

5 km

5 miles

Sentinel Hub, 2019

Greenland's disappearing ice – 50 billion elephants in the room

Everywhere on the planet, ice is disappearing. It is not just glaciers in the Himalayas that are melting. Arctic and Antarctic sea ice is thinning and permafrost in the northern reaches of Canada and Russia is receding, threatening to release massive amounts of carbon and methane.[22] Making matters worse, the Arctic tundra is burning. Massive fires broke out in parts of Siberia in 2019 that released a cloud of soot and ash larger than all the countries that make up the European Union. Researchers predict that because summer temperatures are rising, fires in the boreal forests and Arctic tundra – an area covering about a third of the global land surface – could quadruple by 2100. As the permafrost melts and wetlands burn, the vegetation on which other ecosystems depends is collapsing. These fires could dramatically increase the amount of carbon being released into the atmosphere and undercut even the most inspired attempts to curb emissions.

Left: **Burning arctic fires** – Northwest Territories, Canada, south of the Mackenzie River. Satellite image taken from Sentinel-2 L1C on 27 July 2019.

Right: **Greenland – 2019.**

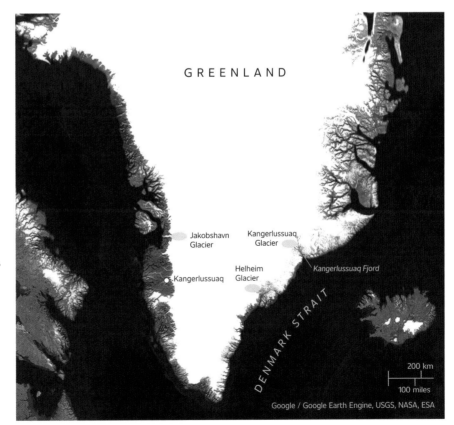

GREENLAND

Jakobshavn Glacier

Kangerlussuaq Glacier

Helheim Glacier

Kangerlussuaq

Kangerlussuaq Fjord

DENMARK STRAIT

200 km
100 miles

Google / Google Earth Engine, USGS, NASA, ESA

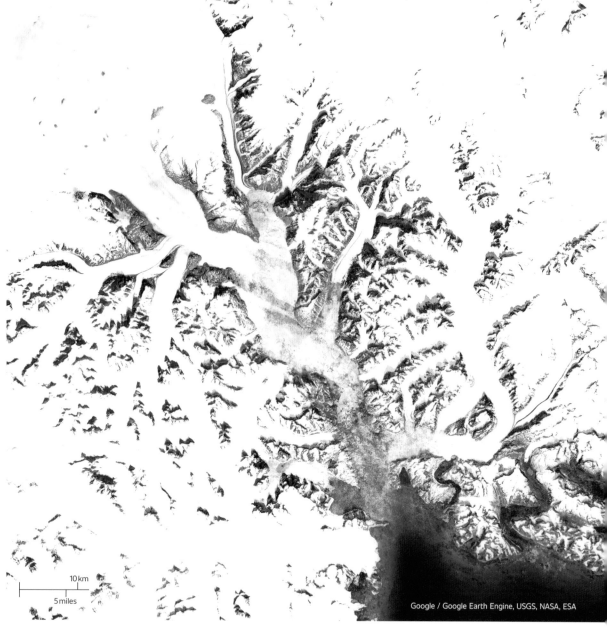

10 km
5 miles

Google / Google Earth Engine, USGS, NASA, ESA

The retreat of the world's glaciers is a wake-up call. In Alaska, glaciers such as Mount Hunter,[23] Mendenhall[24] and Columbia[25] are among the fastest retreating ice fields in the world, losing up to twelve miles since the 1980s. When we climbed Kilimanjaro, Africa's tallest mountain, two decades ago, there were ice fields over 300 feet high. Yet today, 80 per cent of the ancient peak's equatorial ice has all but disappeared. Another place where glaciologists are taking the pulse of climate change is in the land of the midnight sun. Surrounded by huge ice sheets, Kangerlussuaq, in Greenland, may be the most important expanse of ice you've never heard of. Seen from space, the

Kangerlussuaq ice sheet melt – 1984 and 2019.

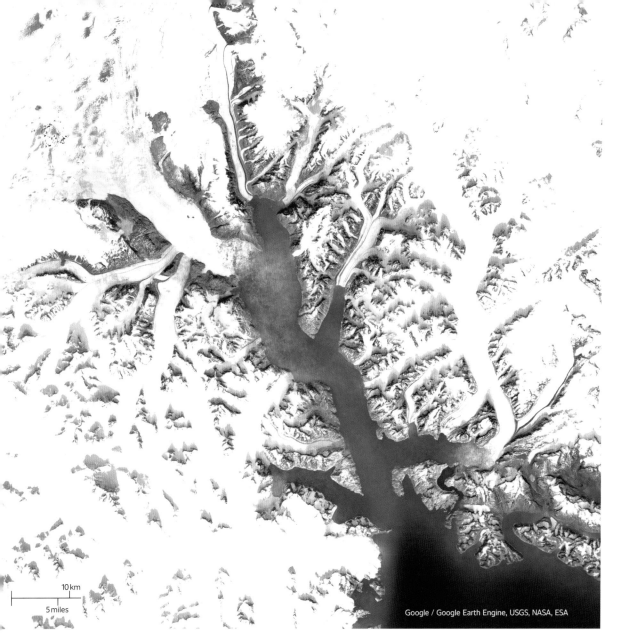

10 km
5 miles

Google / Google Earth Engine, USGS, NASA, ESA

white snow and ice surrounding Kangerlussuaq (which means 'big fjord' in Greenlandic) are turning to grey and blue.

As you can see in this map, there is nothing green about Greenland. Legend has it that mischievous Norse settlers arrived in AD 981 and gave the world's biggest island its name as a marketing ploy, to attract unsuspecting new traders and tourists. Then as now the island is actually covered by snow and ice fields that are more than 1,500 miles long, 450 miles wide and, on average, about 1.2 miles deep. These ancient ice sheets account for almost a tenth of the planet's freshwater supply. If they all melted at once, global sea levels would rise by more than twenty feet.[26] Satellite images of Kangerlussuaq show that it has retreated further than at any time since observations were first recorded in the 1930s.

The latest scientific findings emerging from Greenland are chilling. For the first time in thousands of years, the island's melting seasons are becoming longer and more intense.[27] A study in *Nature* determined that the ice sheets are melting at twice the speed of pre-industrial rates.[28] Rain is also becoming more frequent and hastening the melting of Greenland's ice.[29] When glaciers melt, large pieces can break away, or calve, generating icebergs as high as 3,000 feet.[30] But this is only part of the problem. When permafrost thaws it releases methane from the organic matter stored below.[31] This is deeply worrying because methane is thirty times more potent than carbon dioxide.[32] And the news gets worse. When ice melts, water flows into the world's oceans. The disruptive effects this has on ocean currents and sea levels are the stuff of nightmares.

The melting of Greenland's ice sheets is just the tip of the iceberg. Even a relatively modest melt can have far-reaching effects. As we can see on the map, when white ice on the Kangerlussuaq glacier thaws, it exposes blue seawater below. Sunlight that would ordinarily be reflected back into space is instead absorbed by the water. This heats up the ocean and speeds up the melting of the ice in a vicious loop. Most of the ice caps you see south of the Kangerlussuaq fjord are likely to be reduced to water before the end of the century.[33] In less than two decades, Greenland's ice sheets have already shrunk at a rate of 269 gigatons a year. That's the equivalent of 50 billion large elephants.

1900

Global temperature anomalies – 1900 and 2018

2018

Rising global temperatures – the hottest it's been in a billion years

Ice is receding from the Himalayas to Greenland because global temperatures are rising. These trends persist in 2020, despite the worldwide lockdowns. New remote sensing technologies can help us to map global surface temperature anomalies stretching back over the past century. This map signals global hot spots. The last ten years were the hottest on record according to the National Oceanic and Atmospheric Administration's 139-year registry.[34] Chart-topping temperatures were registered just about everywhere – from East and

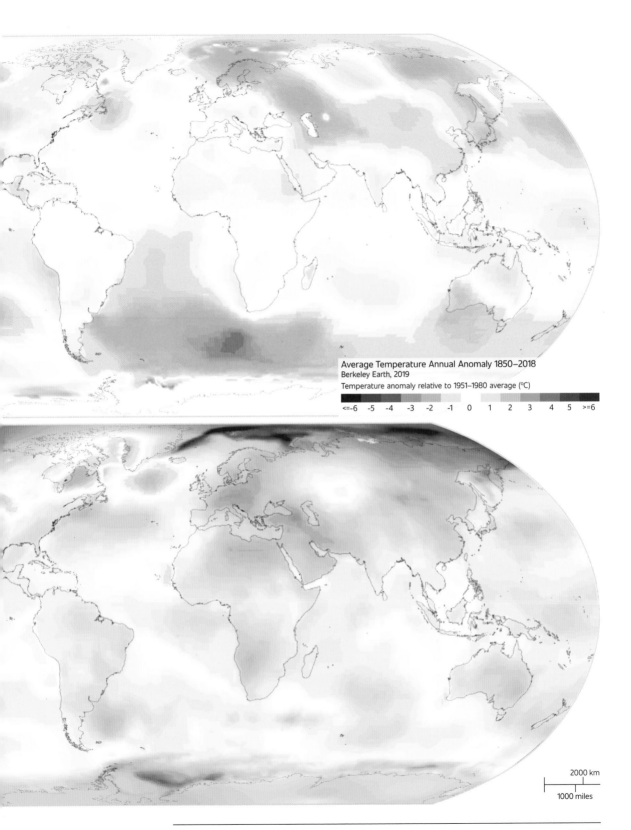

Average Temperature Annual Anomaly 1850–2018
Berkeley Earth, 2019
Temperature anomaly relative to 1951–1980 average (°C)

<=-6 -5 -4 -3 -2 -1 0 1 2 3 4 5 >=6

2000 km
1000 miles

South Asia, the Middle East, North and Sub-Saharan Africa to Western and Eastern Europe, Australia and New Zealand and the surrounding oceans. They gave rise to some of the longest and most severe heat-waves ever recorded. High temperatures can lead to heat exhaustion, heat stroke and ultimately organ failure.

While planetary temperatures have fluctuated over billions of years, the differences today are fourfold: temperatures are hotter than ever, changes are speeding up, they are lasting longer, and humans are mostly to blame. Already, about a third of the world's population is exposed to deadly heat for twenty days or more each year. In India, the world's most populous country, 65 per cent of the population was exposed to heatwaves in 2019. According to research featured in *Nature Climate Change*, if greenhouse gas emissions are not dramatically reduced, extreme heatwaves (defined as two consecutive days of abnormally hot temperatures) could threaten three-quarters of the global population by 2100. Put bluntly, the death toll from extreme heat could rise by more than 2,000 per cent within this century. Hotter days are especially dangerous for the elderly, people with underlying illnesses, and individuals with limited access to cooling. This is why it's not necessarily just high temperatures that are worrying, but how far they deviate from what's considered normal.

Global warming is accelerating because of spiralling greenhouse gas emissions. More carbon dioxide, methane, nitrogen oxide and other noxious gasses were emitted over the past thirty years than in the previous 150.[35] This is because we are burning more fossil fuels for energy, cutting down more forests for food, increasing meat consumption, and using more fertilisers and fluorinated gases than ever. As the map on the previous page shows, global warming is not just heating up land, it is also heating up the oceans. The latest scientific evidence suggests that the world's seas are dramatically warming and becoming less oxygen-rich, threatening marine ecosystems and the coastal communities that depend on them. We have witnessed at first hand the way rising temperatures are hastening coral bleaching and the destruction of marine life from Hawaii and Cozumel to the Maldives and the Great Barrier Reef, where half of all coral has already died.[36] The runaway burning of fossil fuels is also hastening extreme weather events, including unprecedented rainfall and prolonged dry spells,[37] all of which are set to multiply.[38]

The burning of fossil fuels – coal, oil and gas – makes up the largest

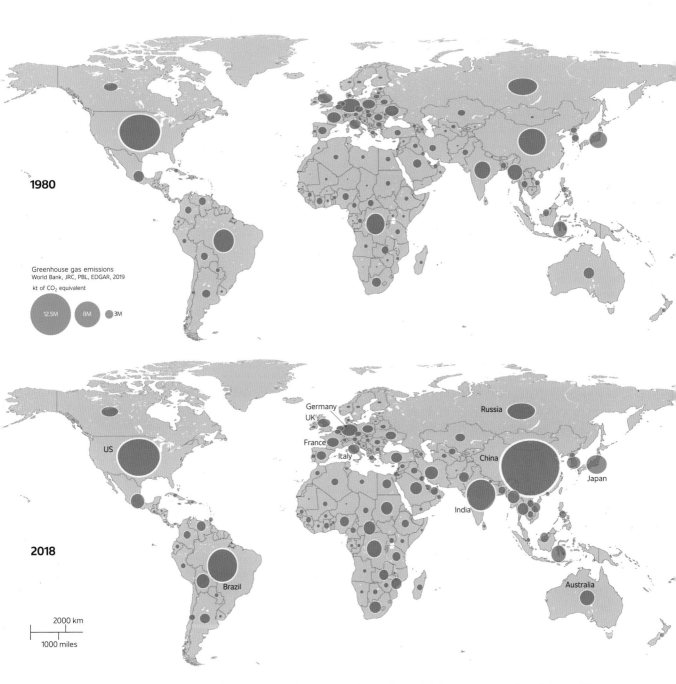

1980

Greenhouse gas emissions
World Bank, JRC, PBL, EDGAR, 2019

kt of CO_2 equivalent

12.5M   8M   3M

Germany
UK
France
Italy
Russia
China
India
Japan
US
Brazil
Australia

2018

2000 km
1000 miles

National contributions to greenhouse gas emissions. [39]

share of emissions because of our insatiable appetites for things like energy, transportation, food, and cement.[40] But it turns out that a handful of countries and companies are responsible for the lion's share of emissions. As the maps show, the US and Europe were the biggest culprits, releasing hundreds of billions of tons of carbon dioxide a year.[41] Today, after decades of staggering economic growth, China releases more greenhouse gases than the US, Western Europe and Russia combined. Alongside these big polluting countries are others like India, Japan, South Korea, Iran, Saudi Arabia, Canada, Indonesia, Mexico, Brazil, South Africa, Turkey and Australia. So, who sources these fossil fuels? The truth is that just 100 companies generate more than 70 per cent of all global emissions.[42] Collectively, these super-emitters produced 635 billion tons of carbon emissions between 1988 and 2016 – the equivalent of 1.8 million Empire State Buildings. Not surprisingly, few of them are inclined to release much information about the extent of their carbon footprints.[43]

Despite the temporary fall in carbon dioxide and nitrogen dioxide levels during the COVID-19 pandemic, global efforts to curb industrial emissions further are moving far too slowly. Since 2000, just twenty countries have successfully reduced their carbon foot-prints, including Denmark, France, Ireland, the UK, Ukraine and the US. While exceedingly high, Chinese emissions seemed to be levelling off after public authorities set a nationwide trading scheme, imposed caps on coal and started transitioning to renewable energy.[44] After the US pulled out of the 2015 Paris climate agreement, China went from international pariah to global leader in calling for mitigation and adaptation. The latter's 'green leadership' may be short-lived since emissions started rising again.[45] Despite announcements by major fund managers like BlackRock (with assets of roughly $7 trillion) to divest from fossil fuels by 2030, too few companies are investing in cleaner energy and greener supply chains. The truth is that we are nowhere near reversing course. Before the COVID-19 pandemic struck, global carbon emissions had surged to record levels, driven by demand for fossil fuels in China, India and Europe.[46] If we are going to turn this dire situation around we need systemic change to occur far faster than at the current pace.

The most energetic calls for greener standards and reductions in emissions are coming not from diplomats or CEOs, but progressive politicians and civic activists. For years, leaders from Pacific Island countries like Kiribati, the Solomon Islands and Vanuatu have pressed major polluters to reduce their emissions. Now they are threatening to sue fossil fuel companies to help meet the costs of their daily battle against rising seas.[47] In the US, some states are taking legal action against massive companies like Exxon, accusing oil giants of knowingly obfuscating their climate change impacts.[48] Big cities like New York[49] and San Francisco[50] are likewise seeking compensation from oil and gas companies for climate change damages. And students are suing the US federal government in a landmark case to stop all new leases for fossil fuel production. Another stunning breakthrough came in 2015, when hundreds of Dutch citizens won a lawsuit against their national government, forcing it to cut greenhouse gas emissions by at least 25 per cent before 2021 (compared to 1990 levels).

If there is any good news it is that public opinion is reaching a tipping point. The Paris Agreement and the 2016 Sustainable Development Goals (especially the goals focused on climate action, life below water and life on land) were key milestones. Climate scientists have also taken off the gloves, and reports issued by the Intergovernmental Panel on Climate Change – a platform created to equip policymakers with regular scientific assessments on climate change – have become more and more urgent. But arguably the most impactful shifts are being precipitated from an unlikely source: children. The most impressive example of this is the 'Fridays for Future' protests initiated by the teenage Swedish climate activist Greta Thunberg, that have encouraged tens of millions of young people to march and demand change. Extinction Rebellion is another political movement launched in more than sixty countries and counting. XR, as it is called, deploys civil disobedience and non-violent resistance to urge governments to acknowledge the 'climate emergency' and take action to prevent ecological collapse. The millennials and Generation Z are just getting started, and not a moment too soon.[51]

World on fire – how San Francisco became the world's most polluted city

Greenhouse gases are released not just by industrial emissions, but also by burning forests and grasslands. A whopping 32 billion tons of carbon dioxide were released by forest fires in 2018 compared to 37 billion tons emitted by fossil fuel companies that same year. The figures for 2019 and 2020 will be even higher. Historically, most fires were naturally occurring: they are a critical feature of the ecosystem since they consume dead vegetation, clear space for new fauna, and reduce the density of plant life. Most of the time, fires are moderated by rain and snow. But the drier the grass and shrubs, the easier they burn. During our travels in the Sahelian countries of Mali, Niger

French Fire
(August 2018)

Camp Fire
(November 2018)

Mendocino Complex Fire
(August 2018)

Carson City

Sacramento

NEVADA

Stockton

South Fork Fire
(August 2018)

San Francisco

CALIFORNIA

Woolsey and Hill Fires
(November 2018)

100 km

Los Angeles

50 miles

NASA-FIRMS, 2018

Concow

Paradise

Parkhill

Big Bend

Butte Valley

2 km

1 mile

Landsat 8, NASA Earth Observatory, 2018

Above: **The Camp Fire, 8 November 2018**
This was the deadliest and most destructive wildfire in Californian history and the sixth-deadliest US wildfire overall.

Fires in California:
All fires in California during 2018.

and Nigeria we saw how dramatically warming temperatures are not only drying out vegetation and reducing forest cover, through extensive slash and burn agriculture, but also exacerbating violent intercommunal tensions between farmers and herders. There, and in other hot spots, arable land is diminishing, dry seasons are becoming longer and wet seasons are shortening.

Today, humans are the chief cause of large-scale wildfires around the world. And many of their fires can be seen from space, appearing in Landsat satellite images prepared by NASA and the US Geological Survey. Forest fires are now a pressing concern, including for big cities. As we can start to see on the map, the US is being roasted by as many as 100,000 forest fires a year. About 80 per cent of the country's 1.5 million forest fires since the 1990s were human-induced.[52] Until recently, fires burned up to 5 million acres a year. In 2018, a record-breaking 9 million acres were reduced to ash.[53] The map reveals how California is one of the worst affected states. In 2018 it tallied up 9,000 separate wildfires that charred 1.9 million acres and shrouded cities and towns in hazardous smoke. We can see that the largest of them – including the Mendocino Complex, the Camp, and the Woolsey and the Hill Fires that collectively covered 500,000 acres – were all visible from space.[54] These fires produced more air pollution in two days than all the vehicles in the state expel in an entire year.[55] For a few weeks in 2018, three Californian cities – San Francisco, Stockton and Sacramento – were considered the most polluted metropolises on the planet.[56]

Gas flares – flaring in North Dakota is the equivalent of 1 million cars on the road

As in the case of forest fires, it is possible to track the tell-tale plumes from coal-fired power stations[57] and the flares generated by hydraulic fracturation, or fracking. As we can see in the map, the US has a lot of gas flares – more than any other country.[58] A closer inspection reveals

Right: **The Bakken Formation in North Dakota, USA**
Gas flares from hydraulic fracturing or fracking.

Gas flares over North Dakota, 2019.

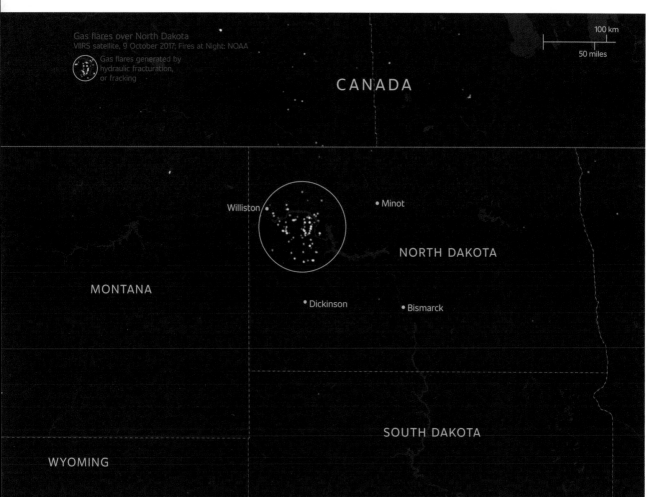

Gas flares over North Dakota
VIIRS satellite, 9 October 2017; Fires at Night: NOAA
Gas flares generated by hydraulic fracturation, or fracking

100 km
50 miles

CANADA

Williston • Minot

NORTH DAKOTA

MONTANA

• Dickinson • Bismarck

SOUTH DAKOTA

WYOMING

Williston

Missouri River

Watford City

Gas flares over North Dakota
Google / Google Earth Engine, USGS, NASA, ESA.
Fires at night: NOAA, VIIRS, 2017

Shale gas generated by hydraulic fracturation,
or fracking

5 km

3 miles

that North Dakota is a fracking hot spot: oil production there reached a record 1.5 million barrels a day by the end of 2019. Owing to the lack of interstate pipelines to bring natural gas to market, producers burn excess oil instead. Oil companies there flared as much as 2.5 billion cubic feet a day – twice the state target – in 2019, an all-time high.[59] This wasted gas could have powered all North and South Dakota's energy needs. Not only are these gas flares a waste (representing over a

Deaths from air pollution vs lethal violence

Air pollution kills several times more people than wars, terrorism, murders and suicide combined.

Death rates from intentional homicides and air pollution – indoor and outdoor
Institute for Health Metrics Evaluation (IHME), 2016

Deaths per 100,000

Self harm and interpersonal violence

245   157   61

Air pollution

245   157   61

$1 billion in needless losses a year) but they are dirty, pumping more than 5 million metric tonnes of carbon dioxide into the atmosphere. This is the equivalent of adding a million cars to the road[60] or heating 4.25 million homes for a year.

Vehicle fumes, gas flares and forest fires are not just bad for the climate, they are bad for public health. Smoke contains tiny particulates that penetrate deep into the lungs, causing inflammation, asthma, respiratory illness and cancer.[61] Ambient and indoor pollution from burning fossil fuels led to over 7 million premature deaths in 2018, making it one of the world's top killers.[62] As the map makes clear, air pollution kills several times more people than wars, terrorism, murders and suicide combined. It is especially dangerous for people

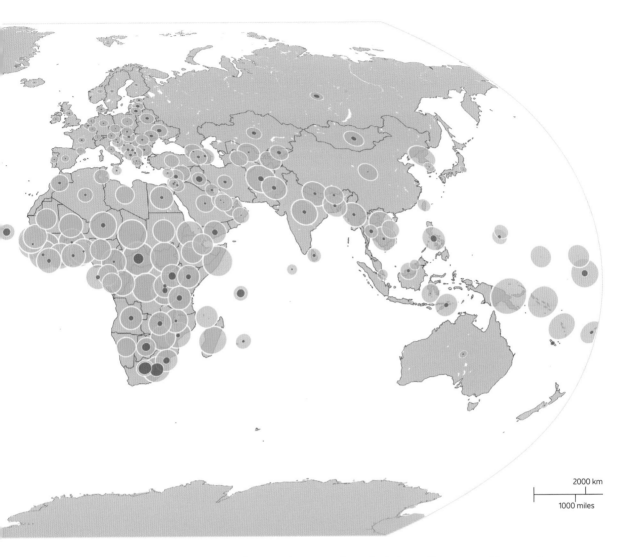

2000 km

1000 miles

over sixty-five, whose risk of dying from cardiovascular causes increases dramatically when they are exposed to smoke, smog and dust.[63] In California, and around the world, governments, businesses and citizens need to fundamentally rethink land use and housing codes and introduce smarter forestry practices.[64] If they do not, they'll literally get burned.

Forest fires have deep roots. For tens of thousands of years, farmers and herders have set fire to the remains of their fields to eliminate grassland and scrub and return nutrients back to the earth. Whether in the African savannah, rural Asia or the forests and plains of South America, people also burn forests to create charcoal, which in the absence of electricity or other fuels is vital for cooking and heat. As you can see from this map, hundreds of thousands of small fires are burning wherever there is an abundance of rainforest and grassland.[65] When viewed from above, Africa's slash and burn fires account for the largest proportion of areas burned globally. Asia is not far behind.

Pollution from global fires
Google / Google Earth Engine, USGS, NASA, ESA.
Fires at night: NOAA, VIIRS, 2019

Thousands of small fires causing global pollution

1000 km

1000 miles

Smoke released by brush fires, cooking stoves and smouldering waste not only generates carbon dioxide and other toxic fumes, but also black carbon, which suppresses rainfall, including in some of the biggest carbon sinks on earth.[66]

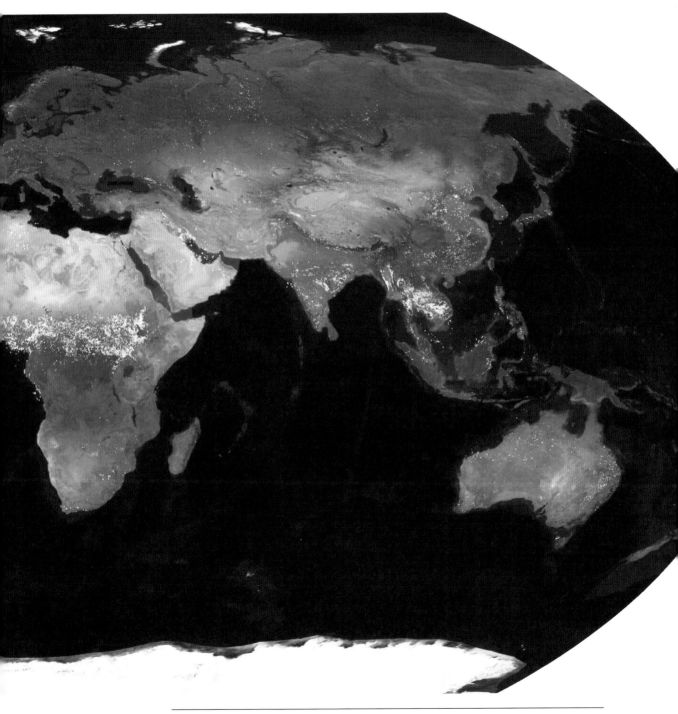

Amazon's flying rivers

The Amazon basin is centre stage in the debate over the causes of and solutions to global warming. It is called the 'planet's lungs' for good reason. Covering about 2.7 million square miles, it accounts for more than 40 per cent of the world's entire stock of tropical forests and regulates rainfall, cloud cover and ocean currents. Over the past few hundred million years it has converted trillions of tons of naturally emitted carbon dioxide into oxygen. But it is turning from a carbon sink into a carbon source.[67] In recent years, these forests have come under increasing threat from fires, relentless deforestation and degradation. Many of these fires were deliberately set by ranchers and farmers. Since the 1980s, hundreds of thousands of square miles have been cleared to make way for pasture, soya beans, sugar, timber and minerals.[68] The deforestation of an area bigger than France translates into billions of tons of carbon dioxide.[69] And since Amazonian rivers

Smoke plumes over the Amazon, 2019
Satellite images taken from Sentinel-2 L1C on 10 August 2019.

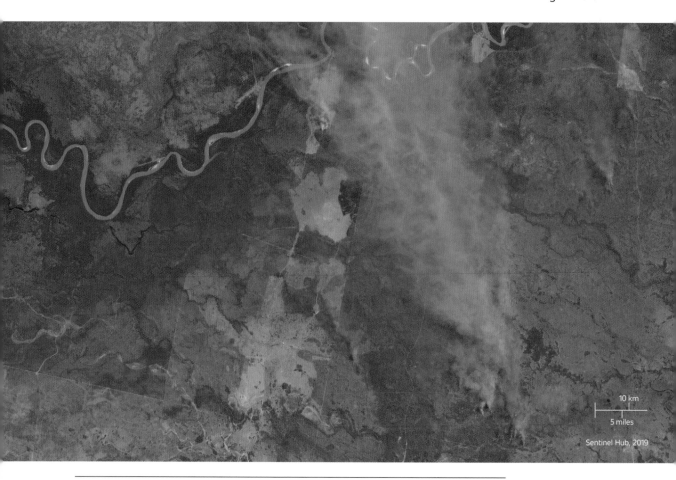

10 km

5 miles

Sentinel Hub, 2019

also emit carbon back into the atmosphere, this means the ability of the world's largest forest to clean the air is declining.[70]

We witnessed first-hand the increase in fires due to slash and burn agriculture and droughts in forests and farms across Brazil, Bolivia, Colombia and Peru. These satellite images only begin to highlight the 80,000-plus forest fires in Brazil during 2019. The more trees that burn, the less carbon is removed from the atmosphere through photosynthesis. And as droughts intensify, the thinner the protective canopy becomes and less moisture is retained. Drier forests also become more prone to fire, and so the fires spread further and faster. And if this isn't bad enough, the warming of the Atlantic and Pacific Oceans is exacerbating dry spells.[71] Scientists are concerned that the Amazon is perilously close to an inflection point, creating conditions so hot and dry that local species will not regenerate. If between 20 and 25 per cent of the tree cover is deforested, the basin's capacity to absorb carbon dioxide might collapse. This could trigger a 'die-back'

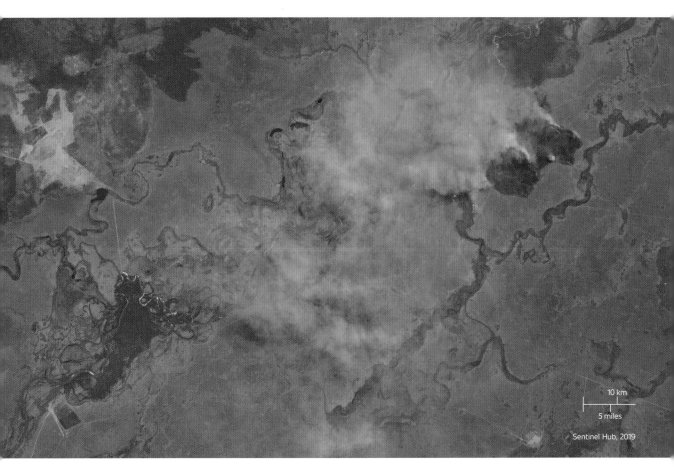

10 km
5 miles
Sentinel Hub, 2019

scenario in which the world's largest tropical forest will become its biggest patch of scrubland. Die-back would not only lead to rapid deterioration of biodiversity, it would profoundly upset the process of evapotranspiration, disrupting cloud cover and the circulation of ocean currents around the world.

The Brazilian state of Rondônia, in the upper Amazon basin, is one of the most intensively deforested places on Earth. It is a huge expanse of land, five times the size of Switzerland.[72] As we can begin to see on

Deforestation and degradation in Rondônia – 1985 and 2019.

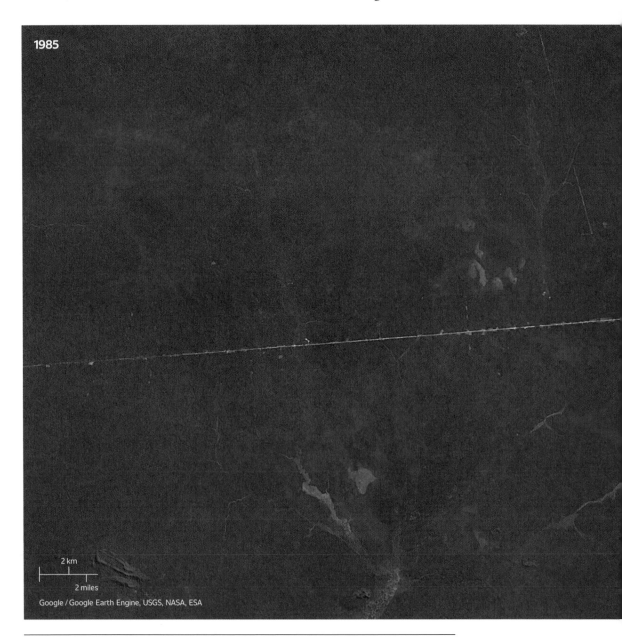

1985

2 km

2 miles

Google / Google Earth Engine, USGS, NASA, ESA

these maps, between 1985 and 2019 roughly a third of its area was converted to pasture and crop-land.[73] The map reveals how deforestation typically starts along roads and then fans out to create a fishbone pattern. Over the past three decades extensive expanses of virgin tropical forests were intensively farmed and poisoned with chemicals to extract precious metals such as gold and coltan. Left-over mercury and nitrogen deposits from fertilisers and livestock effluent are also diminishing biodiversity, polluting rivers and infiltrating the food chain.[74] Today,

2019

2 km

2 miles

Google / Google Earth Engine, USGS, NASA, ESA

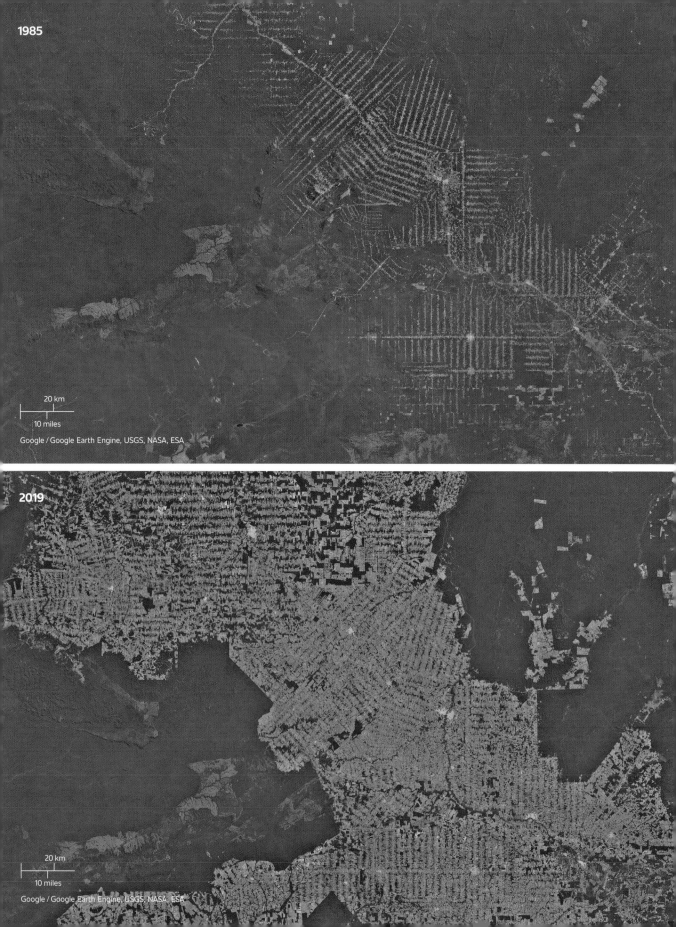

1985

20 km
10 miles

Google / Google Earth Engine, USGS, NASA, ESA

2019

20 km
10 miles

Google / Google Earth Engine, USGS, NASA, ESA

Brazil is losing the equivalent of two football pitches' worth of forest every minute.[75] Deforestation rates sped-up even more dramatically in the wake of the COVID-19 pandemic. While satellite images suggest that parts of the Amazon, Africa and Southeast Asia might be 'greening',[76] this is due to the expansion of monocrops that are steadily replacing the world's rainforests.

One impact of deforestation that has received less attention is rain-fall depletion. Think of forests as fountains, sucking water from the ground and releasing vapour into the atmosphere. A single tree can transpire hundreds of litres of water a day. When billions of trees liberate water molecules, they collectively create enormous aerial 'rivers'. These flying rivers eventually form clouds and generate rain-fall, hundreds and sometimes thousands of miles away.[77] But when these are depleted, precipitation is reduced These problems are amplified because the surface of the earth, depleted of trees, also heats up. In some parts of the Amazon, researchers have detected as much as a three degrees Celsius difference between forested areas and surrounding pastureland.[78] When transpiration falls, then water shortages and drought become more common.[79] This is not a problem 'out there'. Over the past decade, almost 900 Brazilian cities and towns faced severe water shortages – this in a country with 20 per cent of the global fresh water supply.[80]

Deforestation and degradation in Rondônia – 1985 and 2019.

Rising seas and sinking cities

The consequences of climate change are not some remote or distant possibility – they've already arrived, especially in coastal cities.[81] Even in the unlikely event that global temperatures rises are somehow kept from surpassing the two degrees Celsius by 2050, at least 600 cities[82] with more than 800 million residents[83] between them will be ravaged by rising seas and storm surges,[84] water salinisation and an unfathomable financial burden.[85] Yet most of the world's fastest growing coastal cities are not even remotely prepared to deal with rising sea levels. Despite clear and present dangers, many vulnerable coastal cities across North America and Western Europe are still failing to put in place essential mitigation and adaptation measures. All but a handful of cities in Latin America, Africa and Asia have yet to craft basic plans to prepare for the effects of climate change.

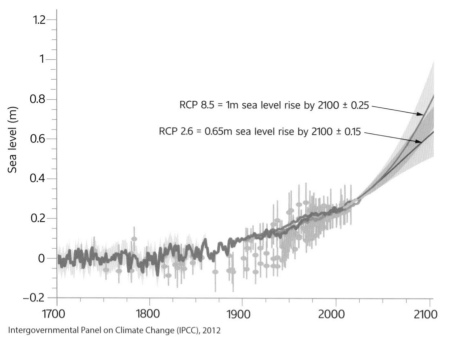

Intergovernmental Panel on Climate Change (IPCC), 2012

Past and projected global sea level rise. The Representative Concentration Pathway (RCP) is a greenhouse gas concentration trajectory employed by the IPCC. The pathways describe different climate futures, all of which are considered possible depending on the volume of greenhouse gases emitted in the years to come. There are four RCPs: 2.6, 4.5, 6 and 8.5. [86]

Sprawling megacities such as Lagos, Shanghai, Miami and Mumbai are facing sea-level rises well above the global average.[87] And because many coastal cities are built on swamps and flood-prone land, they are sinking at the same time as sea-water is seeping in. This is not only because they are heavy (which they are), but because their residents

Projected sea level rise in Jakarta.

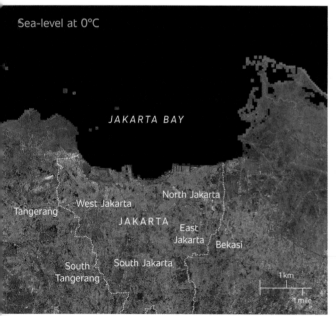
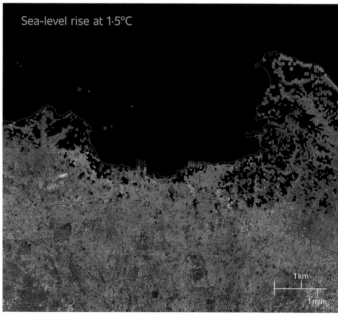

Climate Central. Basemap: Google / Google Earth Engine, USGS, NASA, ESA

are extracting vast quantities of groundwater, sometimes illegally. Consider Jakarta, the current capital of Indonesia and home to almost 10 million people, where some neighbourhoods have sunk by more than eight feet over the past decade.[88] When global temperatures rise by 1.5 degrees Celsius or more, entire neighbourhoods are likely to be swallowed up.

Jakarta is a victim of geography. It is hemmed in by the Java Sea to the north, built on soggy land and is criss-crossed by more than a dozen rivers, many of them intensely polluted. Despite an abundance of surface water, local authorities are struggling to meet even 40 per cent of the city's demands. The metropolitan government's inability to enforce regulations preventing ground-water extraction means that illegal tapping by local vendors and residents is out of control. And despite routine flooding, luxury apartments are still popping up.[89] But make no mistake, the building bonanza will soon go bust: as much as 95 per cent of the city is expected to be underwater by 2050. In a last-ditch attempt to save the city, the government announced that it would build a $40 billion sea wall with Dutch and South Korean support.[90] Yet there are also signs the government is giving up: more than 400,000 people have already been relocated away from overflowing riverbanks and reservoirs,[91] and the nation's capital is moving to the island of Borneo starting in 2024.

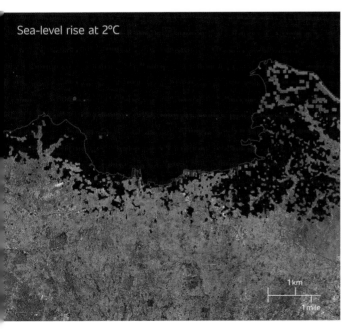

Sea-level rise at 2°C

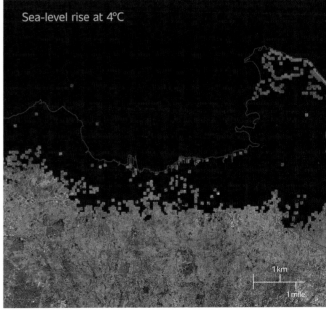

Sea-level rise at 4°C

Jakarta is hardly the only Asian city facing extinction. The number of people living in at-risk flood plains in Asia is expected to double by 2060. In fact, about four of every five people impacted by sea-level rise by 2050 will live in Asia.[92] So-called delta cities will be the first to be submerged.[93] Today, hundreds of millions of people live in or around fifty delta cities such as Dhaka, Guangzhou, Ho Chi Minh City, Hong Kong, Manila and Tokyo. These cities were once ideal places to settle because of their access to the sea and fertile farmland. These character-istics help explain why the Brahmaputra, Ganges, Indus, Nile and Yangtze served as cradles of many of the world's great civilisations. But these ancient virtues are now modern liabilities. Coastal living is becoming dangerous: the costs of sea-level rise will increase to the tens of trillions of dollars by 2100.[94] Not surprisingly, insurers are already pricing in a 'new normal' of higher frequency and more extreme events.[95]

Projected sea level rise in Miami.

Climate Central. Basemap: Google / Google Earth Engine, USGS, NASA, ESA

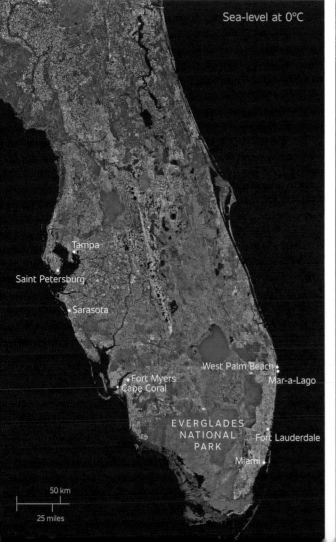
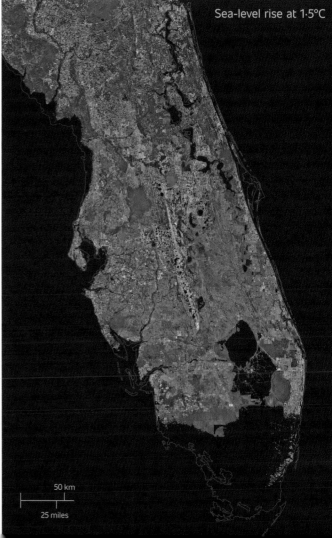

Cities across the Americas, especially those on the North American eastern seaboard and Gulf coasts, are on the front line of sea-level rise. Their vulnerability is due in large part to melting ice in Greenland and the weakening of Atlantic Ocean currents. In the US, more than 90 coastal cities are experiencing chronic flooding – a number that will double by 2030.[96] New York is one of the most at-risk owing to its location, but also its population size.[97] Over the coming decades, twenty-two of the twenty-five cities most vulnerable to coastal flooding in the US will be in Florida. Faced with between ten and thirty feet of sea-level rise by the end of the century, Miami is the poster child for cities on the edge. It was not planned with sea-level rise in mind, which means its real estate and road infrastructure are highly exposed. Making matters worse, southern Florida's underlying geology is like Swiss cheese. It is made up of compressed reef and limestone and is predisposed to heavy flooding.[98] As if that were not enough, hurricanes

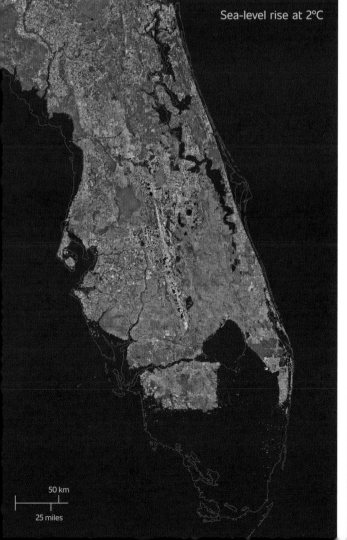

Sea-level rise at 2°C

50 km

25 miles

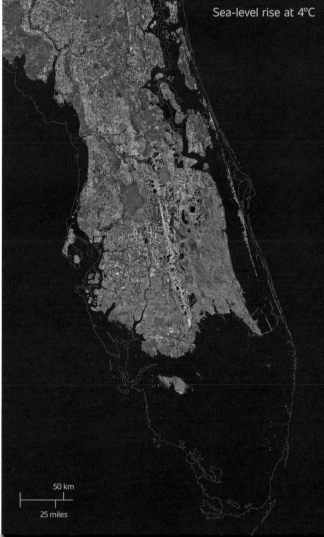

Sea-level rise at 4°C

50 km

25 miles

and extreme weather events are growing in frequency and ferocity, which means more rainfall and storm surges.

Whether they like it or not, Miamians face a floating future. City officials are busily installing engineering fixes like sea walls, water treatment plants, and pumps to force water back into the bay. They are also seeking to make their city more resilient to climate stresses by revamping building codes, raising roads, restoring wetland areas and issuing municipal green bonds to pay for mitigation efforts. As in Jakarta, some neighbourhoods are being surrendered to rising seas, and vacant properties are being converted into water-friendly parks and water retention sinks to absorb overflow.[99] Governments, businesses and civic groups around the world have good reason to plot cities with climate change in mind. Roughly three-quarters of all European urban centres will be hit badly by rising seas, with Dutch, Spanish and Italian cities especially exposed.[100] Rapidly urbanised and crowded coastal cities in Africa and Asia are also in trouble, especially their most vulnerable residents, many of whom are often densely crowded into low-income informal housing at the water's edge.[101]

Faced with prospects of annihilation, a growing number of city councils, business groups and civic groups are busily making their cities climate-proof.[102] For example, within days of the US withdrawing from the 2015 Paris climate agreement, more than 1,000 cities signed a *Mayors Climate Protection Agreement* to meet – and exceed – global targets.[103] Even cities like Austin, Dallas and Fort Worth in the oil state of Texas are striving to be carbon neutral within the next decade.[104] Already far ahead of their US counterparts, European cities are also actively adopting mitigation and adaptation strategies.[105] Take the case of Copenhagen, in Denmark, which should be carbon neutral by 2025.[106] Oslo, Europe's greenest city, is also rushing to reduce 95 per cent of its carbon emissions before 2030 and is well on track. Meanwhile, powerful global city coalitions like the C40 Cities and Carbon Neutral Cities Alliance are helping cities kick their addiction to fossil fuels before 2050.[107] As the map shows, a Global Covenant of Mayors for Climate and Energy has already rallied more than 10,000 cities – home to over 900 million people – to the cause.[108]

At the vanguard are Dutch coastal cities, which are rolling out an array of solutions to manage sea-level rises. They have good reason to be proactive since close to a third of the country is already below sea level (its lowest point is about twenty-two feet below sea level). To

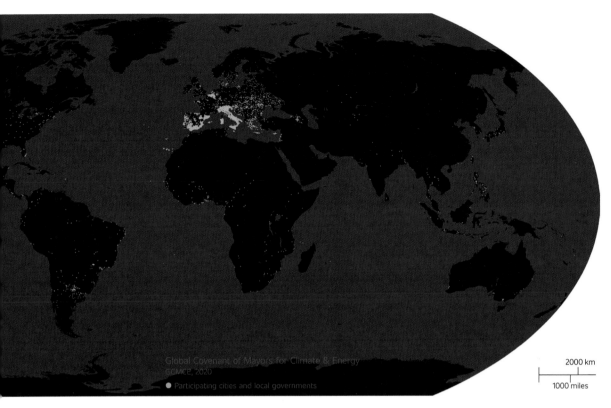

Global Covenant of Mayors for Climate & Energy
GCMCE, 2020
● Participating cities and local governments

2000 km
1000 miles

City signatories of the Global Covenant of Mayors for Climate and Energy, 2020
As of 2020, more than 10,000 cities and local governments from 138 countries have signed up to the Global Covenant of Mayors for Climate and Energy.[109]

accelerate municipal efforts, the national government has decentralised many aspects of water management, delegating cities to take charge. For example, flood protection is the responsibility of regional water management boards and not the central authorities. This is not to suggest that the national government is passing the buck: it has funded hard defences such as a 2,300-mile network of dykes, dams and sea walls, as well as the impressive Maeslant Barrier. Built to protect Rotterdam (which is itself 90 per cent below sea level) from water surges and floods, the barrier is the size of two Eiffel Towers flat on their sides.[110]

Cities like Rotterdam offer an inspiring model to others for how to manage sea-level rise. It is one of the safest delta cities in the world precisely because it has learned to live with water. This mindset can be traced back as far as the thirteenth century, when local merchants and city administrators erected a quarter-mile-long dyke to keep high water from seeping in, but also to facilitate drainage. New canals were built in the 1850s to improve water quality and reduce routine outbreaks of cholera. The Maeslant Barrier was constructed in 1997, after catastrophic floods killed more than 1,800 people in the early

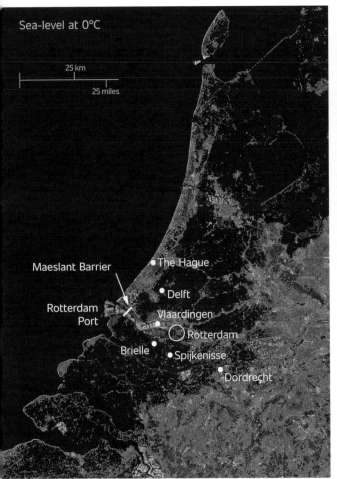

Sea-level at 0°C

25 km
25 miles

Maeslant Barrier
Rotterdam Port
Brielle

The Hague
Delft
Vlaardingen
Rotterdam
Spijkenisse
Dordrecht

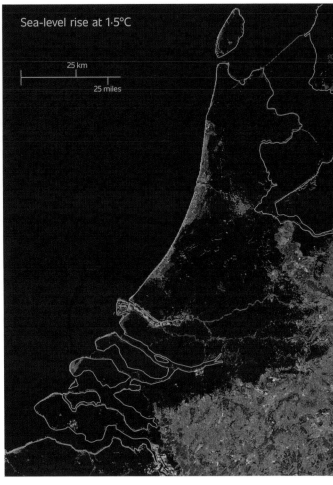

Sea-level rise at 1·5°C

25 km
25 miles

Climate Central. Basemap: Google / Google Earth Engine, USGS, NASA, ESA

1950s. Today these titanic gates protect the city's 1.5 million people from floods with no impediments to sea traffic.

Projected sea-level rise in Rotterdam.

But the critical ingredient in Rotterdam's success isn't engineering, it is attitude. The city's long-serving mayor, Ahmed Aboutaleb, has claimed that local residents 'do not view climate change as a threat, but rather as an opportunity to make the city more resilient, more attractive and economically stronger'. In his view, climate adaptation is a chance to upgrade infrastructure, increase biodiversity and more meaningfully engage citizens in everyday city life. To this end, local authorities adopted a climate change adaptation strategy to 'climate-proof' the city by 2025.[111] Since then it has started converting ponds, garages, parks and plazas into part-time reservoirs. The city government is also working with community groups to revitalise neighbourhoods, reduce inequality and build resilience to future shocks.

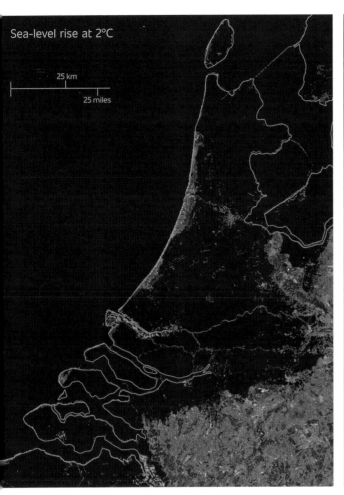

Sea-level rise at 2°C

25 km
25 miles

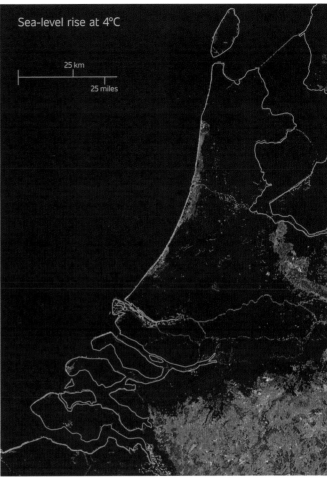

Sea-level rise at 4°C

25 km
25 miles

Few places are more at risk than island nations like Kiribati, the Marshall Islands, Tuvalu or the Maldives. Climate change will engulf these nations. Besieged by rising waters, Kiribati is negotiating to buy 5,000 acres of land in neighbouring Fiji on which to move its 113,000 citizens when necessary.[112] A government website concedes that national survival is unlikely,[113] making it the first nation state likely to cease to exist as a result of climate change.[114] Residents of the Marshall Islands face a similarly bleak choice: leave the country or climb higher. Public authorities there are desperately looking for ways to reclaim land and build islands that are high enough to withstand rising seas.[115] The Maldives is also attempting to reclaim, fortify and build new islands, and relocate when necessary.[116] These countries are the canaries in the mine for what is to come.

Mapping the future of climate action

Climate change is already banging down the front door. As we can see, forest fires, gas flares and industrial pollution are contributing to soaring mortality and morbidity. An astonishing 200 species are disappearing every day and millions are threatened with extinction if temperatures continue rising. Every second, more than two acres of tropical forest are being destroyed or irreversibly degraded around the world.[117] This is contributing to droughts and floods, which are in turn connected to rising food prices, food insecurity, migration, and even communal violence[118] and the outbreak of armed conflict[119] in some of the world's most vulnerable settings.[120] There are signs that things are about to get a whole lot worse. But now is not the time for despondency and inaction.

Radical mitigation is needed, including a determined shift to a zero-carbon world. This can be achieved only by dramatically shifting

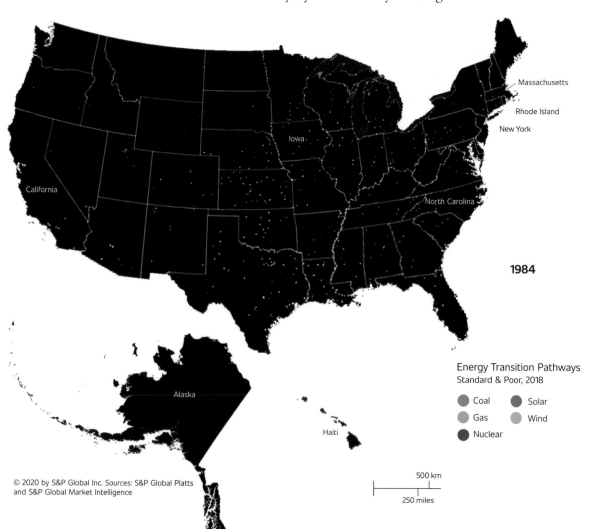

1984

Energy Transition Pathways
Standard & Poor, 2018

- Coal
- Gas
- Nuclear
- Solar
- Wind

© 2020 by S&P Global Inc. *Sources:* S&P Global Platts
and S&P Global Market Intelligence

500 km

250 miles

away from fossil fuels and towards renewable energy. It is that simple. As the maps on pages 88 and 89 show, the world's top polluter has taken steps over the past few decades to diversify its energy matrix and move away from coal and gas to nuclear, wind and solar power. It is entirely possible for all fifty US states to be 100 per cent renewable by 2050.[121] The real challenge is not whether this should be done, but rather if the political will can be mustered to speed up this transition in time. The proposal to craft a 'green new deal' in the US – a sweeping set of legislation to address climate change and economic inequality – is just the first step in a monumental transition. Subsidies for electrical vehicles is another. A global conversion to renewable energy is essential for our collective survival.

The reduction in our carbon footprint must be accompanied by a radical shift from a linear to a circular economy across all major supply chains. The circular economy is one that decouples economic activity from the consumption of finite resources, and designs waste out of the

US energy transition to cleaner energy – 1984 and 2016

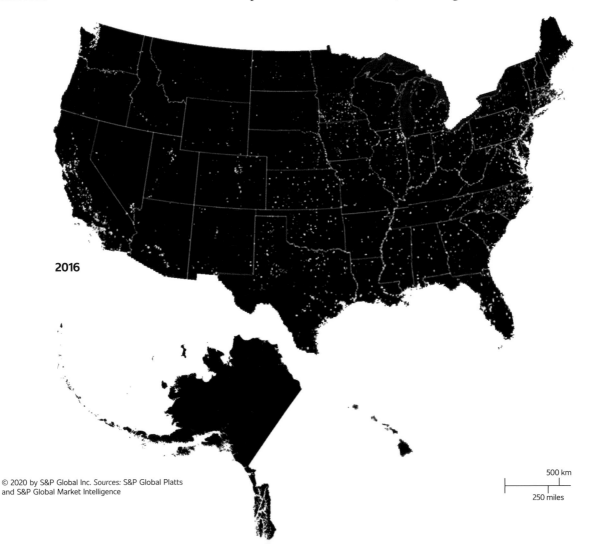

2016

500 km

250 miles

system. This will require a dramatic change in consumption habits, including ensuring that a much greater share of what we consume can be reused, repaired, recycled and converted back into raw materials. Some parts of the world are moving faster on the circular economy agenda than others. Countries like Canada, Denmark, Germany and Japan have already started incentivising companies and nudging citizens to make the leap. Even China, which introduced a *Circular Economy Promotion Law* in 2008, is requiring large factories involved in everything from cement to aluminium production to minimise all waste to close to zero.

Mitigating and adapting to climate change requires a fundamental change of mindset. It means rethinking the very idea of waste from the beginning to the end of a product's life cycle. One way to do this is through something called industrial symbiosis, a collaborative process across companies that can speed up circular economy solutions. This typically involves bringing firms together to find practical ways to convert one company's waste into another company's raw material. In Denmark, for example, governments and businesses have forged partnerships to test out new approaches, launch pilot projects and fund start-ups, from algae production facilities to bioethanol generation.[122] In China there are hundreds of eco-industrial parks putting industrial symbiosis into practice. There, fertiliser factories are using sugar by-products produced by nearby breweries, and paper and pulp plants are trading wood leftovers for green mud for buildings, white sludge for cement factories and hot water for aquaculture.[123]

A combination of carrots and sticks is required if we are to have any hope of meeting the Paris Agreement commitments. At the top of the list must be the immediate phasing out of subsidies and tax breaks to the more than 2,425 coal-fired power plants around the world.[125] Carbon taxes and offset schemes together with emissions reduction targets and scrubbing technologies are all part of the solution. So are targeted enforcement measures focused on the 100 top 'super-polluters' detailed in the figure over the page. In the US, fewer than 100 facilities – just 0.5 per cent of all industrial polluters – produce more than a third of the nation's toxic air pollution and a fifth of all greenhouse gas emissions.[126] We also need to kick our addiction to animal products: avoiding meat and dairy is one of the single biggest ways to reduce our carbon footprint.[127] By 2050, just thirty-five meat producers in Argentina, Brazil, Canada,

China, the EU and US could generate more emissions than all the big fossil fuel firms combined.[128] This will require more conscious consumption, including of meat substitutes. A planet-saving diet is a plant-based diet.

The best news of all is that we know what needs to be done. The largest report on climate change ever assembled says that the only way to keep global temperatures from rising above 1.5 degrees Celsius is to decrease greenhouse gas emissions by 45 per cent from 2010 levels before 2030.[129] This is a tall order. The reality is that despite the Paris Agreement, which committed all countries to keep temperatures from rising by more than two degrees Celsius, emissions are continuing to rise. It is telling that the World Economic Forum's 2020 survey of corporate leaders listed climate change as their number one concern for the first time. But can these preoccupations be turned into action? Further delays will only make a bad problem much worse. This is why it is essential that now, more than ever, we elect political leaders and support corporations who understand the gravity of the situation and are prepared to take the necessary measures to act on the emergency we face. The COVID-19 pandemic is just a warm-up to the kinds of disruption that are ahead if action is not taken on the climate front.

Climate mitigation and adaptation needs to be embraced by the mainstream, not the fringe. It should become part of the core educational curriculum of schools and universities as well as for companies and public agencies. And media outlets have a critical job to do, ending the false equivalency between climate science and climate denial. At the very least, politicians, business leaders and activists need to be focused on solutions and not quibbling over the evidence, which is irrefutable. The decision by some companies to divest from fossil fuels is a step in the right direction. And it is true that there are still a great many disagreements about how to proceed. But if we are to effectively mitigate and adapt, nations, corporations and citizens need to develop new ways to share the burden of climate change. This will require innovative financing models and bold partnerships and, above all, a new mindset primed for a hothouse world. The carnage wrought by COVID-19 might actually help create the foundations for a more resilient future.

The survival of the planet requires compromise and co-operation. Today's wealthiest countries have relied on fossil fuels to power their industries and progress. They now account for about three-quarters

Top 100 polluters around the world – cumulative emissions 1988 to 2015[124]

Producer	Location
China (Coal)	Beijing, China
Saudi Arabian Oil Company (Aramco)	Dhahran, Saudi Arabia
Gazprom OAO	Moscow, Russia
National Iranian Oil Co	Tehran, Iran
ExxonMobil Corp	Irving, TX, US
Coal India	Kolkata, India
Petroleos Mexicanos (Pemex)	Mexico City, Mexico
Russia (Coal)	Moscow, Russia
Royal Dutch Shell PLC	The Hague, Netherlands
China National Petroleum Corp (CNPC)	Beijing, China
BP PLC	London, United Kingdom
Chevron Corp	San Ramon, CA
Petroleos de Venezuela SA (PDVSA)	Caracas, Venezuela
Abu Dhabi National Oil Co	Abu Dhabi, United Arab Emirates
Poland Coal	Poland
Peabody Energy Corp	St. Louis, MO
Sonatrach SPA	Hydra, Algeria
Kuwait Petroleum Corp	Kuwait City, Kuwait
Total SA	Courbevoie, France
BHP Billiton Ltd	Melbourne, Australia
ConocoPhillips	Houston, TX
Petroleo Brasileiro SA (Petrobras)	State of Rio de Janeiro, Brazil
Lukoil OAO	Moscow, Russia
Rio Tinto	London, United Kingdom
Nigerian National Petroleum Corp	Abuja, Nigeria
Petroliam Nasional Berhad (Petronas)	Malaysia
Rosneft OAO	Moscow, Russia
Arch Coal Inc	St. Louis, MO
Iraq National Oil Co	Baghdad, Iraq

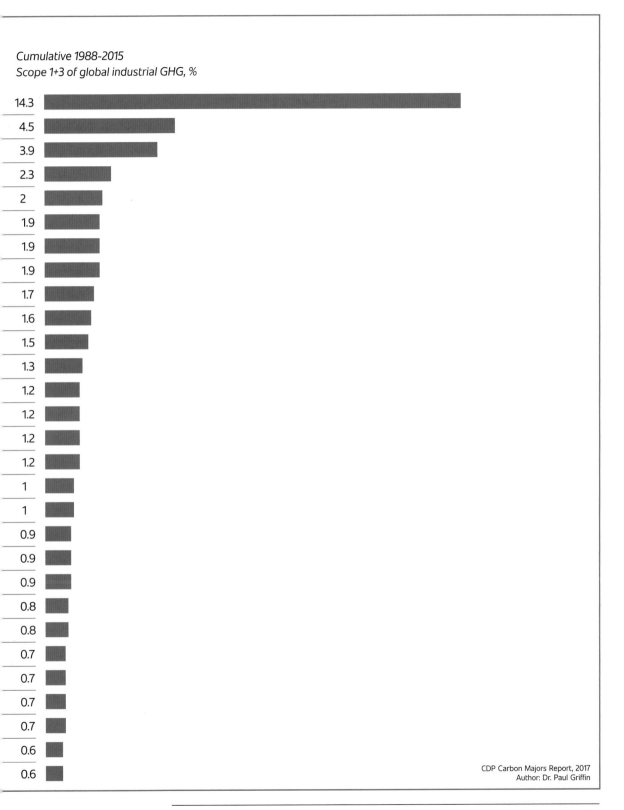

Cumulative 1988-2015
Scope 1+3 of global industrial GHG, %

14.3
4.5
3.9
2.3
2
1.9
1.9
1.9
1.7
1.6
1.5
1.3
1.2
1.2
1.2
1.2
1
1
0.9
0.9
0.9
0.8
0.8
0.7
0.7
0.7
0.7
0.6
0.6

CDP Carbon Majors Report, 2017
Author: Dr. Paul Griffin

Top 100 polluters around the world – cumulative emissions 1988 to 2015[124]

Producer	Location
Eni SPA	Rome, Italy
Anglo American	London, United Kingdom
Surgutneftegas OAO	Surgut, Russia
Alpha Natural Resources Inc	Kingsport, TN
Qatar Petroleum Corp	Doha, Qatar
PT Pertamina	Central Jakarta, Indonesia
Kazakhstan Coal	Kazakhstan
Statoil ASA	Stavanger, Norway
National Oil Corporation of Libya	Tripoli, Libya
Consol Energy Inc	Canonsburg, PA US
Ukraine Coal	Ukraine
RWE AG	Essen, Germany
Oil & Natural Gas Corp Ltd	Vasant Kunj, Delhi, India
Glencore PLC	Baar, Switzerland
TurkmenGaz	Ashgabat, Turkmenistan
Sasol Ltd	Johannesburg, South Africa
Repsol SA	Madrid, Spain
Anadarko Petroleum Corp	The Woodlands, TX
Egyptian General Petroleum Corp	Cairo, Egypt
Petroleum Development Oman LLC	Muscat, Oman
Czech Republic Coal	Czech Republic
Remaining 50 producers in sample	

of the stock of greenhouse gases responsible for climate change. They need to transition to renewables immediately. Also support needs to be given to poorer countries to also transition away from fossil fuels and to leapfrog towards more renewable solutions, while still providing much-needed energy for their citizens, and not least the billion people who still have no access to electricity. Technology alone will not save us. Even with the most powerful carbon sequestration

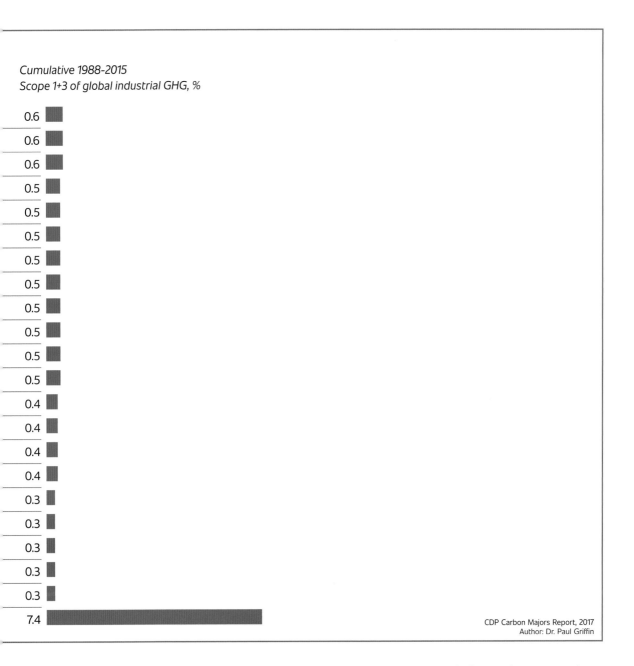

Cumulative 1988-2015
Scope 1+3 of global industrial GHG, %

0.6
0.6
0.6
0.5
0.5
0.5
0.5
0.5
0.5
0.5
0.5
0.5
0.4
0.4
0.4
0.4
0.3
0.3
0.3
0.3
0.3
7.4

CDP Carbon Majors Report, 2017
Author: Dr. Paul Griffin

technologies, emissions cannot be removed from the atmosphere at sufficient scale or speed. To slow climate change we must stop carbon from entering the atmosphere. This means supporting governments and businesses that are actively pursuing zero-carbon solutions[130] and green policies.[131] In this challenge, there is tremendous opportunity. Our collective survival depends on making the right choices right now.

This map uses accumulated light
at night over a 25 year period
to show the location of massive
cities in coastal China.
NOAA, Google, 2018

Urbanisation

The world is urbanising at unprecedented speed

Most future urbanisation will occur in Africa and Asia

Megacities and megaregions are driving the global economy

Urban fragility is deepening, especially in Africa and Asia

City networks are stepping up on climate and migration

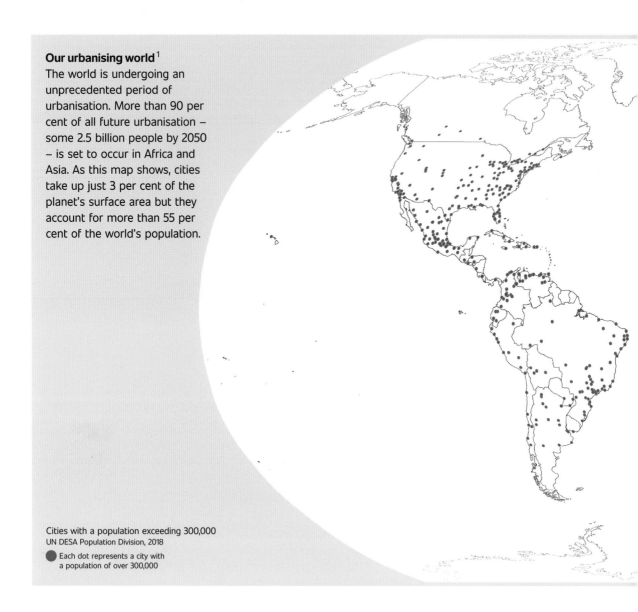

Our urbanising world[1]
The world is undergoing an unprecedented period of urbanisation. More than 90 per cent of all future urbanisation – some 2.5 billion people by 2050 – is set to occur in Africa and Asia. As this map shows, cities take up just 3 per cent of the planet's surface area but they account for more than 55 per cent of the world's population.

Cities with a population exceeding 300,000
UN DESA Population Division, 2018
● Each dot represents a city with a population of over 300,000

Introduction

Cities are the anchors of civilisations. For thousands of years, dense human settlements have concentrated power, capital and ideas. It is easy to forget that nation states are the new kids on the block, dominating global affairs for only the past few centuries. As nation states have risen to prominence since the seventeenth century, once-influential cities have fallen into decline and disrepair. But times are changing. The world is experiencing an unprecedented period of urbanisation which includes the ascendance of powerful global cities. Today, for the first

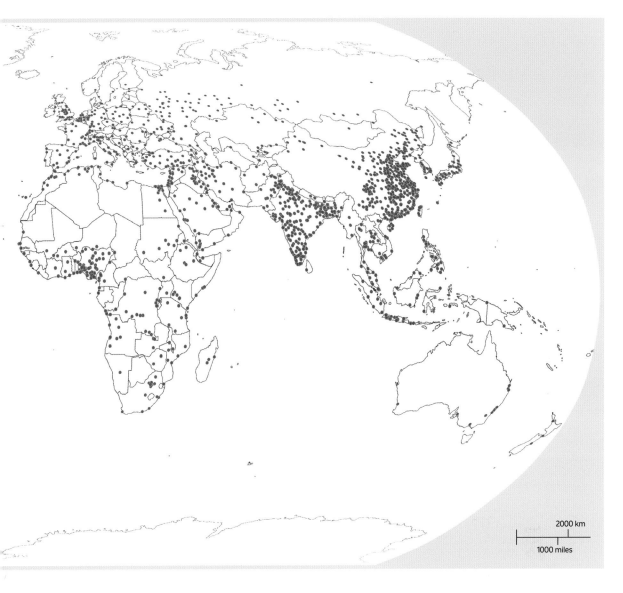

time, more people live in urban areas than in rural ones. Over the next few decades, virtually all urbanisation will occur in Africa and Asia. The numbers are astonishing. In ten years, more than two dozen new cities will cross the 5 million threshold. At least 2.5 billion more people will move to cities by 2050 – an unprecedented demographic transition.[2] While it is true that nation states are still the dominant players and will remain so for the foreseeable future, over the coming years powerful cities will play a key role in mediating global affairs.

There are good reasons why cities are making a comeback. For one, they usually confer benefits to those who are fortunate enough to live

Global urbanisation from
3700 BC to AD 2000
Reba, M., Reitsma, F. & Seto, K., 2016

- AD 1901 – AD 2000
- AD 1801 – AD 1900
- AD 1501 – AD 1800
- AD 1001 – AD 1500
- AD 1 – AD 1000
- 3700 BC – 0

1000 km

1000 miles

Urbanisation through the ages – 3,700 BC to 2000 [4]

Cities have been around far longer than nation states. Yet for the vast majority of human existence, people lived as nomads and not in urban settlements. The first permanent settlements started emerging only 10,000 to 12,000 years ago. Using the earliest recorded information, this map traces the evolution of cities since 3700 BC.

in one. While all cities face problems such as inequality, congestion and pollution, urbanites are on balance healthier, wealthier and longer living than their rural counterparts. The reason why so many people are moving to cities – about 3 million a week – is precisely because they offer so many advantages. But urban life is also creating some problems: the gap in living standards and values between urbanites and rural residents is fuelling tension and resentment. Divisions between the haves and have-nots are also sharpening within many cities. The COVID-19 pandemic exposed the deep fault-lines and inequalities that stratify the world's cities, especially the vulnerability

of elderly, migrant and less affluent populations. As a result, a handful of big metropolises are thriving while many other secondary and tertiary cities are not nearly as prosperous, connected or organised. There are also thousands of fast-growing cities experiencing fragility, struggling to provide even the most rudimentary services and seeing their social contracts unravel.[3]

This chapter explores the dual-edged character of urbanisation. We start by considering the origins and spread of cities and then trace the explosive growth of urban agglomerations during the twentieth and twenty-first centuries, including the rise of megacities and mega-regions. Many of us are mesmerised by a handful of global cities like San Francisco, Stockholm and Singapore. Yet we know virtually nothing about what is happening in the countless other cities that are stalling, being held back and falling behind. To understand some of the reasons why some cities are failing, we review the causes and consequences of their fragility – especially in those parts of the world undergoing turbo-charged urbanisation. A better understanding of what makes today's cities more resilient can potentially help prepare and empower the urban leaders of tomorrow.

Cities are far older than nation states

Cities have been around much longer than nation states. Given their totalising influence over our lives – on our politics, commerce and culture – it is easy to forget that nation states are the newcomers. The birth of the modern international state system can be traced to the mid-seventeenth century, or more specifically, the 1648 Treaty of Westphalia. Nation states were forged out of war – especially the Thirty Years War. Desperate to end a host of bloody armed conflicts spanning the continent, European aristocrats and diplomats gathered in cities like Osnabrück and Münster to hash out a deal. There, in these thousand-year-old cities, they crafted the foundational principles of international relations including ideas like 'national sovereignty' and 'non-interference' in one another's domestic affairs. For the next few centuries, it was nation states – and not cities – that dominated world affairs.

Cities are one of humanity's most successful experiments in social engineering. And while far older than nations, they are also relatively new forms of social organisation. Most of *Homo sapiens'* history involved nomadism. The first large human settlements are believed to have emerged between 10,000 and 12,000 years ago. Their rise is intimately connected to the expansion of agricultural production.[5] Once subsistence farmers learned how to store food and domesticate animals, they could afford to be more sedentary, to specialise their labour and trade with neighbours. Although no one agrees which city holds the title as the world's oldest, Eridu, Uruk and Ur, settled about 9,000 years ago, are all contenders. Other ancient cities like Aleppo, Damascus and Jericho have been continuously inhabited for over 6,000 years.[6] As people, trade and knowledge spread, cities like Beirut, Byblos, Erbil and Jerusalem dominated the Middle East.

Cities are engines of industry and innovation.[7] They have shaped the economic and cultural evolution of civilisations in China and the Indus Valley, the Mediterranean and Mesopotamia, and throughout Mesoamerica and the Andes. Capital cities of faded empires – Alexandria, Angkor, Athens, Cádiz, Chengzhou, Córdoba, Patna, Varanasi, Xi'an and Istanbul – are still standing today. Many of them were the drivers of earlier phases of globalisation, projecting power and influence across vast distances. Rome, for example, was likely the first city to register 1 million residents, at the end of the first century BC. It oversaw a formidable trading system spanning Europe and Persia. Incredibly, it took almost 2,000 years before another city reached the same population size – London by around 1810, then New York in 1875.

One of the world's most awe-inspiring ancient cities was Angkor Wat, the capital of the former Khmer empire in what is today Cambodia. Until recently, however, most of the world was unaware of its true scope and scale. For centuries, foreign archaeologists focused on uncovering and restoring sacred temples and shrines. They were totally unaware of the once-thriving city that sustained these religious tenements. It

Angkor Wat – Asia's ancient urban juggernaut [8]
Satellite imaging and remote sensing techniques are helping researchers to better understand ancient cities. Between 2012 and 2015, laser aerial surveys helped reveal how Cambodia's Angkor Wat was one of the world's largest ancient cities, spreading out over fifteen square miles.

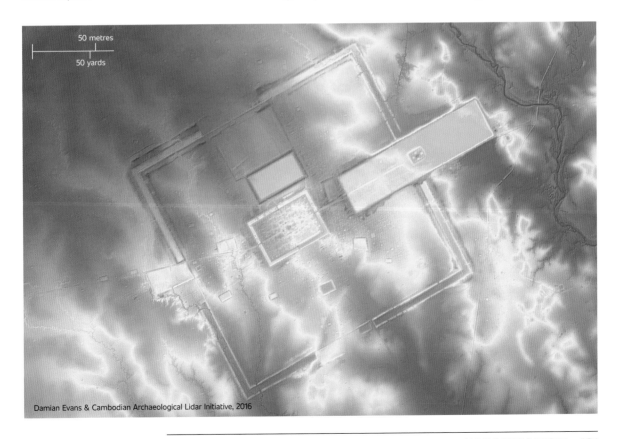

50 metres

50 yards

Damian Evans & Cambodian Archaeological Lidar Initiative, 2016

was only after NASA conducted laser aerial surveys between 2012 and 2015 that we got a glimpse of a vast grid of streets and canals spread across an area larger than present-day Paris, Sydney or Los Angeles.[9] At its peak in the twelfth century, Angkor was home to as many as 750,000 people – and this at a time when London had roughly 18,000 residents. The city had a densely populated core spanning fifteen square miles and a vast agro-urban hinterland, an early prototype of the now familiar suburban sprawl.[10] No one knows exactly why the city fell into decline from the fourteenth century onwards. It is possible that Angkor's rapid expansion had unintentionally destroyed the surrounding watershed, contributing to catastrophic flooding and the collapse of local water systems.

A small number of cities amassed tremendous power from the twelfth to the sixteenth centuries. From East and South Asia to Europe and the Americas, cities grew as they amalgamated and consolidated surrounding areas. This allowed some of them to finance their terrestrial and maritime forces. Despite suffering recurring plagues, feuds and outright wars, Italian city-states like Venice and Genoa accumulated spectacular wealth by trading in spices, silk, textiles and slaves. Some cities found strength in numbers. One of the first inter-city networks – the Hanseatic League – included over one hundred members stretching from London to Novgorod. The league developed its own legal system, tax code and police force.[11] Although a handful of urban behemoths like Constantinople, Beijing and Vijayanagar grew to half a million people or more, cities were typically quite small. And for most of human existence, city living was a comparatively rare event. In the sixteenth century there were around two dozen cities with more than 100,000 people.[12] Today, there are at least 4,000 cities exceeding this threshold.[13]

The era of turbo-urbanisation – 1950 and 2050[14]
These maps describe the transition from predominantly rural to urban societies. In 1950, most people lived outside urban settings. Only North America, most European countries, some Gulf states, Australia, Argentina, Japan, New Zealand and Venezuela were mostly urban. In 2050, virtually every country in the world will be majority urban, with some exceptions in Africa and Asia. This is the most significant urban transition in history.

1950

2050

Rural vs urban majority
UN DESA Population Division, 2018

majority rural
majority urban

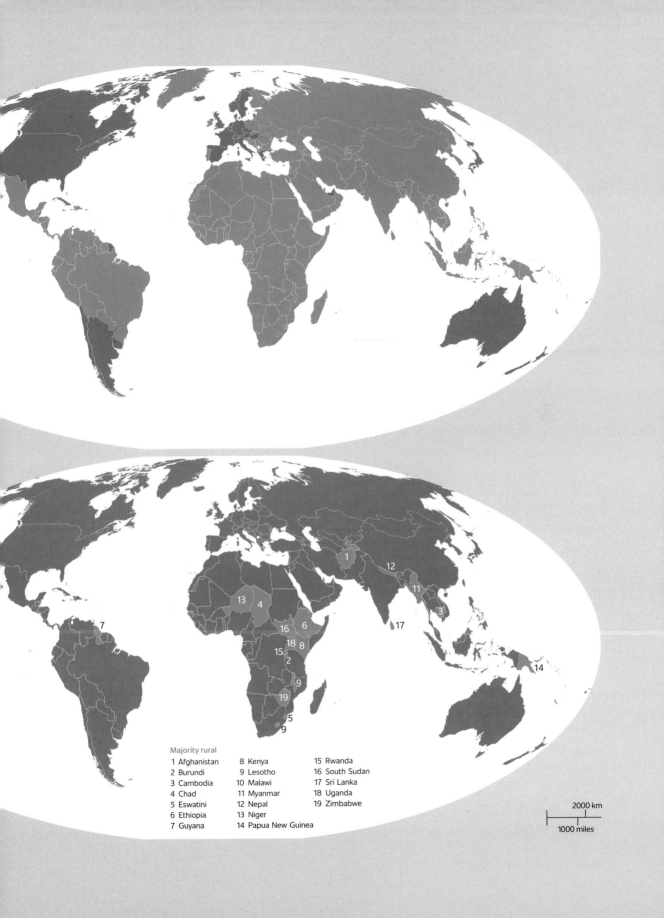

Majority rural
1 Afghanistan   8 Kenya   15 Rwanda
2 Burundi   9 Lesotho   16 South Sudan
3 Cambodia   10 Malawi   17 Sri Lanka
4 Chad   11 Myanmar   18 Uganda
5 Eswatini   12 Nepal   19 Zimbabwe
6 Ethiopia   13 Niger
7 Guyana   14 Papua New Guinea

2000 km
1000 miles

Urbanisation gathered steam from the late nineteenth century onwards. The industrial revolution, advent of railways and emergence of manufacturing centres encouraged Europeans and people living in the colonies, in particular, to move to cities in ever greater numbers.[15] In the US, for example, less than 2 per cent of the population lived in cities in 1800. By 1920, more than half of the country's residents had moved to one.[16] While generating work opportunities and driving economic growth, mass migration into European and North American cities had its downsides, including the explosion of poverty, disease and other social challenges. Central and South American countries urbanised even more rapidly than the US or those in Europe did – doubling the proportion of urban dwellers between 1950 and 2000. Their cities also contain some of the highest rates of inequality and crime in the world, as the chapter on violence shows.[17] Asia urbanised more slowly, although that started changing half a century ago. Controlled and coercive urbanisation such as China's Great Leap Forward (1958–1961) was devastating, leading to the deaths of around 45 million people.[18] But reforms in the 1970s and a sustained economic boom since the 1990s changed the game. In 1980, just one in five Chinese citizens lived in a city; by 2017, the ratio had shrunk to one in two.[19]

Cities may be key to our collective survival

Today, cities are economic powerhouses, generating over 80 per cent of global GDP.[20] They are also pivotal to the knowledge economy and are the source of more than 90 per cent of all patents. But what makes cities so incredibly important is that they are probably one of humanity's most realistic hopes to renew democracy, reverse inequality, prepare for and respond to infectious diseases, and mitigate and adapt to climate change, including the transition to a zero-carbon world. Cities are already taking actions without asking permission. They are busily setting targets to reduce greenhouse gas emissions and testing new models of governance and economic thinking in stark contrast to nation states, many of which are stagnating and becoming polarised.

The world is rapidly changing, but the basic units of global affairs are stubbornly fixed in time. For all sorts of reasons, many of us are wedded to

1960

2018

Share of total population living in urban areas – 1960 and 2018[21]
For the first time in history, more people live in urban areas than rural ones. The maps show how virtually every country saw a growing share of its population living in cities between 1960 and 2018. The next big urban boom will occur in Africa and Asia. Just three countries – China, India and Nigeria – will account for 40 per cent of all future urban population growth between 2020 and 2050.

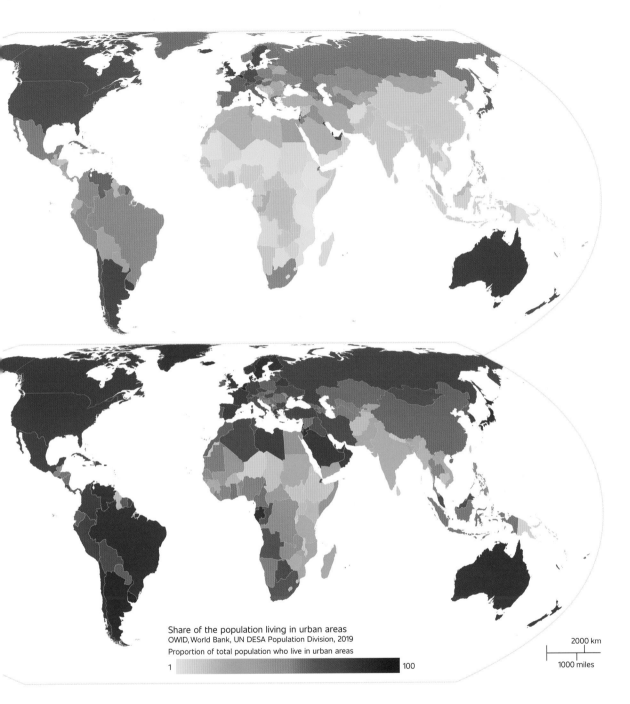

Share of the population living in urban areas
OWID, World Bank, UN DESA Population Division, 2019
Proportion of total population who live in urban areas

1 ▮▮▮▮▮▮▮▮ 100

2000 km

1000 miles

the inter-*national* system of 193 nation states. The way our governments manage global affairs – on issues of defence, migration, trade and aid – are still determined by the principles of national sovereignty and non-interference. This is because the global multilateral system – including the United Nations and the Bretton Woods institutions

– was built to service the needs and interests of nations. Even though nation states successfully scaled up essential public goods, they are proving to be insufficiently nimble at addressing some of the most urgent existential threats facing humanity, like massive pandemics, climate change and the rise of artificial intelligence (AI). We need new mental maps, including ones that put cities at their centre. This makes sense. We are, in the words of economist Edward Glaeser, an 'urban species'.[22]

Are there 2 billion more people living in urban areas than we thought?

Cities are the talk of the town. Municipal authorities, businesses, service providers and scholars are striving to build happier, more creative and smarter cities. Dozens of urban indices now catalogue everything from city liveability and openness, to business, digital preparedness, safety and security. But there are also big knowledge gaps when it comes to cities. While everyone from scholars to investors are busily studying what the sociologist Saskia Sassen calls 'global cities', much less is known about what is happening in most cities where the majority of the world's population actually live.[23] Complicating matters, there are also basic disagreements about what urbanisation is and how it ought to be defined. Incredibly, we still do not know how many cities there are in the world.[24] Depending on the definition used, there could be anywhere between 50,000 and one million of them.

In the absence of a universal definition, it is hard to distinguish the city core[25] from the sprawl that surrounds it[26] – what together is often referred to as the metropolitan area.[27] The truth is that cities are much more than just a dense clustering of people and buildings. They are administrative zones overlaid with complex networks of physical and digital infrastructure.[28] Cities are political, social and economic communities where individuals live, exchange and work.[29] But how are cities different from, say, large towns?[30] It turns out that size doesn't (always) matter. In the UK, for example, it all comes down to what the monarch thinks. She can designate any inhabited space as a city. Today, there are sixty-six cities in the UK, including one with just 1,800 inhabitants.[31]

Notwithstanding these semantic disagreements, no one doubts that urbanisation is accelerating and that the number of urban agglomerations is multiplying.[32] The United Nations' *World Urbanization Prospects* is arguably the most authoritative source on urbanisation trends.[33] In 2018, it noted that 55 per cent of the world – more than 4.2 billion people – lived in urban areas, up from only 751 million people in 1950. The report claimed that 80 per cent of North Americans and Latin Americans and 74 per cent Europeans were living in urban areas as compared to roughly half of all Asians and Africans. It also predicted that the share of the world living in urban areas will grow to 68 per cent – more than 6.7 billion people – by 2050, with close to 90 per cent of that growth occurring in Asia and Africa.[34] Meanwhile, some areas are actually seeing their overall populations – including their urban ones – shrink. The most dramatic declines are occurring in Japan, South Korea, Poland, Russia and Ukraine, where ageing populations, smaller families and, in some cases, industrial decline are to blame.[35]

A radical revision of urbanisation

A new estimate of urban living.[36]

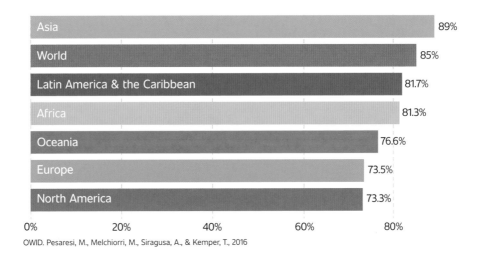

OWID. Pesaresi, M., Melchiorri, M., Siragusa, A., & Kemper, T., 2016

New findings could turn accepted understandings of urbanisation trends on their head. In 2018, a group of geographers from the European Commission came up with an entirely new way to measure urbanisation.[37] They deployed high-resolution satellite images and a new methodology reclassifying built-up areas into 'urban centres' (50,000 people or more and a density of at least 1,500 people per square kilometre) and 'urban clusters' (5,000 people or more and a population density of 300 people per square kilometre).[38] In this way they determined that 52 per cent of the world lives in urban centres,

33 per cent in urban clusters and just 15 per cent in rural areas (fewer than 5,000 people). Seen this way, the planet is already 85 per cent 'urban'. The chart depicts how Asia – and not Latin America – is possibly the most urbanised part of the world and Europe and North America the least urbanised. According to the European Commission estimates, there are already 6.4 billion people currently living in 'urban areas', which means that the UN estimate may be off by at least 2 billion people.[39] These estimates have come under criticism (especially from the UN)[40] but, if shown to be correct, could be a big deal. After all, where people live shapes everything from tax collection and budget allocations to urban planning and service delivery. Not surprisingly, the UN has convened an expert group to try to get to the bottom of things.[41]

Despite the best efforts of the UN and the European Commission to define urbanisation, governments still have different ways to classify the urban hierarchy.[42] Thresholds for what is 'urban' range from a low of 200 people in Iceland to 10,000 in Portugal and to a high of 30,000 residents in Japan.[43] In Australia and Canada, for example, populated areas with 1,000 people or more can be designated urban.[44] In China, only spaces with at least 1,500 people per square kilometre are designated urban. Meanwhile, in Western Europe, planners distinguish between 'urban areas' and 'commuting zones'.[45] Applying these kinds of parsimonious classification systems, just 40 per cent of Europeans live in an urban centre and 20 per cent reside in community zones, not three-quarters, as described by the UN, or over 85 per cent by the European Commission. We will not resolve these taxonomic conundrums in this chapter.[46] One issue that is not heavily disputed is the relationship between controlled urbanisation, economic growth and improved living standards. In most parts of the world, urban areas tend to benefit more from greater access to services – including electricity, drinking water and sanitation – than rural areas.[47]

The megacities arrive – one in five city dwellers will be Chinese by 2030

The sheer size and scale of today's cities is unprecedented.[49] In 1900, there were thirteen cities with one million inhabitants or more. By 1950, there were 83 cities with one million people and three

Los Angeles

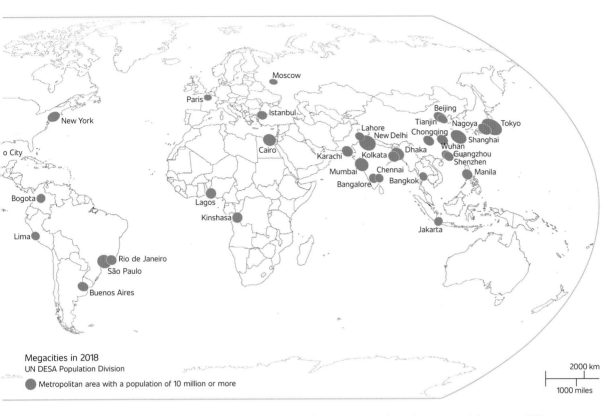

Moscow
Paris
Istanbul
New York
o City
Cairo
Lahore
New Delhi
Karachi
Kolkata
Dhaka
Mumbai
Chennai
Bangalore
Bangkok
Beijing
Tianjin
Chongqing
Nagoya
Tokyo
Wuhan
Shanghai
Guangzhou
Shenzhen
Manila
Bogota
Lagos
Kinshasa
Lima
Rio de Janeiro
São Paulo
Buenos Aires
Jakarta

Megacities in 2018
UN DESA Population Division

● Metropolitan area with a population of 10 million or more

2000 km
1000 miles

The rise of the megacity – 2018 [48]

The size of the blue circles represents the number of people living in the world's 34 megacities, each with a population over 10 million.

megacities – New York, Paris and Tokyo – with 10 million. Now there are more than 500 cities with populations exceeding one million people, including at least 34 megacities. Tokyo-Yokohama, with 38 million people, is the undisputed urban behemoth. The next in line is Delhi, with more than 28 million people, followed by Shanghai, with 24 million. Other urban giants include São Paulo, Mexico City, Lagos and Wuhan.[50] These cities are central nodes in international financial networks and global supply chains. The most successful of them have moved away from a dependence on manufacturing into services and high-tech sectors. Just as the industrial revolution drove urbanisation across Europe and North America in the 1800s, the intensification of globalisation since the 1970s has triggered an urbanisation boom in Asia and Africa.[51] But are these next generation cities ready to compete?

Some megacities are denser than others. Take the case of downtown Manila, considered the world's densest city. It is home to at least 1.6 million people[52] with 46,000 people packed into every square kilometre.[53] But if all metropolitan Manila is included, the density drops to 'just' 13,600 people per square kilometre. By way of comparison,

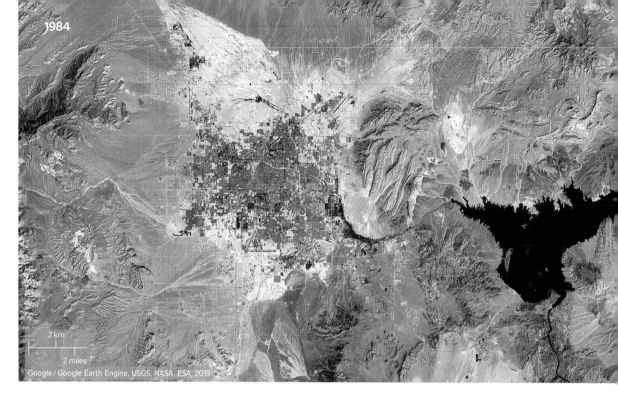

1984

2 km

2 miles

Google / Google Earth Engine, USGS, NASA, ESA, 2019

downtown Paris is home to more than 2 million inhabitants[54] and 21,000 people per square kilometre. The city of New York, with more than 8.6 million residents, has the highest density in the US, with 10,000 people per square kilometre. By contrast, cities like Sydney, with 4.4 million people, are among the least densely populated, with only 1,900 people per square kilometre.[55] One of the least dense cities in the world is Atlanta with fewer than 600 people per square kilometre.[56]

North American cities dominate the low-density sweepstakes. As the map demonstrates, Las Vegas is the quintessential low-density city, with only 1,875 people per square kilometre. Despite its extreme desert heat and limited water supplies, the city is one of the fastest-growing metropolitan areas in the US, expanding from 540,000 residents in 1984 to 2.2 million by 2018. Low density living translates into urban sprawl.[58] The city grew precisely because of unsustainable zoning and an over-reliance on car commuting. The combination of rapid population growth, sustained residential demand and climate change is contributing to risky water shortages.[59] Nearby Lake Mead, the largest man-made reservoir in the US, is at dangerously low levels.[60] The lake lost over half its water reserves in the past two decades. Las Vegas is a reminder of how people with means can temporarily escape climate change by living in air-conditioned cocoons. Yet like

The Las Vegas sprawl, 1984 and 2018[57]
Las Vegas is one of the lowest density cities in the world. It grew from 540,000 to 2.2. million people in just three decades.

The car is king in Las Vegas, 2012–2016
The reason Las Vegas has expanded so rapidly is due to an over-reliance on cars and lack of public transport infrastructure. As the yellow stain shows, most people depend on cars to get around. Less than 5 per cent of Las Vegans use public transport or other modes of transport to get to work, school or play.

2018

Summerlin

North Las Vegas

Las Vegas

Summerlin South

Winchester

Paradise

Whitney

Lake Mead

Enterprise

Henderson

Boulder City

2 km

2 miles

Google / Google Earth Engine, USGS, NASA, ESA, 2019

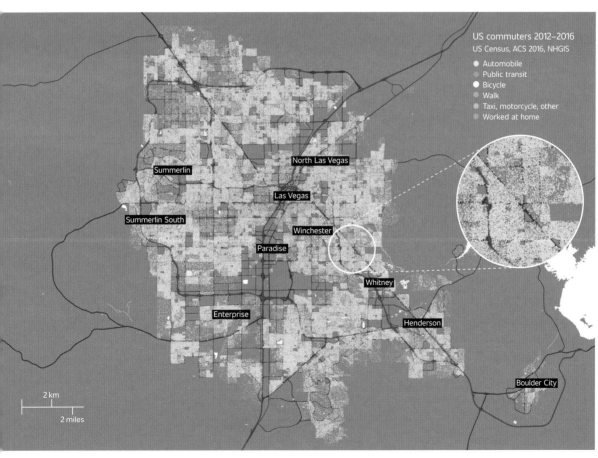

US commuters 2012–2016
US Census, ACS 2016, NHGIS

- Automobile
- Public transit
- Bicycle
- Walk
- Taxi, motorcycle, other
- Worked at home

Summerlin

North Las Vegas

Las Vegas

Summerlin South

Winchester

Paradise

Whitney

Enterprise

Henderson

Boulder City

2 km

2 miles

other desert cities – including Dubai and Doha in the Gulf that rely on fossil fuels to pump water and cool their residents – it is both a contributor to and victim of the climate crisis.

Ground zero for turbo urbanisation is Asia and Africa. By 2050, approximately 292 million Chinese people will have flocked to cities, along with 404 million Indians and 212 million Nigerians. Together these three countries will account for 40 per cent of all future global urban growth. The scale of China's urban trajectory is mind-boggling. Today, more than 58 per cent of its population already live in cities, compared to just 18 per cent in 1980. There are officially 662 Chinese cities, including at least 160 with a million people or more.[61] Compare this to the US, which has just ten cities with more than one million inhabitants, or the UK, with two (depending on how you count).[62] Since the 1980s, China has focused on building special urban growth clusters, or economic zones – Shantou, Shenzhen, Xiamen and Zhuhai are prominent examples. Since bigger cities are associated with higher productivity and faster growth, Chinese authorities are busily building nineteen gigantic urban clusters that amalgamate sprawling cities. As the map of lights at night shows, three of them are well advanced, including the Pearl River Delta near Hong Kong, the Yangtze River Delta surrounding Shanghai, and Jing-Jin-Ji, encircling Beijing.

Urbanisation proceeds in different ways and at different speeds around the world.[64] In China, city growth is typically centrally planned, with strict controls on where people live and who can access public services. In spite of tough rules and regulations, the Chinese authorities have been unable to avoid sprawling cities, soaring emissions and rising inequality.[65] The urbanisation process is more disorderly in India, now the world's most populous country.[66] According to the political scientist Reuben Abraham, most Indian cities 'extend well beyond their administrative limits . . . growth is unregulated and unplanned, marred by narrow roads . . . limited open space and haphazardly divided plots'.[67] Although India's urban explosion is just getting underway,[68] it is already home to 2,500 cities, forty of them exceeding a million people.[69] At least eight of the world's fastest-growing cities (by population) are Indian,[70] as are eighteen of the twenty most polluted.[71] Indian authorities believe that at least $1.2 trillion will be required to accommodate the hundreds of millions more Indians who will move to cities over the next two decades.[72] Like China, Indian officials have proposed building dozens of high-tech cities from

Changchun

Shenyang

Beijing

Tianjing

Jinan

Qingdao

Seoul

SOUTH
KOREA

Busan

X'ian

Zhengzhou

CHINA

Nanjing

JAPAN

Shanghai

Chengdu

Wuhan

Chongqing

Changsha

Taipei

Kunming

TAIWAN

Hong Kong

Macau

VIETNAM

Hanoi

HAINAN

Lights at night over 25 years
NOAA, Google, 2018

Decrease   Increase

China's urban boom [63]
It is possible to map urbanisation by examining light emissions projected into
space. This map highlights aggregate light emissions above China, Taiwan, South
Korea and parts of Japan and Vietnam over a twenty-five-year period. Areas that
are red or yellow registered net increases in light emissions, while those in green
are neutral and those in blue represent a decline.

200 km

100 miles

scratch.[73] In 2015, the federal government announced a multibillion-dollar plan to build 100 'smart cities' by 2020 – but progress is much slower than advertised.[74]

Meanwhile in Nigeria, Africa's most populous country, urbanisation is even more disorganised than in China and India.[75] Already more than half of Nigeria's 150 million citizens live in cities, compared to just a third in 1990.[76] The country has 248 cities, including seven with at least 1 million people.[77] Consider the situation of Lagos, a city of just 200,000 residents in 1960 and an estimated 20 million inhabitants today (though no one really knows its exact size).[78] Should Lagos's population continue expanding at its present rate, demographers predict that it could hit the 100 million mark by 2100.[79] It is hard to fully comprehend the pace and scale of urbanisation in Lagos. The city can expect roughly 77 new arrivals *every hour* between now and 2030.[80]

The legendary Lagos sprawl, 1984 and 2019[81]
One of the world's fastest-growing cities population-wise is Lagos, in Nigeria. As the map shows, the urban footprint expanded dramatically between 1984 and the present. The city had roughly 3.3 million residents in 1984 and has around 20 million today. About 77 new Lagosians will arrive in the city every hour between now and 2030.

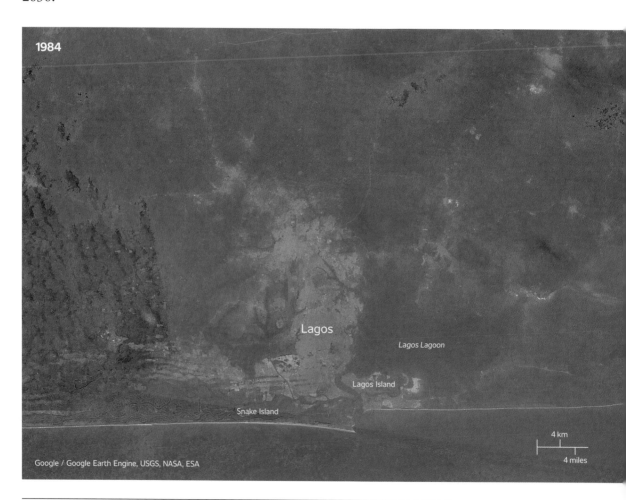

1984

Lagos

Lagos Lagoon

Lagos Island

Snake Island

4 km
4 miles

Google / Google Earth Engine, USGS, NASA, ESA

The world's largest cities exert more economic clout than most nation states. In 2016, McKinsey singled out 600 of the most powerful metropolitan areas – the so-called C600 – responsible for powering 60 per cent of the global economy.[82] These cities accounted for more than a fifth of the global population and stunning economic productivity. Take the case of metropolitan Tokyo and New York, with nominal GDPs that are equivalent to those of South Korea and Russia, respectively.[83] Or consider Los Angeles, which has a larger GDP than that of Australia; Paris, a bigger economy than South Africa; and London, Moscow, Shanghai and Delhi, which outstrip the Netherlands, Austria, Norway and Israel.[84] Today, most of the C600 are restricted to wealthy developed countries. But by 2025, the list will include an additional 136 new cities, most of them from South and East Asia and Latin America. As the economic clout of big cities grows, some are demanding more political power and autonomy.[85]

1 Cascadia (Seattle to Portland)
2 NorCal (San Francisco, Palo Alto, and San Jose)
3 SoCal (Los Angeles to San Diego)
4 Tor-Buff-Chester (Toronto, Buffalo and Rochester)
5 Chi-Pitts (Chicago to Pittsburgh)
6 Char-Lanta (Charlotte and Atlanta)
7 Bos-Wash (Boston, New York, Washington DC)
8 SoFlo (Miami to Tampa)
9 Texas Triangle (Dallas, Houston, San Antonio and Austin)
10 Mexico City
11 São Paulo
12 Lon-Leeds-Chester (London, Leeds, Manchester)
13 Par-Am-Mun (Paris, Amsterdam, Brussels and Munich)
14 Vienna-Budapest
15 Ista-Burs (Istanbul to Bursa)
16 Barcelona-Lyon
17 Rome-Mil-Tur (Rome, Milan and Turin)
18 Cairo-Aviv (Cairo to Tel Aviv)
19 Abu-Dubai (Abu Dhabi and Dubai)
20 Delhi-Lahore
21 Singa-Lumpur (Singapore, Kuala Lumpur)
22 Shandong (Jinan, Zibo and Dongying)
23 Jing-Jin-Ji (Beijing, Tianjin and cities in Hebei province)
24 Shang-zou (Shanghai to Hangzhou)
25 Greater Tokyo
26 Osaka-Nagoya (Osaka to Nagoya)
27 Hong-Shen (Hong Kong and Shenzhen)
28 Seoul-San (Seoul to Busan)
29 Tapei

2000 km
1000 miles

Cities are living laboratories. Around the world, smart, digital, sensory and intelligent cities are harnessing vast troves of data to increase their energy efficiency, improve transportation flows and better target basic services.[86] Sensing opportunity, technology vendors and management consultancies are busily selling 'solutions' to municipal authorities.[87] The spread of 5G infrastructure and the Internet of Things (IoT), detailed in the technology chapter, are giving rise to 'networked urbanism' which, for better and worse, is changing city landscapes. Not only will it speed up the digital economy and

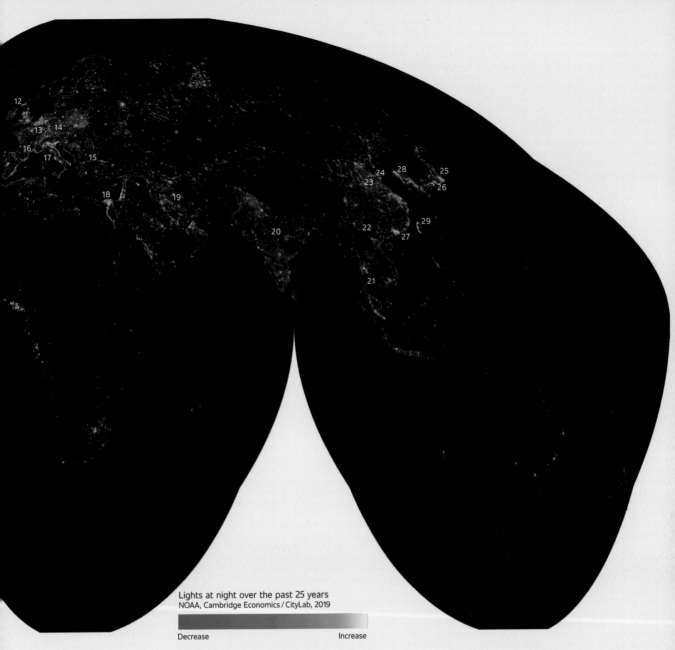

Lights at night over the past 25 years
NOAA, Cambridge Economics / CityLab, 2019

Decrease   Increase

The rise of mega-regions [89]
It is possible to acquire new insights about urbanisation by examining light
emissions reflected into space. This map highlights the spread of mega-regions
together with aggregate light emissions over a twenty-five-year period. Areas that
are red and yellow have registered net increases in light emissions, while those in
green are neutral and those in blue represent a decline.

digitisation of services, it is already generating huge vulnerabilities to
ransomware and other cyber threats since everything from nuclear
plants to home security devices will be digitally connected.[88]

Twenty-nine city clusters account for more than 60 per cent of the global economy

The twenty-first century is giving rise to entirely new categories of urbanity. For the first time, the world witnessed the emergence of 'super cities', with more than 40 million residents. These are the outgrowth of 'hyper-cities' – metropolitan areas, with over 20 million inhabitants – and 'megacities' – with at least 10 million residents. The spectacular speed at which cities are growing is unlike anything we have ever experienced.[90] Take the case of Pakistan's largest city, Karachi, that expanded from roughly 1 million people in 1950 to more than 20 million today.[91] As they absorb surrounding municipalities, fast-growing cities like Karachi are coming to the people rather than the other way around. While these huge sprawling metropolises are rarely the result of centralised or calculated planning, their sheer size may well give them a competitive edge in the global economy.

Some of the world's fastest-growing metropolitan areas are gigantic, stretching across hundreds of miles.[93] Back in the 1960s, the geographer Jean Gottmann saw them coming – describing them simply as 'megalopolises'.[94] Today, megalopolises – also known as 'megapolises', or 'conurbations'[95] – are everywhere, from the cities spanning the Yangtze River Delta in China to the Boston-New York-Washington DC corridor in the US. One of the most spectacular examples is Jing-Jin-Ji, a truly massive mega-region that connects three enormous cities – Beijing, Tianjin and several large cities in Hebei province. All told, it includes as many as 130 million people in its catchment.[96] The areas included within the Jing-Jin-Ji jurisdiction account for a tenth of China's economy, with an output of roughly $1.2 trillion a year, on a par with that of Mexico.[97]

Chinese planners view Jing-Jin-Ji as a possible template for future urban and rural development at home and abroad.[98] As the map shows, the sheer scale and dimensions of the megaregion are difficult to fathom. As part of the 'integration' of the three megacities, the authorities built more than a thousand major inter-city highways, bus lanes and rail projects. Thousands more public and commercial offices, universities and hospital buildings are being purpose-built about sixty miles to the south in Xiong'an New Area, China's 'city of the future', established in 2017. Another 1,200 manufacturing and logistics companies were shuttered in Beijing and hastily relocated to

China's extraordinary megacity – Jing-Jin-Ji – lights at night over the past 25 years [92] It brings together several massive cities – Beijing, Tianjin, Baoding, Shijiazhuang, Tanghsan, Cangzhou, Langfang, Zhangjiakou, Qinhuangdo, Dongyin and Chengde – and their combined 130 million inhabitants.

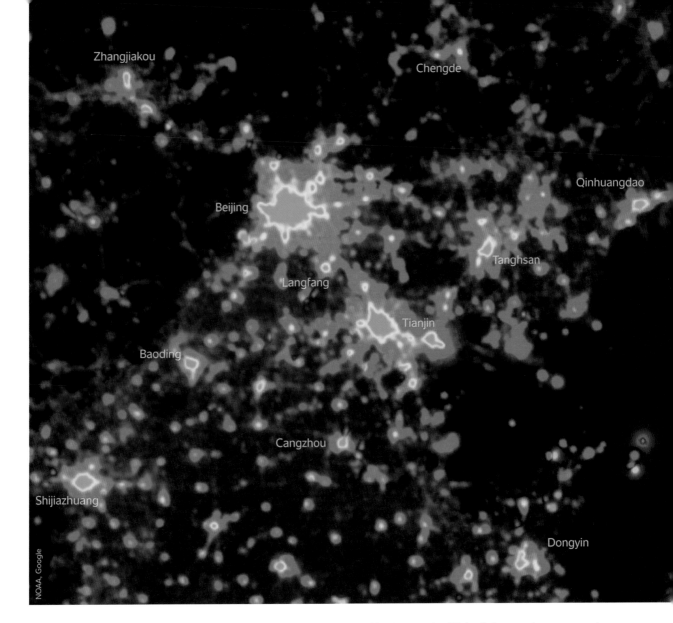

other cities in order to reduce pollution. Mindful of the environmental consequences of urban growth in other Chinese cities, the new mega-region's authorities have pressed the private sector to plant millions of trees. These and other efforts have reportedly reduced air and water pollution and cut fine particulate matter (PM2.5) densities by half since 2013.[99]

Satellite maps can help reveal the wiring of these interconnected mega-regions.[100] According to urban specialists Richard Florida, Charlotta Mellander and Tim Gulden, mega-regions are made up of at least two adjacent metropolitan areas with populations over 1 million and combined economic outputs of at least $300 billion.

At least twenty-nine mega-regions meet this definition, including eleven in Asia, ten in North America, six in Europe and one each in Latin America and Africa.[101] The wealthiest of them – the Boston-New York-Washington DC corridor, or BoWash – has a combined population of 47.6 million people and an economic output of over $3.6 trillion, making it the fifth largest GDP on the planet after the US, China, Japan and Germany. Europe's largest mega-region is made up of Paris-Amsterdam-Brussels-Munich – called Par-Am-Mun – and generates $2.5 trillion combined output. Other examples include Seoul-San (Seoul and Busan), the Texas Triangle (Dallas, Houston, San Antonio and Austin) and Lon-Leeds-Chester, which encompasses London, Leeds and Manchester.[102] While mega-regions can generate prosperity and reduce environmental stress through greater connectivity, not all cities are benefiting. Many of the fastest-growing cities population-wise are becoming ungovernable, even fragile.

The total number of slum dwellers could rise to 2 billion by 2030

A handful of global cities are concentrating wealth, talent and opportunity. Meanwhile thousands of fast-growing cities in South and Southeast Asia, North and Sub-Saharan Africa and Central and South America are falling behind. High-tech cities of the future – like Chongqing, Gandhinagar and Songdo – are diverting our gaze from cities in middle- and lower-income countries where most future population growth is set to occur.[103] Many of them – with names you've never heard of – are struggling to attract investment, coping with extreme poverty, and battling crime along the lines described by historian Mike Davis in his book *Planet of Slums*.[104] Unlike in North America and Europe, cities in Africa and Asia are urbanising before they industrialise. And while they have the potential to leapfrog legacy systems and take advantage of new technologies, many of them are giving rise to vast slums and shanty towns.

Slums are still widespread across Asia, Africa and the Americas. In India and Brazil, for example, as many as one-third of all city residents live in slums while in some African cities the proportion rises as high as nine in ten.[105] Today there are roughly a billion people living in slums – but by 2030, the number will double.[106] Shanty towns like

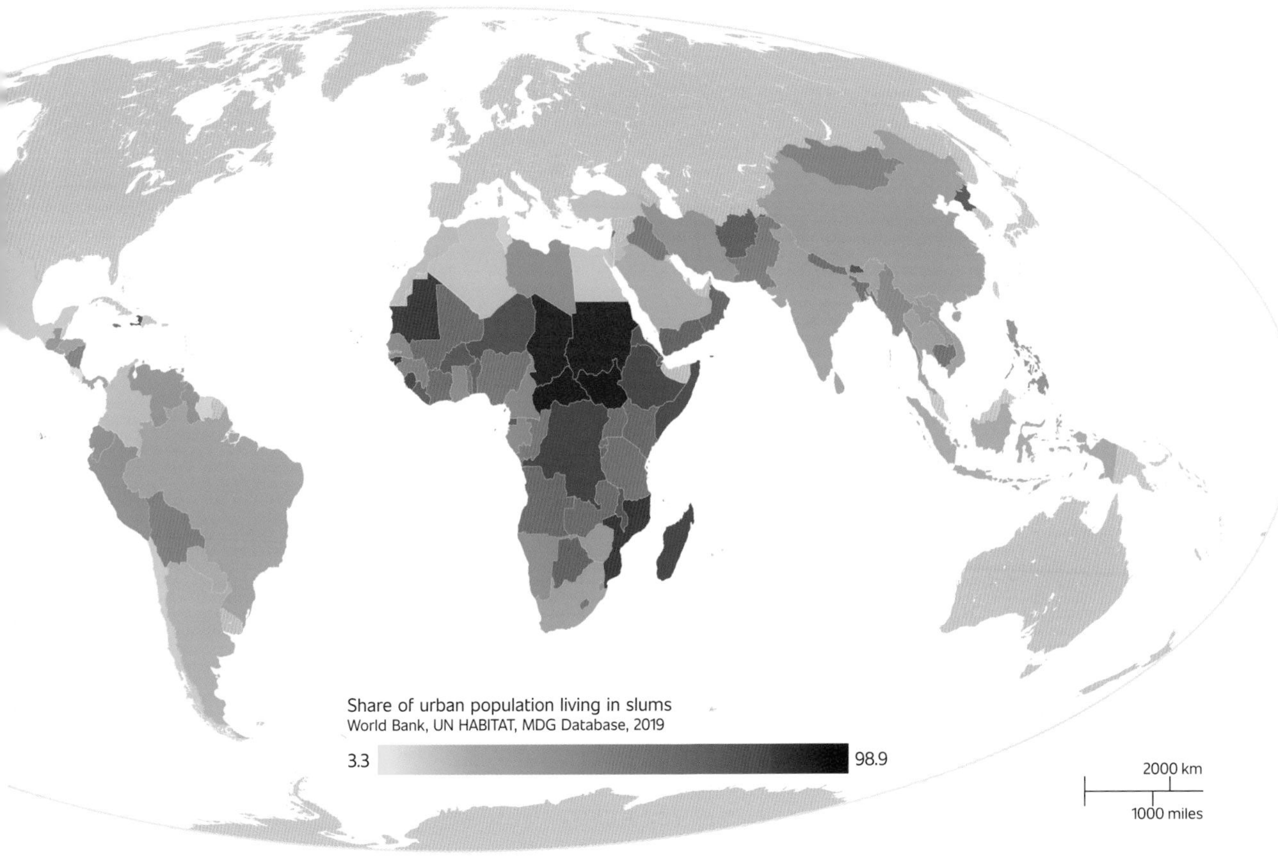
Share of urban population living in slums
World Bank, UN HABITAT, MDG Database, 2019
3.3
98.9
2000 km
1000 miles

Slums in the southern hemisphere – slum populations as a share of urban population, 2014 [107]

The world's slum population will grow from 1 billion in 2014 to 2 billion by 2030. This map highlights how slum populations are especially concentrated in Sub-Saharan Africa, Central and South America, and parts of South Asia and Southeast Asia. While the aggregate number of slum dwellers is growing, their overall proportion is decreasing due to overall population growth.

Dharavi (Mumbai), Khayelitsha (Cape Town), Orangi Town (Karachi) and Neza (Mexico) are cities in their own right, with as many as a million residents. High-density slums like Petare (Venezuela), Rocinha (Rio de Janeiro) and Kibera (Nairobi)[108] can literally be seen from space.[109] While they have different histories and spatial forms, they share several common characteristics, including an abundance of informal housing; unreliable and poor quality electricity and social services; mountains of rubbish and waste; and heightened exposure to communicable diseases.[110] Even so, ingenious solutions to these challenges are emerging from slum dwellers themselves. Many of them

Orangi Town, Karachi, 2019
Google / Google Earth Engine, USGS, NASA, ESA

2 km
1 mile

New Karachi Town
North Karachi Township
Orangi Town
North Nazimabad Town
Sindh Industrial Trading Estate
Nazimabad
Karachi

are fighting for land and tenure rights, affordable housing, improved infrastructure and better services. Residents are launching profitable businesses, innovative mobility solutions and off-grid electricity options. The truth is that the answers to many of tomorrow's biggest urban challenges probably reside in the world's lower-income informal settlements.[111]

The overall slum population may be growing, but the proportion of people living in slum households is falling in most countries. For example, in China the percentage of households classified as slum dwellers plummeted from 44 to 25 per cent between 1990 and 2014. In India the ratio dropped from 55 to 24 per cent over the same period, and from 77 to 50 per cent in Nigeria.[112] Yet because urban populations are still rapidly expanding and cities are unable to absorb new residents, the absolute number of people living in slums will keep rising.[113] Some researchers believe that slums are an inevitable if regrettable outcome of economic development. They are a form of 'affordable housing' in lower- and middle-income cities. Others contend that slums are an avoidable symptom of depressed investment in basic public goods and are self-reinforcing poverty traps.[114]

Satellite images of Orangi Town, Neza, and Dharavi – 2019.

Neza, Mexico, 2019
Google / Google Earth Engine, USGS, NASA, ESA

Texcoco

Chimalhuacán

Neza
(Nezahualcóyotl)

Mexico City

La Paz

4 km

2 miles

Mumbai

Bandra

Mithi River

Dharavi

Antop Hill

*Mahim
Bay*

Matunga

Dharavi, Mumbai, 2019
Google / Google Earth Engine, USGS, NASA, ESA

2 km

1 mile

Whatever one's view, the countries experiencing the fastest economic growth are those that have also significantly reduced the proportion of urban residents living in slum conditions.[115]

Deepening city fragility – more than 90 per cent of African cities are fragile

Clearly, not all fast-growing cities are moving in the same direction. In richer and poorer countries alike, some cities have entered a post-industrial phase while others are still dominated by heavy manufacturing and other sunset industries. There are cities that are struggling in democratic states just as there are cities thriving in countries that are more authoritarian.[116] Cities like Surat in India and Zunyi in China are experiencing double-digit economic growth,[117] while Mosul in Iraq and Mogadishu, the capital of Somalia, are on life support at the bottom of the rankings. In those cities battered by political upheaval and economic instability, the social contract binding municipal governments to urban residents is unravelling altogether.[118]

What makes some cities more fragile than others? Ancient cities like Corinth or Pompeii didn't stand a chance – they were obliterated by great upheavals like war, earthquakes and volcanoes. Other low-density cities such as Anuradhapura and Tikal simply ran out of resources owing to poor planning and bad luck.[119] Urban fragility cannot be boiled down to a single factor. It is a result of the cumulative impact of multiple stresses such as high levels of inequality and poverty, unregulated population growth, soaring unemployment, congestion and pollution, violent crime and exposure to natural disasters.[120] Nor is it a fixed condition – urban fragility fluctuates over time. But some risk factors stand out. For example, cities whose populations grow at 3 per cent a year or more, which exhibit severe income inequality, and have policing and criminal justice deficits, tend to exhibit more

The surprising spread of city fragility around the world – 2000 and 2015[121]
Some cities are more fragile than others. This map charts the evolution of urban fragility in more than 2,100 cities with populations of 250,000 or more between 2000 and 2015. City fragility occurs when there is a convergence of multiple risks – high levels of inequality, unemployment, crime and exposure to natural disasters. There is evidence of rising fragility in most parts of the world, especially in Africa, the Middle East, South Asia and parts of Southeast Asia.

2000

2015

Urban fragility score
Igarapé Institute, 2015

Less fragile   More fragile

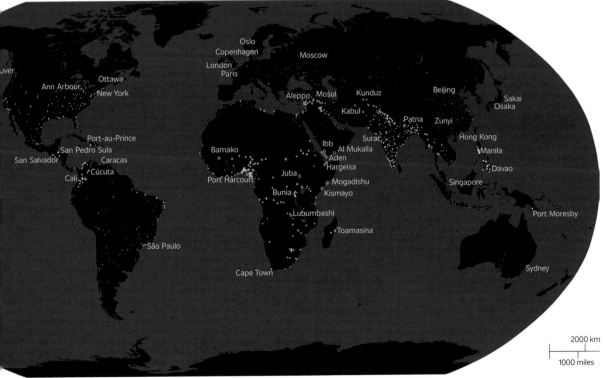

Oslo
Copenhagen
Moscow
London
Paris
uver
Ann Arbour
Ottawa
New York
Beijing
Aleppo
Mosul
Kunduz
Sakai
Osaka
Kabul
Patna
Zunyi
Port-au-Prince
Ibb
Surat
San Pedro Sula
Bamako
Al Mukalla
Hong Kong
San Salvador
Caracas
Aden
Manila
Cúcuta
Hargeisa
Cali
Juba
Davao
Port Harcourt
Mogadishu
Bunia
Kismayo
Singapore
Lubumbashi
Port Moresby
Toamasina
São Paulo
Cape Town
Sydney

2000 km
1000 miles

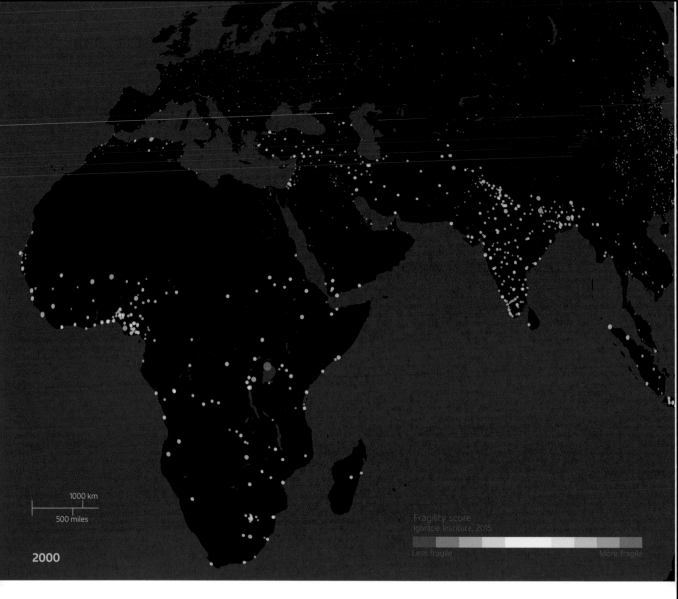

Fragility score
Igarapé Institute, 2015

Less fragile   More fragile

2000

fragility than others.[122] The most at-risk cities are found in poorer countries like Somalia and South Sudan, and also in war-torn middle-income ones like Syria and the Philippines.

City fragility is spreading. On these maps there are more than 2,100 cities with populations of 250,000 inhabitants or more. The larger and redder the city, the more fragile it is, while the smaller and bluer the city, the more resilient. All told, just over one-tenth of all cities fall into the 'high' fragility category while two-thirds are experiencing 'medium' fragility. Surprisingly, less than one-fifth registered 'low fragility'. Roughly 90 per cent of all African cities fell into the high and medium fragility categories.[123] At the top of the global list are Somalian cities like Mogadishu, Kismayo and Merca followed by

The concentration of urban fragility in Africa, the Middle East and Asia – 2000 and 2015 [124]
Africa and Asia experienced the most alarming rise of urban fragility between 2000 and 2015. These maps show how cities across Sub-Saharan Africa, the Middle East and central and southern Asia are especially at risk.

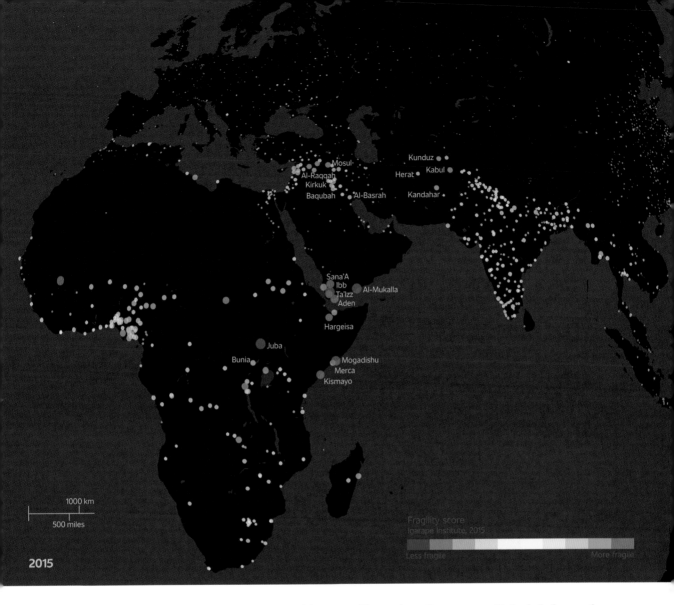

Mosul
Al-Raqqah
Kirkuk
Baqubah   Al-Basrah
Kunduz
Herat   Kabul
Kandahar

Sana'A
Ibb
Ta'Izz   Al-Mukalla
Aden

Hargeisa

Juba
Bunia
Mogadishu
Merca
Kismayo

1000 km
500 miles

Fragility score
Igarapé Institute, 2015

Less fragile   More fragile

2015

cities recently ravaged by war like Aden, Damascus, Kabul, Juba and Mosul. By contrast, the most resilient cities in our list are Ann Arbor, Canberra, Oslo, Ottawa and Sakai.

Looking to the future, the most significant urban security and development challenges are in Africa. The continent is already home to the world's youngest and fastest-growing population and is urbanising more rapidly than any other part of the planet. Africa's population of roughly 1.1 billion citizens is likely to double by 2050 and more than 80 per cent of that increase will occur in cities and their slums.[125] Turbo-charged urbanisation and a bulging youth population (with most young people lacking meaningful job prospects), is a ticking time bomb. Already around 70 per cent of Africans are under thirty.

One of the biggest challenges facing African cities is how to convert their youth bulge into a dividend.

African cities are also confronting a major infrastructure gap. Annual national public spending on infrastructure is extremely low by international standards: an average of just 2 per cent of GDP between 2009 and 2015,[126] compared to 5.2 per cent in India and 8.8 per cent in China.[127] The deficits in urban infrastructure are daunting. Africans need to spend at least $130 billion annually on infrastructure just to meet the continent's basic urban needs.[128] Yet Africa as a whole is already facing annual financing shortfalls of more than $68 billion.[129] Roughly two-thirds of the investments in urban infrastructure needed by 2050 have yet to be made.[130]

Making matters worse, African cities are expanding during a period of rising climate stress. Africa's urban areas will most likely suffer disproportionately from climate change because the continent as a whole is warming 1.5 times faster than the global average.[131] More than eighty-five of Africa's 100 major coastal cities have yet to develop basic climate mitigation and adaptation strategies. The strain on basic services and natural resources, as Cape Town's water crises reveal, will only increase. In 2017, Cape Town came perilously close to Day Zero – the day when the water level of the major dams supplying the city falls below 13.5 per cent. But this date with destiny was merely postponed, and not avoided.[132] If Africans do not find a way to build more sustainable and inhabitable cities then they face a future of deepening fragility.

Los Angeles

The 100 largest cities by GDP in 2035
Oxford Economics, 2018

- China (34)
- North America (28)
- Rest of Asia (15)
- Europe (12)
- Australia (4)
- Middle East (4)
- India (3)
- Latin America (3)
- Africa (1)

The urban future is Asian – a million people flock to Asian cities every week

The future looks somewhat brighter in the supercharged cities of Asia. The world's ten fastest-growing urban economies are in India, with tech hubs like Agra, Hyderabad and Chennai growing at 8 per cent a year. Chinese cities like Beijing, Guangzhou, Hangzhou, Quanzhou and Suzhou are not far behind. If their blistering growth rates persist – and this is far from certain given the crushing economic effects of the COVID-19 pandemic – then Asia's largest cities will have a larger GDP than all North American and European cities combined. But it's worth underlining that most Asians are not moving to

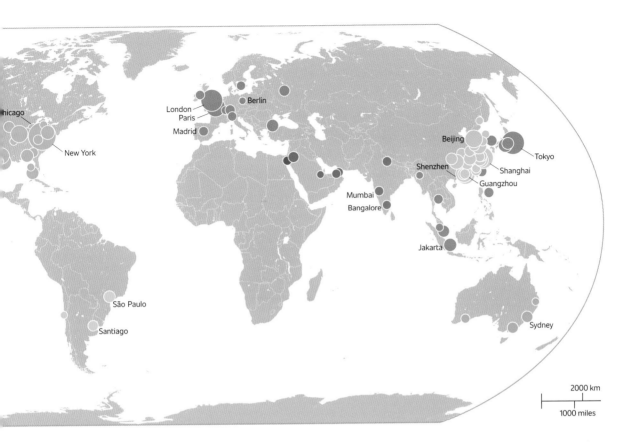

The 100 largest cities in the world by GDP – 2035 [133]

By 2035, most of the world's largest cities (measured by nominal GDP) will be in East and South Asia. Big cities in North America and Europe will remain competitive, but the fastest-growing urban centres will be elsewhere. Notably, Africa and Latin America combined will have just four of their cities in the top 100.

megacities. On the contrary, the populations of the largest megacities are stabilising and, in some cases, declining. Rural-to-urban migration is increasing, but for the most part to medium and large cities, and not the biggest ones. [134]

Asia's spectacular city growth is one of the success stories of the past few decades. The UN estimates that well over half of Asia's 4.5 billion residents will live in cities by 2026, and the continent will be home to two-thirds of the world's megacities. [135] Large Asian cities are consolidating their position as the world's new economic centre of gravity. With notable exceptions, the overall quality of life for most Asian city dwellers is improving. China and Japan, in particular, have amassed the necessary wealth and capacity to deal with half a century of accelerated urbanisation. To varying degrees, they have decent infrastructure, housing, utilities and services to show for it. The rapid response of Asian cities to the COVID-19 pandemic has been testament to this.

All is not well in Asian cities, of course. Chinese cities are fast approaching their peak size[136] and the populations of most Japanese cities have flat-lined or are in steady decline.[137] While comparatively better off than before, urban populations in these and other countries are facing serious challenges stemming from ambient air pollution, water shortages, extreme and even fatal heat-waves, faltering service delivery, and social and economic segregation. There are signs of rising frustration and dissent in some big urban centres. Meanwhile, secondary cities across India, as well as in Indonesia, the Philippines, Thailand and Vietnam, are humming. These two regions – South and Southeast Asia – with their young populations will unleash the world's next growth wave. Although they are busily investing in infrastructure

Surat – the world's fastest growing economy, 1984 and 2019 [139]
Surat has one of the fastest-growing economies in the world. The city's GDP is ticking along at over 9 per cent a year, a rate that could continue for the next quarter of a century. While the city is emerging as a twenty-first century technology hub, it also has more than a thousand years of history.

1984

2 km
1 mile

and new technologies, with boundless investment on the way, social and environmental risks are mounting.[138]

One place to catch a glimpse of the future of urbanisation is Surat. The Indian port city is a global hub for the diamond trade: eight out of ten stones sold in the world are cut and polished there. It is also undergoing a renaissance having been awarded prizes for everything from smart tech to cleanliness. Surat, a place that few people outside India have heard of, had a red-hot economy before the onset of COVID-19. Its GDP has grown at an eye-watering 9.1 per cent over the past few years – the fastest in the world. According to the forecasting group Oxford Economics, its economy is set to continue growing rapidly until 2035.[140]

What makes Surat especially interesting is that it's been around for centuries. Over a thousand years ago the city was a gold and textile superpower. During the fifteenth and sixteenth centuries it was also

2019

Google / Google Earth Engine, USGS, NASA, ESA

2 km

1 mile

one of the world's great 'welcoming cities', with religious groups offering free health services for newcomers, as well as cows, horses and even insects. Surat's fortunes started declining because of the rise of Bombay (now Mumbai) in the eighteenth century. It started recovering after the opening of India's railways, and began growing again in the nineteenth century. In 2019, it was the tenth wealthiest city in India with a nominal GDP of some $57 billion, about the same as that of Costa Rica or Lebanon.

How inter-city diplomacy can change the game

In contrast to nation-states, cities and mayors are stepping up to the global challenges of the twenty-first century. Sub-national governments – including metropolitan ones – were among the most effective first responders to the COVID-19 pandemic. Growing numbers of city leaders are taking action to reduce their carbon footprints, absorb and protect migrants, and reduce inequality. In the process, cities are busily redefining the parameters of international statecraft, pursuing ever more entrepreneurial forms of city diplomacy. From China and India to Nigeria and Brazil, cities are setting-up trade and investment promotion offices and commercial ventures to attract capital, people and ideas. They are also forging municipal foreign policies, strengthening bilateral partnerships, and launching city networks on everything from cultural and scientific exchanges to welcoming refugees and providing humanitarian relief overseas.

Diplomacy comes naturally to cities. Mayors have long since dispatched diplomatic emissaries around the world to strengthen opportunities for everything from tourism to trade. In the process, cities have discovered they often have more in common with one another than with their own nations. One area of radical consensus is on climate mitigation and adaptation. As this book was going to press, over 10,000 cities representing more than 900 million people from every continent were committed to a global pact on climate and energy.[141] United by shared values and a sense of urgency, their stated goal is to accelerate action that can meet and ultimately exceed the 2015 Paris climate agreement targets.

Many cities are also resisting reactionary nationalism and helping reinvigorate democratic politics. For example, when the US President signed an executive order in January 2017 to cut funding to so-called 'sanctuary cities' protecting undocumented immigrants, hundreds of states, counties and municipalities refused to enact the legislation. After the White House decided to boycott negotiations on a new global compact to manage migration in December 2017, the largest US cities moved ahead regardless. US cities are not the only ones standing up to national authorities and their policies. Throughout his tenure, Sadiq Khan, the mayor of London, repeatedly called out reactionary politicians in Europe.[142] Cities like Barcelona and Madrid have publicly opposed Spain's restrictive asylum policies.[143] Some Italian cities have also awarded new migrants and asylum seekers 'local citizenship' even after they have been denied such rights by national authorities.[144]

Growing numbers of cities are also calling on international organisations to take urban issues more seriously. Perhaps the largest of these inter-city networks is the United Cities and Local Governments, or UCLG, which was established in 1913. A newer coalition is the Global Parliament of Mayors, a city rights movement calling for a seat at the multilateral table since 2016. City leaders are also engaging directly with UN organisations such as the International Organization for Migration and the UN High Commissioner for Refugees, particularly on issues related to protecting and caring for migrants and refugees. After intensive lobbying, cities influenced the shape of the Sustainable Development Goals, a fifteen-year agenda announced in 2015. This was the first time an international agreement explicitly vouched for the interests of cities alongside their national counterparts. Emboldened by these successes, city advocates are appealing to everything from the G20 to the G7 for a new urban agenda that promotes the rights of cities.

At a time when multilateralism and international co-operation is strained, cities are mobilising new forms of diplomacy and exchange. Today there are more than 300 inter-city coalitions – more networks for cities than for nation states. By forging agreements and standards to decarbonise, welcome refugees and counter radicalisation and extremism, cities are demonstrating their power and influence in an urban world.

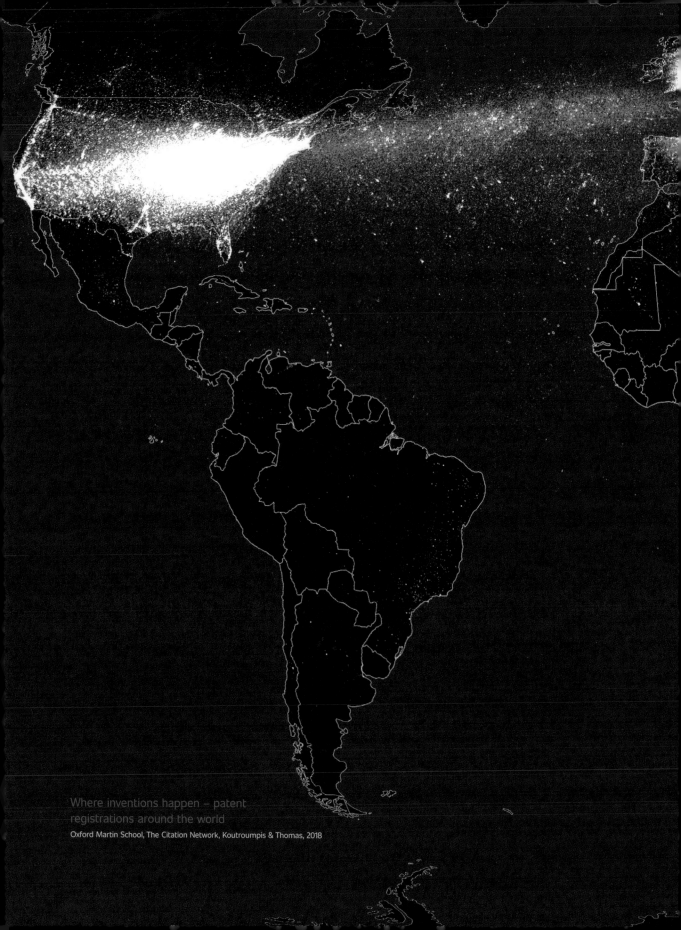

Where inventions happen – patent
registrations around the world

Oxford Martin School, The Citation Network, Koutroumpis & Thomas, 2018

Technology

Technological innovations underpin progress

Health technologies increase longevity sharply

Technologies can be a force for good or for bad

Artificial intelligence and robots will eliminate jobs

Where inventions happen – patent registrations around the world

Each white dot is a patent, with dots across oceans and continents depicting collaboration between the different poles of invention, mainly in the US, Europe, China, Japan, Taiwan and South Korea.

Canada

United States

Patent heat map – 2014
Oxford Martin School, The Citation Network, Koutroumpis & Thomas, 2018

Each white dot represents a patent registration

Introduction

One reason humans excel is because we collaborate in ingenious ways. This map of global patents shows the extent of cooperation between inventors. Each dot represents an individual patent. Where patents have more than one address, we have located the dot at the mid-point between the innovators' addresses. Dots above the Atlantic Ocean are indicative of collaboration between inventors located in the US and Europe, whereas those above central Asia reflect partnerships between innovators in China and Europe. The extent of collaboration between China, Taiwan, Japan and the US is evident in the stream of dots flowing across Asia and the Mediterranean. Within the US, the

extent of collaboration between the Northeast corridor and Silicon Valley is reflected in the solid white amalgamation of dots across the US, and similar patterns are visible across Western Europe and within China, reflecting collaboration between innovators located in various national research centres.

Technological change is accelerating not simply because many more people from diverse backgrounds are innovating but because they are increasingly assisted by digital connectivity and powerful supercomputers. It's as if all the chefs in the world are suddenly in the same giant kitchen with a constant delivery of new ingredients, cooking methods and baking devices. This chapter considers how technologies have changed our lives and asks whether the pace of change is accelerating or slowing down. We review whether technologies widen or reduce inequalities and how new innovations may solve critical problems we are facing now and into the future. Of course, technologies are dual-edged and can cause harm as well as hasten progress. Some of the most pressing existential concerns relate to how technologies are curbing privacy and taking away people's jobs.

What is technology?

Technologies are so seamlessly integrated into our everyday lives that we routinely take them for granted. It is only when they fail – when our internet goes down or there are delays in producing vaccines – that we remember just how heavily we rely on them. Technology, according to the science writer Brian Arthur, is a 'means to a purpose', a tool to help find solutions to problems.[1] When we think of technology, many of us think of computers and other digital gadgets: smartphones, smartwatches and self-driving cars. The truth is that virtually everything we use in our daily routines is a product of technological progress. The book you are holding. The computers we used to write this chapter. Pens, paper, clothes, running water and toilets. Technology permeates literally every aspect of our lives.

The capability to develop sophisticated tools, including the most innovative technologies that dazzle us today, has given our species an edge and provides the foundation for every human achievement. The greatest impacts on our well-being have come from breakthrough technologies, starting around 2.6 million years ago[2] with stone tools and at least a million years ago with fire.[3] It was a very long time before our ancestors started domesticating animals and developing crops, only about 10,000 years ago. More recently, as the chapter on health shows, advances in population health, including the control of diseases through water purification, sanitation and vaccinations, have improved the quality and length of human life.

Humans are also the most destructive of species, and weapons – from arrows to ammunition – have injured and killed millions of people. As we explain in the chapter on violence, they can cause mass destruction and devastate lives. In addition to the weapons designed to maximise terror and harm, there have been many unintended harmful consequences of technologies that were developed with peaceful intent. We have belatedly discovered that fossil fuels, aerosols, plastics, fertilisers, herbicides, asbestos and a widening range of innovations are destroying human health, biodiversity and the ecological systems that sustain life on earth.

Until the late seventeenth and early eighteenth centuries, almost everyone lived on the modern equivalent of $400 to $1,000 a year, just above subsistence level.[4] The industrial revolution, which started in England around 1760, provided a trigger for dramatic improvements. Fossil fuels like coal provided the means for advances in propulsion, heating and electrification. Only two and a half centuries later are we finally realising to what extent this energy source is damaging the global climate and endangering our future survival. Fortunately, alternative forms of energy are available, providing the foundation for an energy revolution, which together with the genomics and artificial intelligence (AI) revolutions currently underway, promises to be even more profound than previous periods of transformative technological change.

As the potential of technology grows, so too does our need to understand and manage it. AI has the potential, in combination with social media, to undermine democracies and to create new means to control human behaviour and thought, with Orwellian possibilities. Some types of innovations – what we call 'general purpose technologies' – were genuinely transformative. They introduced new ways of producing a very wide range of products and spurred many previously unknown opportunities for adaptation and invention. Examples include the printing press, the steam engine, electricity, the internal combustion engine and, more recently, computing and information technology.[5]

The development of computers is arguably the most significant technological development of the post-World War II era. In 1965, Gordon Moore predicted that the number of components on an integrated computer circuit would double every year, which he later revised to every two years. His observation became known as Moore's Law and has defined a historical trend that has lasted for fifty years. The impact of the exponential improvement in computing power is pervasive. Our smartphones now have more than 100,000 times the

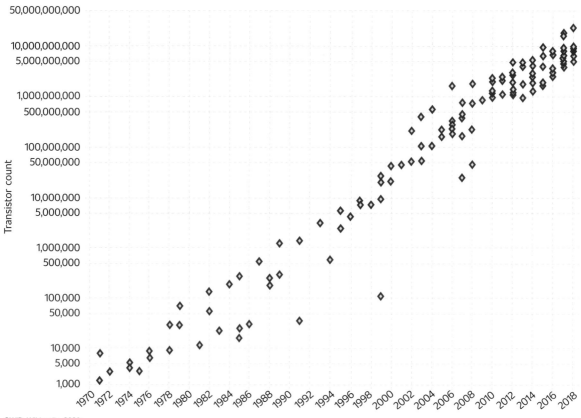

OWID, Wikipedia, 2020

processing power and 7 million times the memory of the computer that guided Apollo 11 to the moon over five decades ago.[6] In twenty years, they will be 100,000 to a million times more powerful, with the difference in estimates reflecting whether one is an optimist or pessimist in the debate which has raged for decades around claims that Moore's Law is running out of steam.[7]

Improvements in computing power are the engine of the present technological revolution. For the first time, we can observe objects ranging in scale from subatomic particles to the universe. Entirely new disciplines and approaches have emerged, such as biotechnology, genomics and quantum computing, with each of these providing new springboards for further scientific advances. In medicine, for example, these hold out promise for major breakthroughs in many areas, including in cures for cancer, and through DNA sequencing and associated methods many crippling hereditary and other diseases could be addressed. In surgery robotic augmentation is already leading to more accurate treatments and faster recovery times.

Exponential power: rising processing speed 1970–2018
The figure shows the exponential rise in processing power over time as measured by the number of transistors on integrated circuit chips; these have risen from about 1,000 in 1965 to over 50 billion today.

Turbocharged technological change

While computing power provides the hardware for accelerating change, education and connectivity provides the software. Technological change is accelerating because of the increase in the number of connected brains generating new ideas. In 1990 the world's population stood at around 5.3 billion, of which almost 4 billion were literate.[8] Today, of the 7.7 billion people on earth, about 6.7 billion are literate.[9] Not only are there 2.7 billion more people able to read and write, but they are also increasingly interconnected, eager to discover new ideas and share their own in ways that were totally unimaginable half a century ago. Advances periodically occur as a result of individual brilliance, but also they are often down to teamwork. Individuals spark ideas in each other, which also explains why dense cities are hubs of creativity and productivity.

The rapid exchange of new ideas and innovations is dramatically scaled up by the internet, which allows, for the most part, free, open and continuous communication. We are mindful of the fact that too often this is undermined by censorship and that fake news and bad ideas are also spread on the internet. But at its best, the exchange of ideas and the ability to store vast amounts of data in the cloud have dramatically augmented the capacity of individuals, companies, labs and start-ups in distant locations, often on different continents, to exchange ideas, establish global collaborations and churn out ideas across time zones. In general, the more diverse the team, including in relation to nationality, gender, race and age, the more innovative its output.[10] Migrants in particular are often high-achievers and it is no accident that they are disproportionately represented in patent citations, Nobel Prizes and Academy Awards.[11] Albert Einstein, whose name is synonymous with invention, was a refugee, as was Sergey Brin, the co-founder of Google, and the father of Steve Jobs, the founder of Apple.

Not only are there more ideas and technologies being produced, but they are being adopted more rapidly and in more places than ever before. This is due to rapid communication and the globalisation of supply chains, including digital ones. It is also because the costs of new technologies are lower and therefore easier to adopt than ever before. Indonesia only managed to acquire one of the steamships it invented 160 years afterwards.[12] Electricity took sixty years to reach Kenya.[13]

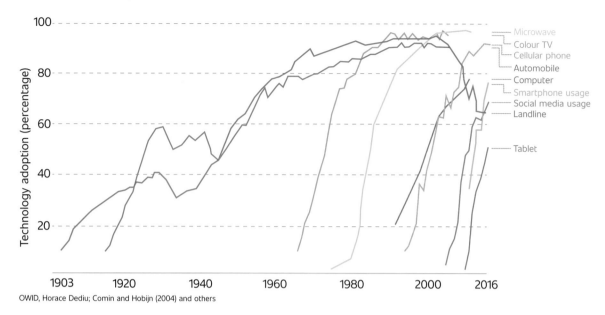

OWID, Horace Dediu; Comin and Hobijn (2004) and others

Desktop computers were widely available in Vietnam fifteen years after they were invented in the US.[14] The latest iPhone was on sale in more than 100 countries within a week and was available in China before we could purchase one in Oxford or Rio de Janeiro.[15] Meanwhile, in Rwanda, the pioneering delivery of blood products and life-saving vaccines by drones is occurring long before it happens elsewhere.[16] In fact, today's technologists designing solutions in some of the world's most challenging settings may change the game. This is because more pressing needs, combined with access to the latest technologies, create better conditions for radical innovation and rapid scale.[17]

Accelerating adoption rates are due in part to the declining costs of these technologies, relative to rising incomes. This also happens because many of the breakthroughs in information and communication technologies (ICT) do not require the scale of investment in infrastructure and services that accompanied the rollout of railways or electricity.

It's getting steeper: Technology adoption in US households from 1903
Adoption rates of new technologies have accelerated over the past 120 years. Contrast, for example, the slow rise in the proportion of US households that used landlines in any one year, to the almost vertical increase since the 1990s in the use of mobile phones.

Leapfrogging

Techno-evangelists argue that low-income countries in parts of Africa, Asia and Latin America could use technology to leapfrog over the failures that currently characterise their education, health, energy and other sectors. They could do this by side-stepping capital, intensive infrastructure and investing in lower-cost and digitally-enabled pathways instead.[18] At a minimum, new technologies are helping some societies overcome the constraints of expensive legacy systems. For example, digital health, mobile banking and remote education have already given millions of poor people in low-income countries access to services at a greatly reduced cost. A widely celebrated example of leapfrogging in ICT is that of Kenya, where 80 per cent of adults use mobile phones for banking. Compare this to the UK, where barely 40 per cent use mobile banking, or the US, where the proportion is closer to 30 per cent.[19] M-Pesa (M for mobile, and pesa, meaning money in Swahili) was launched in 2007 in Kenya and has rapidly become the most widely used system for payments and transfers, with more than 33 million accounts in a poor country of 50 million people.[20] The success of M-Pesa has given rise to other platforms, including M-Tiba, and M-Health, which allows subscribers to set funds aside for health care. Funds stored on M-Tiba are used for services and medication at health centres. Specialised maternal and infant care products are particularly popular, with local clinics in the Kibera shanty town offering ultrasound scans at a fraction of the price and with much greater accessibility than in public hospitals.[21]

Mobile banking took off in Kenya as mobile phone penetration was high and the relatively poorly developed incumbent banking system allowed newcomers to penetrate the market, whereas entrenched banks

Leapfrogging: Mobile and landline telephone subscriptions in the UK and Gambia

The figures show that Gambia, which has a per capita income of $716,[24] and the UK, with a per capita income of $42,962,[25] have about the same level of mobile phone penetration, with Gambia having leapfrogged fixed lines, which never really penetrated.

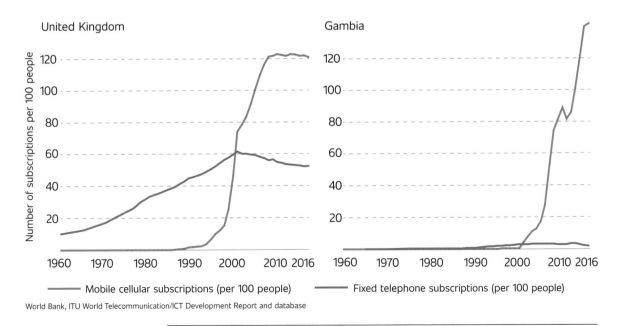

World Bank, ITU World Telecommunication/ICT Development Report and database

and regulators prevented this in the UK and Europe.[22] Until recently over half of the mobile bank accounts in the world were in Africa, but China is now by far the biggest mobile banking market, with well over 10 billion transactions a year through Alipay, WeChat and other platforms.[23] The development of mobile banking has allowed millions of people, many of whom previously did not have bank accounts, to access credit and transfer money in a secure, efficient and low cost manner.

Technological 'leapfrogging' has enabled some new-comers to benefit from the absence of major prior invest-ments and the associated inertia. The idea that latecomer countries can catch-up or even leapfrog is an attractive one but in practice only applies to a comparatively small set of technologies that can be introduced independently of the institutional, educational or economic environment.

From dirty to clean energy

New technologies offer the potential to overcome some of humanity's greatest challenges, such as reducing our fatal dependence on fossil fuels. As we noted in the climate chapter, many countries signed up to the global commitment in 2016 in Paris in order to keep global temperatures from rising above 2 degrees Celsius. This means that fossil fuels must be phased out by 2050 and societies will have to shift to renewable and other non-fossil forms of energy. The rapid decline in the cost of wind and solar energy is the result of successive improvements in these technologies. The maps on pages 148 and 149 show the remarkable growth between 2010 and 2018 in the renewable solar and wind capacity of many countries, in gigawatt hours, with purple representing wind and red solar capacity. By 2018, China's renewable capacity had outstripped other countries, with wind capacity (211GWh) exceeding its solar capacity (175GWh).[26] In the European Union, wind in purple (178GWh) similarly

Mobile phone penetration: Rapidly rising connectivity
Whereas in 2000 mobile phones were largely confined to a minority of people in rich countries, by 2018 they were almost universal, with only the poorest countries still disconnected, depicted in light green. North Korea and Eritrea, which prohibit connectivity, are depicted in yellow.

2000

2018

Mobile cellular telephone subscriptions – per 100 inhabitants
World Bank, ITU World Telecommunication / ICT Development Report and database, 2019

0   100

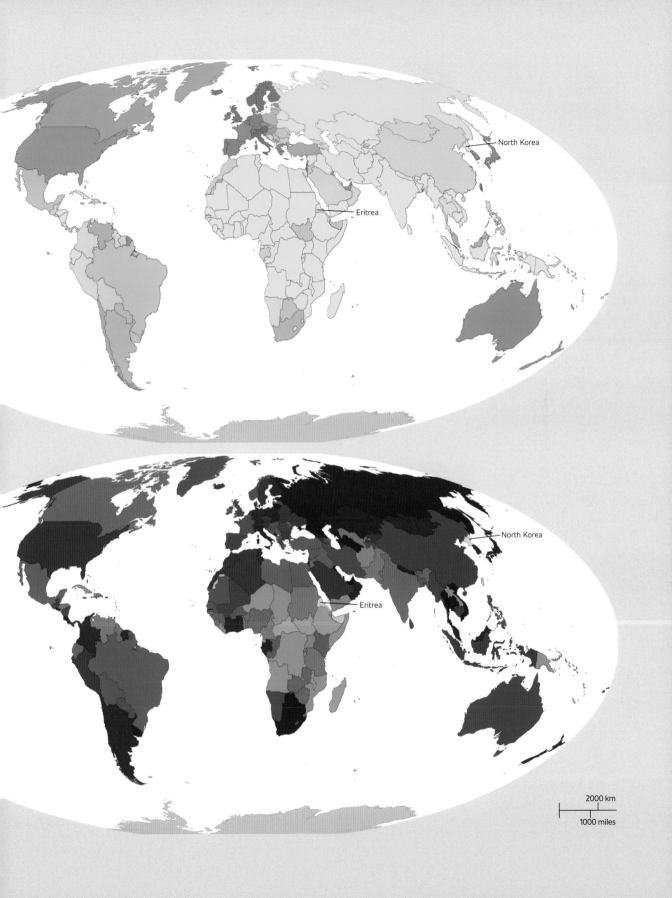

North Korea

Eritrea

North Korea

Eritrea

2000 km
1000 miles

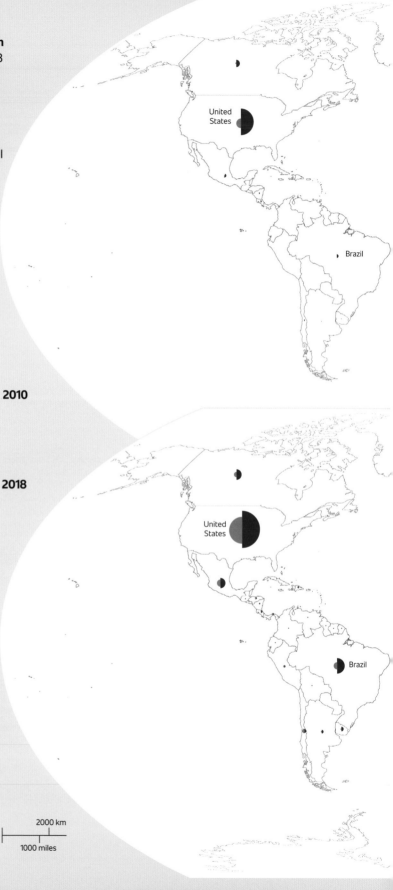

Hot and windy: Renewable power growth
The rapid growth between 2010 and 2018 in the renewable solar and wind capacity of many countries, in gigawatt hours, is apparent, with purple representing wind and red solar capacity. By 2018, China's renewable capacity had outstripped other countries' capacities, followed by all the EU countries' capacities combined.

Wind and solar electric capacity per country
International Renewable Energy Agency (IRENA), 2020

● Wind power (GWh)

● Solar power (GWh)

2010

2018

2000 km

1000 miles

provided more power than solar (115GWh) as it did in the UK, where wind generated 21GWh and solar 13GWh.[27] In the US, wind (89GWh) also exceeded solar (62GWh), while Japan came fourth in terms of generating solar power (55GWh) but nineteenth in terms of wind generation (4GWh), which is why the red circle is so much larger than the purple one.[28]

The distribution of solar power installations across the US is also evident in the map below. While the concentration of photovoltaic capacity in sun-drenched and renewable-friendly California is not surprising, the difference between Nevada and Arizona is striking, as is the relative absence in Florida and concentration in the Northeast. In part this reflects population density. These differences in solar installations also reflect the extent to which the regulatory and taxation environment encourages or discriminates against solar power. The importance of regulation is clear when we zoom in on the Northeast of the US, as we can then clearly see that the state of Rhode Island, which failed to provide incentives for solar power, has a far lower rate of adoption than other states in the Northeast corridor, such as neighbouring Connecticut or Massachusetts.

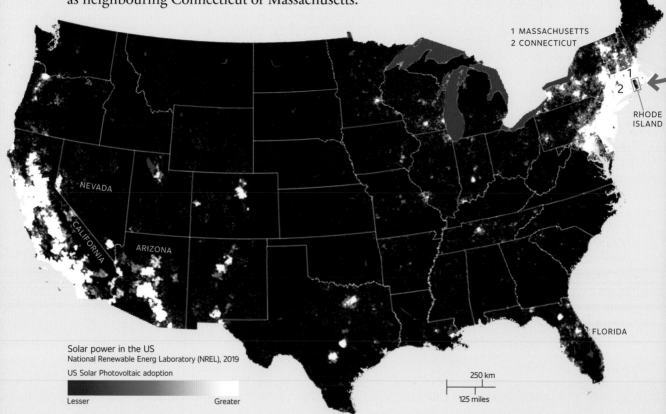

1 MASSACHUSETTS
2 CONNECTICUT

RHODE ISLAND

NEVADA

CALIFORNIA

ARIZONA

FLORIDA

Solar power in the US
National Renewable Energ Laboratory (NREL), 2019

US Solar Photovoltaic adoption

Lesser   Greater

250 km
125 miles

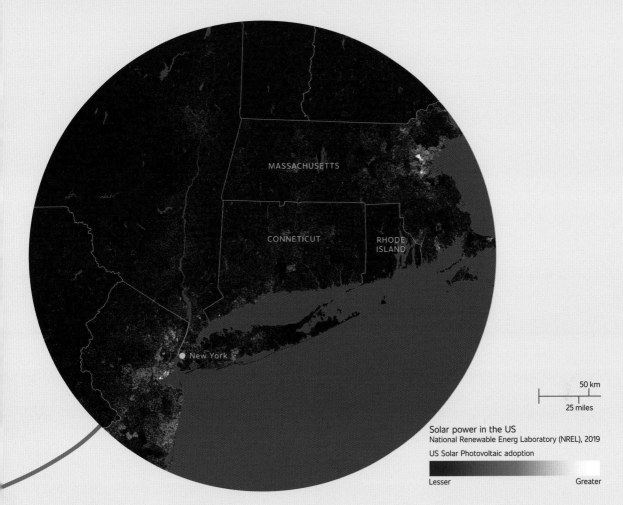

MASSACHUSETTS

CONNETICUT

RHODE
ISLAND

● New York

50 km

25 miles

Solar power in the US
National Renewable Energ Laboratory (NREL), 2019

US Solar Photovoltaic adoption

Lesser   Greater

Solar power in the US is growing quickly, but regulation matters
The differences in solar power identified in white reflect the regulatory and taxation environment that either encourages or discourages investment in renewable energy. Zooming in on the northeast we see that Rhode Island has a far lower rate of adoption due to its lack of incentives for solar power technology.

While not moving fast enough, the world is transitioning to more renewable energy solutions. The energy transition needs to be fast, but also depends on the widespread adoption of electric vehicles. Despite major investments from all the leading car manufacturers, at current rates of adoption electric vehicles (EV) are unlikely to account for more than 10 per cent of all vehicles on the road globally by 2025.[29] In China however, it may well exceed this, due to the determined support from the Chinese government which includes offering subsidies on electric vehicles and regulatory limits on the use of internal combustion vehicles in inner cities, as well as highly visible support for EV manufacturers, not least the establishment of Tesla's Gigafactory in Shanghai. Norway is even more advanced in its transition, with about half of the new cars sold now being electric.[30] Regulatory changes and subsidies could rapidly accelerate the transition to EVs globally – as they have in Scandinavian countries – as could developments in battery life and the rollout of charging facilities. Motivated by concern over rising pollution, electric car sales are booming in China. The map of sales of EVs shows that China accounts for about half of global demand, with more than 1.2 million sold in 2019. Although this remains a small share of the 26 million vehicles sold in China in 2019, the Chinese market for electric vehicles is growing twice as fast as that in the US, and Chinese purchases now account for one out of every two such cars sold globally.[31]

Global digital nervous system

The internet is the world's nervous system. This means that individuals who are not connected will find it increasingly difficult to hear and see what is happening, much less make use of public, private and other services that are powering the digital economy. As more and more people go online, the risks of misinformation also rise. The rapid improvement in connectivity over the decade to 2018 is immediately evident when comparing the two maps of internet access on page 155.

In rich countries, over 85 per cent of the population now have access to the internet, as is depicted in purple, whereas the average for developing countries is 43 per cent and for Africa just 35 per cent. The countries where less than 10 per cent of the population is connected are shown in yellow. In Latin America and Asia, the situation improved markedly by 2018, by which time China had 50–60 per cent coverage across the country, indicated in pink. This remains well below the high level of connectivity in Europe and Australia and the almost total access in Canada, Norway and the parts of the UK that are depicted in darker purple. Among middle-income countries, Chile in Latin America stands out as exceptionally well connected, as does Malaysia in South East Asia.

Electric vehicle sales are growing fastest in China
China accounts for about half of global demand, with more than 1.2 million sold in 2019.

New electric car sales – 2014 and 2018
International Energy Agency (IEA), 2019

>1 million   500,000   100,000

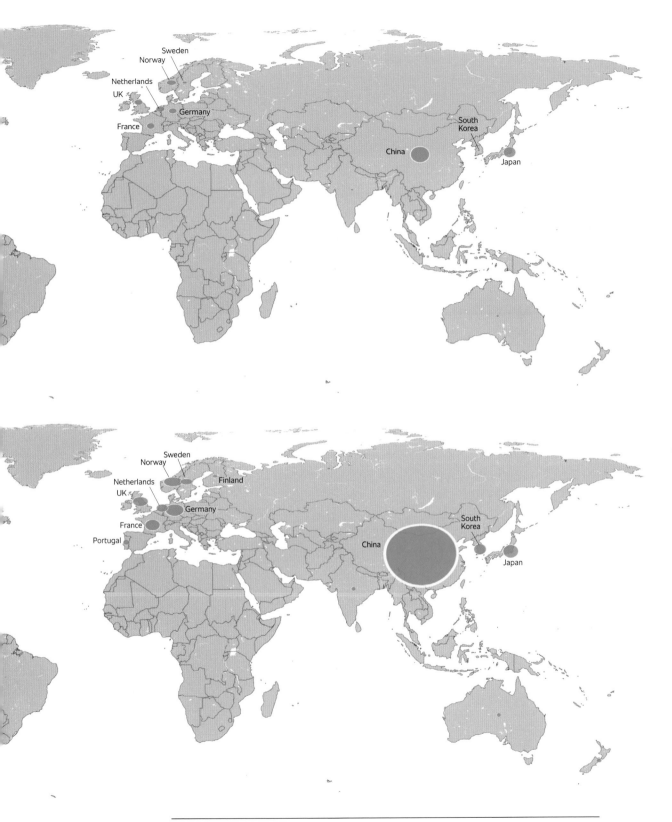

If the internet is the nervous system, then servers are the muscles keeping everything running smoothly. The map overleaf shows the estimated distribution of internet servers in the world, taken from a one-off rarely published analysis following a major botnet attack in 2012.[32] The intensity of the dots reflects the number of servers and although the number of servers has increased markedly since this image was created, the global distribution remains broadly as reflected in the map. Africa has a bigger population than the Americas and Europe combined, but as is apparent, far less internet connectivity. This digital divide undermines the growth potential of Africa and the ability for its vast informal and youthful workforce to benefit from online education and employment opportunities.

Download and upload speeds have increased tenfold every five years since the early 1990s.[33] Older readers will recall the modems that took all night to download large documents or a single song. In 2000, just 4 per cent of households in the US had broadband connections and around 40 per cent were using dial-up modems with a speed of 1Mbps at a cost of around $250 per year.[34] By 2010 speeds in the US averaged 10Mbps and by 2019 while download speeds had increased to around 34 Mbps, average upload speeds remained stuck around 10 Mbps, placing the US in 10th place globally for download and a poor 94th rank for upload, comparable with Angola or Poland.[35] Globally, download speeds are accelerating at an average of 15 per cent per year.[36] With this rate of change, late adopters often have faster speeds than earlier adopters, as new fibre-optic cables are far superior to older copper wires. Singapore currently has the fastest fixed download speed, at 175Mbps, compared to 34Mbps in the US and 30Mbps in China.[37] As more and more devices become connected, through the Internet of Things (IoT), and data increasingly is stored in the cloud,

Global internet access over the period 2000 to 2018 has improved, but remains uneven
The dramatic growth in internet access is evident from the comparison of the 2000 and 2018 maps. In rich countries, over 85 per cent of the population now have access to the internet, as is depicted in purple whereas the average for developing countries is 43 per cent and for Africa just 35 per cent. The countries where less than 10 per cent of the population is connected are shown in yellow.

2000

2018

Share of the population using the internet
World Bank, ITU World Telecommunication /
ICT Development Report and database, 2019

% of individuals who have used the internet in the last 3 months

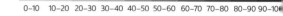

0–10   10–20   20–30   30–40   40–50   50–60   60–70   70–80   80–90   90–100

2000 km

1000 miles

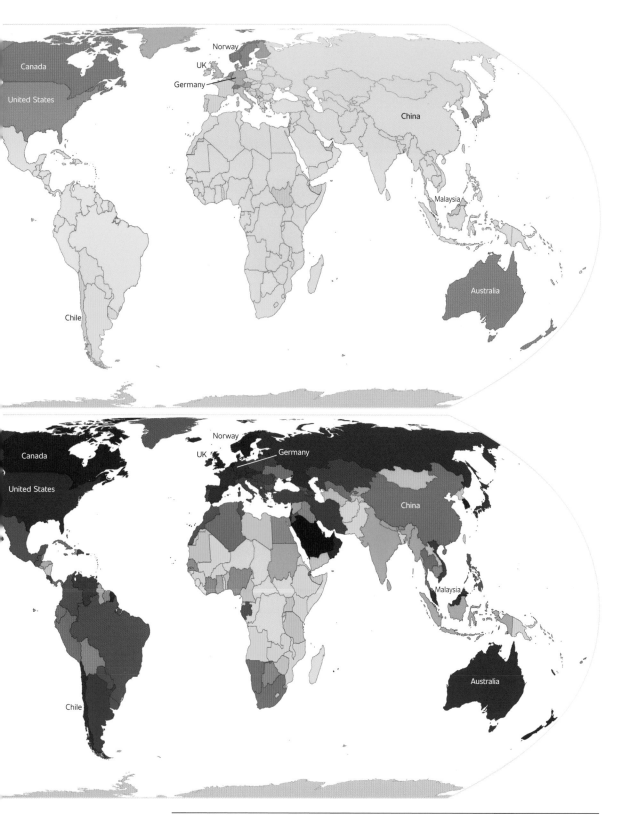

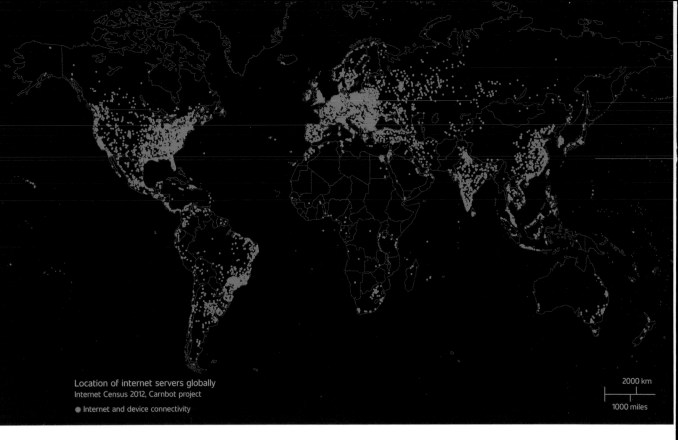

Location of internet servers globally
Internet Census 2012, Carnbot project
● Internet and device connectivity

2000 km

1000 miles

business processes and efficiency increasingly will require a fast internet. Countries and companies left behind will find that they are simply unable to compete and adopt new technologies. Extremely low latency, which is a measure of the delay in time in receiving or sending messages, is critical to the adoption, for example, of cloud computing and for the safe operation of autonomous vehicles.

The rollout of 5G networks promises to be a game-changer. 5G networks have an average download speed of 1Gbps and, as the map shows, are already being adopted by mobile operators in more than ninety countries. These will become the industry standard for private networks with applications for the IoT, local networking and critical communications.[38] The rollout of vastly improved data speeds will enable the development of a digital system that is 100 times faster and also more reliable than the present one. As the chapter on geopolitics shows, the dominance of Chinese-based Huawei in the development of key hardware has become a lightning rod for concerns regarding dependence on China.[39]

The US and some Western governments have applied considerable pressure to ban the installation and use of Huawei equipment. With almost 60 per cent of Huawei's 5G contracts in Europe, the

Location of internet servers globally
The dots reflect internet and device connectivity. Africa has a bigger population than the Americas and Europe combined, but as is apparent, far less internet connectivity.

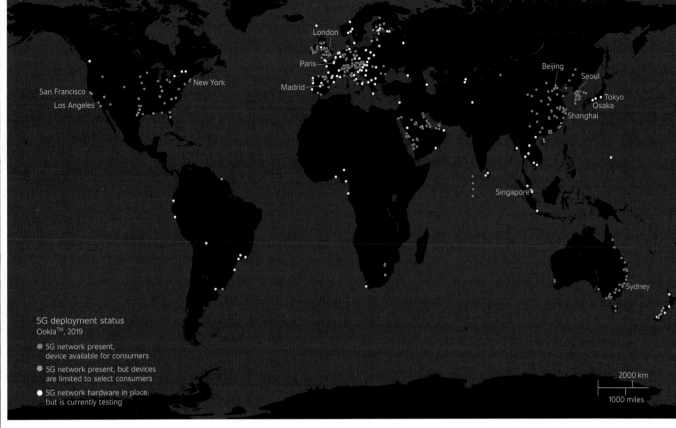

5G deployment status
Ookla™, 2019

● 5G network present,
   device available for consumers
● 5G network present, but devices
   are limited to select consumers
● 5G network hardware in place,
   but is currently testing

2000 km
1000 miles

Rollout of 5G networks globally

The map shows the deployment status of 5G networks globally in 2019, highlighting the extent to which this at present has been concentrated in China, the east coast of Australia and in Europe and the Gulf states.

response of governments, not only poses a threat to the profitability of the company but also threatens to slow the rollout of 5G around the world.[40] The technological race between the US and China is at the top of the geopolitical agenda. The battle over Huawei reflects a much wider anxiety about the seemingly unstoppable capacity of China to compete with the US on AI, not least because of the former's unfettered access to massive datasets and the involvement of the state in home-grown technology giants.[41] As Russia's President Vladimir Putin observed: 'Whoever becomes the leader in [AI] will become the ruler of the world.'[42]

Modern electronic devices are among the most complex things that humans produce. Their components are shuttled around the world in a relay race, crossing oceans and borders several times before appearing as a final manufactured product.[43] One country's ban on a global company can have repercussions that ripple all around the world. The 2019 decision by the US government to prohibit American companies from supplying parts to Huawei and seventy of its affiliates cost the top suppliers $14 billion in just three months.[44] Although Huawei has its own semiconductor subsidiary, it imports some $11 billion of chips a year from Qualcomm, which is based in

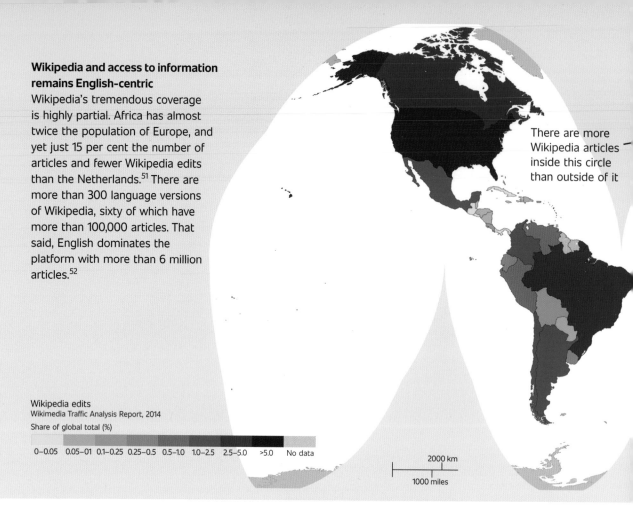

Wikipedia and access to information remains English-centric

Wikipedia's tremendous coverage is highly partial. Africa has almost twice the population of Europe, and yet just 15 per cent the number of articles and fewer Wikipedia edits than the Netherlands.[51] There are more than 300 language versions of Wikipedia, sixty of which have more than 100,000 articles. That said, English dominates the platform with more than 6 million articles.[52]

There are more Wikipedia articles inside this circle than outside of it

Wikipedia edits
Wikimedia Traffic Analysis Report, 2014

Share of global total (%)

| 0–0.05 | 0.05–01 | 0.1–0.25 | 0.25–0.5 | 0.5–1.0 | 1.0–2.5 | 2.5–5.0 | >5.0 | No data |

2000 km
1000 miles

San Diego and makes about half of the world's broadband processors. The main semiconductor manufacturers in turn buy their micro-processors from Taiwan and South Korea, with components and raw materials sourced from more than a dozen other countries.[45] It is not difficult to see that while the US ban may have been designed to slow AI development in China, owing to the integrated nature of supply chains, it will have a chilling effect on the rollout of 5G everywhere.

Meanwhile, the internet has created a platform for the provision of services and is transforming the nature of production. Examples include the largest ride-hailing service, Uber, which owns no cars, and Airbnb, which lists more short-term accommodation than the world's top five hotel companies put together yet owns no properties. It has also given rise to Wikipedia, an open platform that has replaced *Encyclopaedia Britannica*, a venerable repository of knowledge that was in continuous production from 1768 to 2012 and had about 100 full-time editors.[46] In truth, *Encyclopaedia Britannica* didn't stand a chance.

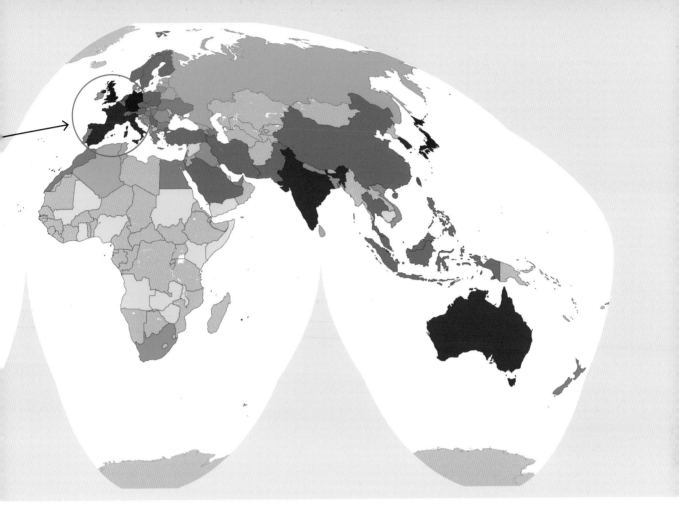

It was up against hundreds of thousands of volunteers, whose daily inputs revise and expand on Wikipedia's ever-increasing content. On any given day, it is accessed by 15 per cent of all internet users.

While more people have access to more content than ever before, this does not mean there are not biases or gaps in the information that is available. The map displays the number of average monthly Wikipedia edits that emanated from each country in 2014, the latest year for which data are available. All of Africa combined reported fewer edits than the Netherlands.[47] There are more than 300 language versions of Wikipedia, sixty of which have more than 100,000 articles. That said, English dominates the platform with more than 6 million articles.[48]Africa has almost twice the population of Europe, and yet just 15 per cent the number of articles.[49] Indeed, there are more articles written about Antarctica – which has no permanent residents – than most countries in Africa, and many in Latin America and Asia.[50]

IoT is the acronym given to the extension of the internet to physical devices and everyday objects, enabling them to send and receive data. It is expected to lead to an even greater dependence on the web in the coming decades, with an exponential growth in connectivity of devices reflecting the equally rapid decline in the cost of sensors and Wi-Fi or Bluetooth connectivity. It will be built on hundreds of billions of sensors and will provide a rapidly growing mountain of data, which each year will be greater than all that created in previous years. Such information is a treasure trove for the companies that can access it and provides the raw material for the rapidly growing sophistication of AI.

The IoT is projected to grow 30 per cent annually over the coming years, with more than 75 billion automated devices estimated to be linked to virtually every person, place and company by 2025.[53] There is no doubt they will generate significant dividends in everything from energy consumption to health and well-being.[54] Even so, the interdependence of a growing range of our activities on interconnected systems creates new risks and greatly amplifies existing privacy and security concerns. One concern is what Harvard scholar Shoshana Zuboff has called 'surveillance capitalism'. Her book shows how this ubiquitous system threatens to track and manipulate our actions, with profound impacts on our behaviour and politics. Another concern is the vulnerability of everything from door locks, vehicle control systems, heart pacemakers and bank accounts to identity theft and hacking as well as to sudden system failures. Already smart cities are being hacked around the world. In just the past few years, ransomware has been used to disrupt the municipal tram system in Dublin, to jam air traffic control and railway ticketing systems in Stockholm, and to shake down power plants from Johannesburg to Hyderabad.[55]

The development of cryptocurrencies such as bitcoin has led to an explosive growth in demand for computer servers. Bitcoin 'mining' is performed by high-powered computers designed to solve complex mathematical problems involving the correct sequence of a mind-boggling random array of sixty-four-digit numbers, which have a one in 13 trillion chance of being solved.[56] Each time the problem is solved a new bitcoin is issued, with the unique solution providing the proof that the bitcoin has been produced. The verified bitcoin becomes part of a public record known as a 'blockchain', with the various blocks that compose the chain kept on the different sets of computer servers, known as 'nodes'.[57] The growth in server demand for increasingly energy-intensive cryptocurrency mining and the spread of server farms is already producing more than twenty-two megatons of greenhouse gases annually, equivalent to Las Vegas or Hamburg, and in recent times this has quadrupled every year, to far exceed Ireland's annual demand for energy.[58]

What is artificial intelligence?

AI is a buzzword, but there is also confusion over what it means. AI is a term that has been around since the 1970s and refers to the ability of computers to be programmed with mathematical formulas, known as algorithms, to recognise patterns in vast amounts of data. Writing the sequence of events into code means that the machines can draw conclusions and take actions when they identify similar patterns.[59] Advances in computing power and the exponential growth in data resulting from connected devices are leading to very rapid increases in the capabilities of AI. Machine learning (ML) builds on AI by using statistical probability functions to detect patterns and to constantly update them to make better predictions. These can include everything from the types of advertisements Facebook includes on your newsfeed and movies recommended by Netflix to the spelling of specific words you are typing on your new Samsung or Apple phone.

Although they might disagree on the balance of positive consequences and risks, both Elon Musk, the founder of SpaceX and CEO of Tesla, and Jack Ma, the co-founder of Alibaba, along with many others believe that AI will fundamentally transform our societies.[60] The development of AI is still in its infancy, but already by 2017 a leading AI company, DeepMind, had beaten the world Go champion, with this regarded as a watershed in AI development since Go is among the most complex games for computers to solve.[61] Some of the most exciting applications of AI involve the creation of virtual experiments to resolve hitherto intractable problems. For example, in medicine millions of different combinations of molecules are tested to try to identify possible cures for cancer and other diseases.[62] Pioneers of AI, including DeepMind and OpenAI, aim to create highly intelligent devices that can solve problems independently, without programming.

The exponential increase in computing power means that machines will acquire perfect recall, multitask in extraordinary ways, and become more intelligent than humans, acquiring what Oxford philosopher Nick Bostrom calls in his book of the same title, 'superintelligence'.[63] Machines and robots can already do many tasks more effectively than humans, such as arithmetic, navigation or welding, in a fraction of the time. However, in our view, machines are unlikely to perform a range of creative and highly dexterous functions, such as are required in caring for infants or the elderly, or many skilled jobs, at least not in the next couple of decades.

Over 2.5 million industrial robots operate around the world, with about 30 per cent in auto and 25 per cent in electronics and the remainder in a wide range of factories, warehouses and other sites. Over 400,000 per year are being added and three-quarters of the total installed are in five countries, China, Japan, Korea, Germany and the US.[64] In these countries robots have already replaced well over half of the jobs in the automotive and related industries, such as vehicle insurers and petrol pump attendants.[65] The highly uneven distribution of robots within countries, as well as between them, is evident from the maps. In the US, the concentration of robots in the Midwestern manufacturing states reflects their historical role as the centre of automotive production lines, not least in Michigan. The economist Carl Frey has shown that raised anxieties regarding automation led workers in the US to support the election of President Trump in 2016.[66]

There are growing apprehensions that AI, together with powerful robots, will soon replace humans in the workforce. While it is true that AI is advancing at breakneck speed, developing capabilities that previously only humans could perform, it is not the case that humans are about to become redundant. Carl Frey has pointed out the importance of distinguishing between two types of technology: *enabling* technology, which helps skilled workers increase their productivity, for example through the use of software or smart tools, and *replacing* technology, which does a job *instead* of a worker.[67] The former pushes up wages, employment and the labour share of income. Technologies that allow machines to perform a widening range of current jobs do the opposite. AI is replacing many work tasks that are rules-based and repetitive, and which do not require great agility or empathy.

Robots per 10 thousand workers
International Federation for Robotics, 2018

Number of robots/labour force
per 10 thousand workers

| 0–1 | 1–5 | 5–25 | 25–50 | >50 |

Robots are taking jobs
The highly uneven distribution of robots within countries, as well as between them, is evident from the maps. In the US, the concentration of robots in the Midwestern manufacturing states reflects their historical role as the centre of automotive production lines, not least in Michigan and Indiana.

Industrial robots in the workforce – USA
International Federation for Robotics, 2018

Number of industrial robots per 1,000 workers

0–1   2–3   4–5   6–7   >7

2000 km
1000 miles

250 km
125 miles

It is not only manufacturing jobs that are threatened by automation. Services now account for more than 80 per cent of employment in many advanced economies. Automated call centres are already receiving higher customer satisfaction ratings than those staffed by experienced phone operators.[68] And computerised systems have already replaced much of the paper processing that formerly characterised the back offices of financial, legal, accounting and retail operations. Improvements in AI mean that translators, lecturers, store assistants, receptionists and even journalists, actors and musicians are among the next wave of workers to be replaced. As we highlight in the chapter on inequality, the question is not only what jobs will be available, but also where the new jobs will be located. This is a vital question, as many countries are experiencing the rise of a new dual economy, where pockets of displaced workers appear unable to move and find work in the dynamic cities that are experiencing record high employment and income levels.

There is a wide range of views on how the latest wave of technological change will impact employment. In a widely publicised study, Carl Frey and Michael Osborne found that, assisted by AI, 47 per cent of all jobs in the US could be automated over the coming decades.[69] Drawing on this work, the World Bank estimated that two-thirds of jobs in developing countries will soon be at risk.[70] The Organisation for Economic Co-operation and Development is more optimistic, predicting that just 14 per cent of jobs in the OECD countries are likely to be threatened.[71] Meanwhile, the consultancy group McKinsey is even more sanguine, predicting that fewer than 10 per cent of all jobs are at risk.[72] In part the different projections reflect the range of assumptions regarding not only technological possibilities but also the political and regulatory responses adopted by governments. None of the studies are able to provide clarity on new jobs or on whether people displaced from existing jobs will be able to find new jobs.

Technology is impacting not just the number and location of jobs, but also their quality. It is facilitating changes in employment from sustained contractual commitments to purely transactional interactions, in which job security, training, conditions of work and loyalty are no longer considered part of the package. Although the outsourcing of work and the undermining of the bargaining power of workers has long been a feature of capitalism, the ability to fragment work into discrete tasks that can be undertaken by a combination

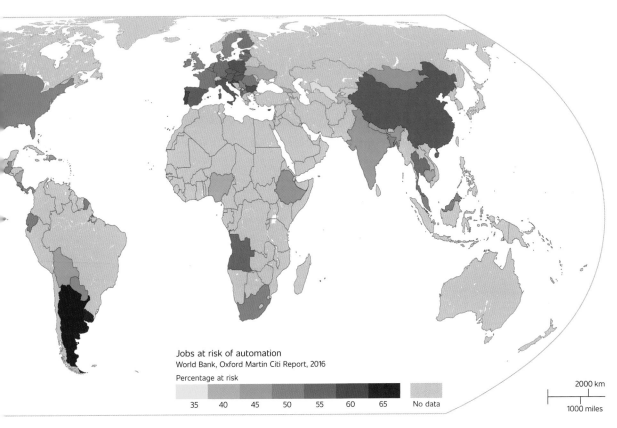

Jobs at risk of automation
World Bank, Oxford Martin Citi Report, 2016

Percentage at risk

| 35 | 40 | 45 | 50 | 55 | 60 | 65 | No data |

2000 km
1000 miles

Rising risk globally of job losses due to automation

A third to two-thirds of jobs around the world are vulnerable to automation over the coming decades. Lower wages and slower technology adoption may slow the impact on developing countries, but machines rather than people are likely to be engaged in mass production manufacturing and repetitive service jobs in the future.[73]

of machines and people is being taken to a new level by automated systems enhanced with AI. As the economist Richard Baldwin discusses in his book *The Globotics Upheaval*, this new phase of outsourcing involves professional services that can just as easily be done by lawyers, architects, accountants or others at far lower cost. Platforms such as Upwork and TaskRabbit allow foreign-based freelancers to replace local employers with global telecommuting.

AI threatens jobs in poor countries

What developing countries have in abundance is low-cost, semi-skilled, rather than skilled, labour. Their ability to leverage this to transition from low- to middle- and then higher-income status has provided the foundation for all previous examples of countries escaping poverty, giving rise to the famed East Asian Tigers, led by South Korea, Singapore, Hong Kong and Taiwan, and later followed by China, Thailand and Vietnam. The worry is that many of the low-skill repetitive jobs that previously provided the middle rungs that allowed countries to climb the development ladder are being removed as these are now being undertaken by automated processes.[74] As we show in the chapter on demography, in Africa, an estimated 100 million young people will be coming into the job market over the coming ten years. The combination of AI and robotics with 3D printing and other new technologies is likely to make it more difficult for young African people to find decent jobs.[75]

Automation and robotic processes require skilled operatives for installation, maintenance and other tasks. But skilled labour is in short supply in many low-income countries. In poorer countries, sophisticated imported machines also tend to be more expensive and it is more difficult to secure finance on favourable terms than in richer countries. Automated factories are likely to be located nearer wealthy consumer markets, where finance is cheaper, the skills and spare parts required to maintain the machines are easier to source and reliable power supplies are taken for granted.

Automation and new industrial techniques, such as 3D printing, allow consumers to order individually customised clothes, footwear, prescription drugs and other products and have them delivered to their home within days or even hours. With the cost advantages of locating manufacturing facilities in far-off locations that previously offered cheaper labour disappearing, and the customer service advantages of locating closer to home rising, the age of outsourcing production to developing countries is fast drawing to a close. The politics of protectionism will accelerate this. The demand to repatriate manufacturing to advanced countries has never been higher, although it is not jobs but automated and AI-assisted robotic processes that are coming home.[76] The 'America First' politics prevalent in the US, and rising nationalism in China, Europe and the UK, in combination with the technological potential to bring manufacturing closer to customers and quickly service their needs at lower costs, means we may well have reached peak supply chain fragmentation globally. The COVID-19 pandemic has not only exposed the vulnerabilities of the system, but is also hastening the decoupling of supply chains. Whereas technological change in the twentieth century facilitated globalisation and the fragmentation of supply chains, in the twenty-first century it is making local production more profitable. Taking into account the carbon footprint embedded in products may further reinforce this localisation, raising even deeper questions about the future of export-orientated jobs in developing countries.

Technology can deepen inequality

Automation can cause wages to fall in the short term, so although it may be good for growth it is bad for equality.[77] This is because automation exercises downward pressure on wages in vulnerable sectors such as routine factory and services work, including secretarial, call-centre and data-processing jobs.[78] Meanwhile, automation increases the profits for owners of the technology platforms and machines, which increases inequality. Highly educated workers, in particular, also see their incomes rise as their skills are put to use in skill-intensive, knowledge-based areas, in a process that economists call 'skill-biased technical change'. All of this results in a continuing improvement in the returns to education and a persistent decline in the real wages of semi-skilled and unskilled workers.[79]

Several years ago, the White House Council of Economic Advisers reported that in the US, automation will hit low-pay workers the hardest. The council projected that 83 per cent of jobs paying $20 an hour or less would be negatively affected in the period to 2035, and those keeping their jobs would probably see their real (after adjusting for inflation) wages fall by at least a quarter.[80] In the medium term, the share of national income of these workers was expected to fall from the current level of around a third to well under 20 per cent over a couple of decades.[81] Meanwhile the wages of skilled workers are projected to increase by at least 56 per cent over the same period.[82]

Technology can increase gender bias

The gender implications of technological change are often poorly appreciated. Take the case of one of the world's most powerful technologies – contraception – and the way it has given women the means to control how many children they decide to have. As the chapter on demography shows, this has led to dramatic reductions in fertility rates around the world. Or consider how medical innovations and practices have reduced the risk of dying during childbirth a hundredfold over the past century, a topic discussed in the chapter on health. Even the rise of advanced housing appliances – from the fridge to the vacuum cleaner – has transformed the nature of housework. Meanwhile, greater access to information and the means to transmit it has helped to raise awareness of the rights of women and of gender equality. For the first time, as is evident in #MeToo and the action of young women like Malala Yousafzai and Greta Thunberg, people acting in isolation can inspire global audiences.

Despite real progress, the World Bank has identified several stubborn obstacles to the potential of information technologies to overcome gender discrimination. In many countries, the weakness of digital infrastructure and political controls are major barriers. For example, women in low- and middle-income countries on average are 14 per cent less likely to own a mobile phone than men.[83] In Egypt and in India, 12 per cent of women say that they did not access the internet more often because of 'social pressures'.[84] Women are globally much less likely to work in the technology sector owing to comparatively lower participation in science, technology, engineering and maths (STEM) education – a topic addressed in the education chapter.[85] Because fewer women are able to access skilled jobs, in most, but not all, developing countries, they are more vulnerable to automation and are concentrated in poorly paid, lower-skilled jobs.[86]

Technology needs better management

If technological change is accelerating so quickly, why can't we see its impacts in terms of higher levels of economic output or efficiency? Productivity captures how much output each person creates every hour. If machines are substituting for, or augmenting, the jobs of workers, we should expect higher productivity. But we don't get it. On the contrary, since the turn of the millennium we have seen a slump in productivity across the world.[87] This creates a tricky puzzle for economists. Three decades after Robert Solow's famous quip that 'you can see the computer age everywhere but in the productivity statistics', it remains the case that technological progress has not led to productivity improvements.[88]

None of the current explanations for the disconnect between accelerating technological change and slumping productivity is satisfactory. Mismeasurement is the most common explanation, with the argument being that the new digital economy reflects a 'dematerialising economy'.[89] Productivity statistics track physical outputs and services that are sold, leaving out free digital services. This increasingly underestimates the real extent of economic activity. Our mobile phones, for example, have completely disrupted camera, GPS, music, telephone switchboard and other forms of production, and although they leave us better off, these efficiencies are reflected in a contraction in output.[90] Since the digital economy accounts for only a modest fraction of total economic activity, even though there is undoubtedly significant mismeasurement of some products that are now cheaper but give us a great deal of consumer satisfaction, this is insufficient to account for the bulk of the slowdown in productivity.[91]

One of the more plausible explanations for the lower than expected productivity gains of technological transformation relates to the speed of change. Because technological change is occurring so quickly, we need to adapt more rapidly and, unless we do, our systems are more rapidly out of date.[92] More investment is required to keep our systems and infrastructure up to date, and yet investment rates are slowing down in many countries and political gridlock means that our rules and regulatory structures are getting more and more out of date. Rapid changes in the structure of work mean that we should be moving to jobs in new places, but the cost of housing and transport and restrictions on migration mean that mobility is declining, both within and between countries. The ominous decline in productivity tells us that we need concerted reform and investments from governments and companies to keep pace with technological change. Yet the failure to set future-looking agendas and the resistance of ageing populations to structural change means that we are falling ever faster behind, resulting in slower growth and a decline in living standards.

Our second renaissance

Klaus Schwab, the founder of the World Economic Forum, has promoted the idea that we are living in a fourth industrial revolution, or 4IR. He argues that whereas previous industrial revolutions liberated humankind from a dependence on animal power, made mass production possible and brought information technologies to billions of people, the current waves of technological change are fundamentally different in their scope and impact.[93] The advent of ubiquitous mobile supercomputing (with machine-learning capabilities, intelligent robots and blockchain), autonomous vehicles, and genomic editing and neuro-technological enhancements are among the many marvels that are currently being developed, many of them at an exponential speed. These are fundamentally changing the way we live, work and relate to each other.[94] COVID-19 has given birth to an unprecedented digitally-enabled global collaboration of scientists who have shared the sequence of the virus and worked together on treatments and vaccines, going well beyond what any one nation could have achieved.

The current pace and scale of change is different to anything humanity has ever experienced before. The previous industrial revolutions percolated relatively slowly. In fact, there are still parts of the world where people till their fields with ploughs pulled behind oxen, untouched by the first industrial or subsequent revolutions. The first industrial revolution began around 260 years ago and brought us the steam engine and mechanical power. The second industrial revolution of the early twentieth century ushered in the modern world, bringing mass production techniques and scientific breakthroughs associated with internal combustion engines, aeroplanes and new chemical products, including fertilisers. Beginning in the 1950s, the third industrial revolution brought semi-conductors, mainframe computing and, from the 1990s, the internet and the digital revolution. All of this helped internationalise production and trade, as discussed in the chapter on globalisation.

Framing the current era as a fourth industrial revolution is optimistic. It suggests that like the previous revolutions we can anticipate more and better jobs, replacing drudgery and danger with cleaner and safer technologies. The first industrial revolution was not, however, a happy time for most who experienced it as it was characterised by great turmoil, unemployment, upheaval and wars.[95] It brought not only new production methods, but also prompted a fierce backlash, which included Karl Marx and communism, the French Revolution and the American Civil War. It may well be the case that in the long term we are all better off, but as John Maynard Keynes noted, in the long run we are also dead.[96]

In their book *Age of Discovery: Navigating the Storms of Our Second Renaissance*, Chris Kutarna and Ian argue that our current technological revolution has more in common with the Renaissance of 500 years ago than an industrial revolution.[97] The Renaissance led to fundamental changes in perspective and to the first age of global commerce. Those left behind felt increasingly anxious and, with the help of the printing press, spread ideas that challenged change. The firebrand Savonarola deposed the Medicis and ignited the Bonfire of the Vanities, burning books. The reaction of the despotic and deeply corrupt Catholic Church was to enforce a reign of terror, which history knows as the inquisitions, in which religious tolerance and scientific experts were outlawed.

The Renaissance demonstrated the power of an information revolution to promote rapid technological change and scientific and artistic progress. But the lesson of the Renaissance is also one of how this challenges authorities and can be profoundly destabilising for societies. Today as in the past, when changes occur rapidly, people get left behind more quickly leading to populist backlashes and rising social tensions. It is vital in periods of accelerating change to ensure that we build inclusive and tolerant societies. Growing inequality, as we discuss in the chapter on that subject, is not a necessary outcome of rapid technological change, but active policies are required to ensure that this change benefits the whole of society. Technology may not be a foe, but it is a fickle friend.

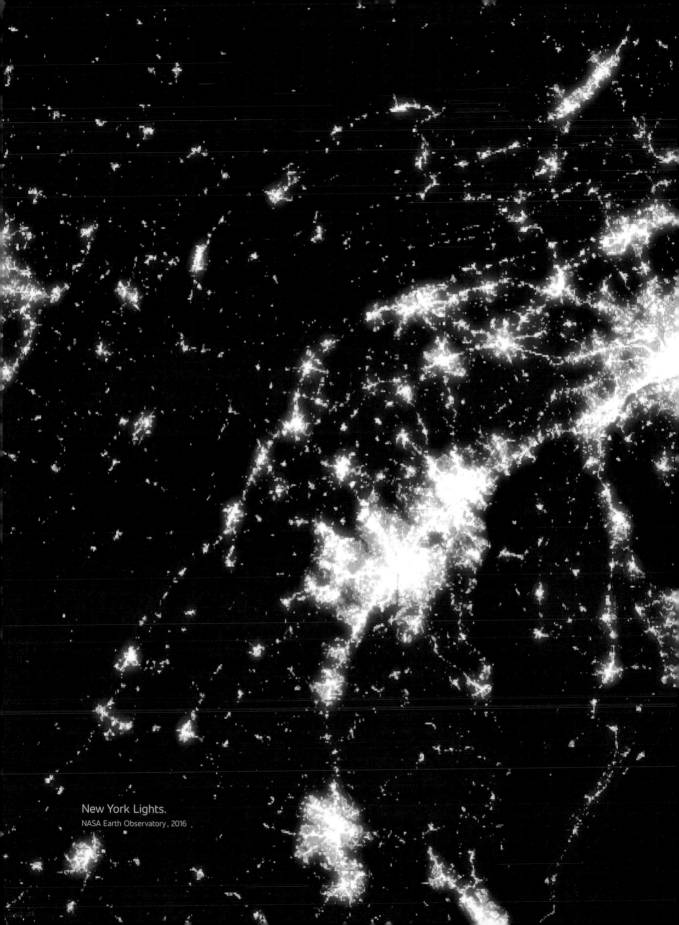

New York Lights.
NASA Earth Observatory, 2016

Inequality

Inequality within countries is growing

Inequality between countries is reducing

Where people live defines their prospects

Extreme inequality is rising, benefiting the richest

Light and darkness
This time lapse of light at night starkly displays a critical dimension of inequality, which is access to energy and light. Africa and Latin America are almost totally in the dark, with spots of light above São Paolo, Johannesburg, Lagos and other megacities, in contrast to Asia, Europe and North America.

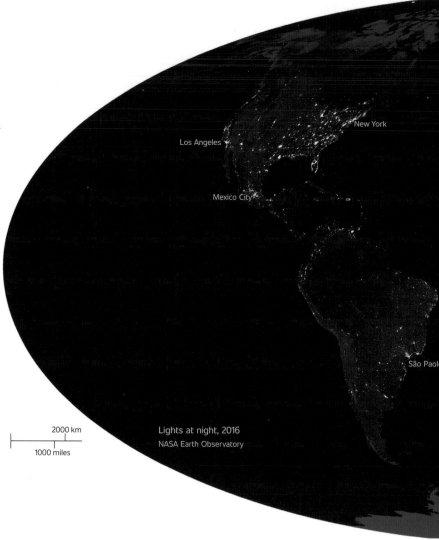

New York

Los Angeles

Mexico City

São Paolo

2000 km

1000 miles

Lights at night, 2016
NASA Earth Observatory

Introduction

Inequality is one of humanity's most persistent economic, political and social challenges. It is also one of its most misunderstood. Much of the debate focuses on skewed income distribution. Another way of visualising inequality is by studying lights at night. As the map shows, the darkness over much of South America and Africa is in stark contrast to the luminosity of the prosperous cities in North America and Western Europe and the incandescence of urban areas in India, China and Southeast Asia. The image is a vivid reminder of the gross inequalities associated with accessing power. Electricity and

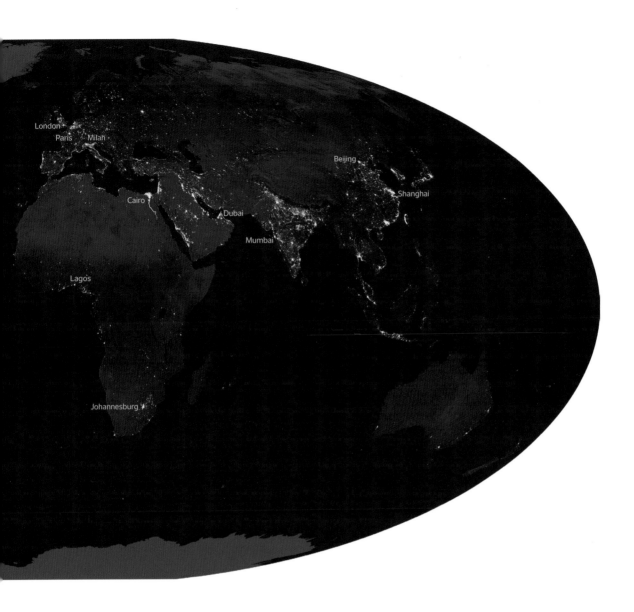

light are essential for households, businesses and cities to survive. The arithmetic is straightforward: no energy, no development.

The conventional way to study inequality is by examining how income and wealth are distributed across societies. Today, the world's elite, the so-called '1 per cent', are under scrutiny. And for good reason. The 1 per cent – the roughly 42 million people who have a net worth in excess of $10 million – control almost 48 per cent of global wealth.[1] What is more, the situation seems to be getting worse.[2] But does inequality matter? Why should we care if Jeff Bezos, the founder of Amazon, has more money than the combined wealth of more than fifty

African and Caribbean countries?[3] Every day we hear that inequality is growing. But is it? And is it responsible for rising populism and anger, as many commentators claim? What do we even mean by inequality anyway?

In this chapter we explore the question of inequality, and why it matters. We offer a very short history of inequality, and then look at how it is distributed between and within states. In the process, we encounter several dimensions of inequality, of income, wealth, health, education, race and gender, to name a few. There's good and bad news. On the one hand, inequality between countries has narrowed considerably. But inequality within countries is rising. We use a series of maps to examine the spatial and temporal characteristics of inequality in order to help think through concrete solutions.

Inequality in the dark

There are multiple ways to define inequality. Many economists use the concept of the Gini coefficient. In a perfectly equal world, where everyone has the same income, the Gini is zero. If one person had all the income it would be one (or 100 per cent).[4] While Gini is straight-forward to measure, the truth is that no two people experience inequality the same way. Nor is inequality easily reduced to dollars and cents. Some people experience harassment and threats, power inequalities. Others experience inequality before the law, or inequality based on their gender, sexual preferences, age, education, origin, nationality, dialect, religion, occupations, disabilities, height, looks or other factors that go into defining what others may think about us and our life chances.

The visualisation of light emissions from space helps illuminate the geography of inequality. The two images included here highlight the striking energy inequalities between New York state, with a population just shy of 20 million people, and Lagos, the biggest city in Nigeria, with just over 20 million people and covering a similar geographical area.[5] Even though Nigeria is one of the biggest oil exporters in the world, poor governance and chronic mismanagement means that its 200 million citizens rely on less than five gigawatts of power supply on an average day, while New York state's 20 million inhabitants consume 392 gigawatts.[6]

Lagos and New York lights – 2016
Lagos and New York state have similar populations but as the satellite images show, New Yorkers have much more energy and light. Indeed, the state of New York consumes more energy than the whole of sub-Saharan Africa.

Lagos lights – 2016

Ibadan

Porto-Novo

Contonou

Lagos

Benin City

Lome

Accra

50 km

25 miles

NASA Earth Observatory

New York lights – 2016

New York

Philadelphia

Baltimore

Washington

50 km

25 miles

NASA Earth Observatory

Not surprisingly, Nigeria's National Electric Power Authority (NEPA PLC) is widely derided by locals, who say the acronym stands for Never Expect Power Always Please Light Candles. The result is that many Nigerians rely on generators, but these are costly to run, which is why small businesses complain that 'when the generator comes on, the profits fly away'.[7] Having a small generator, to operate a TV, refrigerator or charge a mobile phone, is a sign of being middle-class in Nigeria. The poorest people cannot afford power and rely on kerosene and candles instead. This means that they are less able to cook and clean, can study less and are literally kept in the dark. Unequal access to energy reinforces other forms of inequality and, ultimately, poverty.

Energy inequality is a solvable problem. When Ian worked at the Development Bank of Southern Africa, it financed the electrification of more than 500 towns and 1 million homes a year. The sheer range of benefits was astonishing. Streetlights not only improved mobility, they reduced property crime and sexual violence. Home lighting allowed for longer periods of study and improved education results. Electric power permitted refrigeration and the boiling of water, improving nutrition and hygiene and reducing water-borne diseases. And electrification for barbers, butchers, bakers and thousands of other professionals and retailers allowed enterprises to flourish inside homes or in retail outlets. Not surprisingly, incomes and employment improved dramatically. Light matters.

A brief history of inequality

To fully understand how inequality is distributed and why it matters, we need to go back in time. Our earliest ancestors lived very tough lives, with everyone working extremely hard to contribute to the survival of their families and communities. As the chapter on cities shows, the advent of agriculture and food surpluses allowed societies to specialise and develop increasingly elaborate social, religious, political and economic hierarchies. Inequality is not a recent phenomenon: it can be traced to archaeological sites dating back at least 7,000 years. Ancient settlements and cities reveal evidence of inherited land and wealth. And skeletal remains exhibit significantly different diets and health between and within Neolithic communities.[8]

Inequality began increasing with the extravagant accumulation of wealth of the ancient Egyptian pharaohs, Mayan rulers, and Chinese and Roman emperors. Later, the castles, priceless jewels and slaves collected by monarchs, tsars, popes and imperial traders reflected how a tiny minority of powerful people became fabulously wealthy while the vast majority were condemned to live in extreme poverty. The industrial revolution accelerated the concentration of wealth. This occurred first in the UK, and then in Western European countries and the US, with the advent of steam-powered and mechanised technologies.[9] The industrial revolution benefited the middle-class, but not the urban or rural poor, whose incomes and life expectancy showed comparatively limited signs of improvement.[10]

By the time of the French revolution in 1789, the top fifth of the world's population had about three times as much income as the bottom fifth.[11] The technological advances that had started in the UK improved the growth prospects and income of countries that adopted new technologies. For example, in 1800 the average purchasing power (what consumers can afford of local goods and services) of people in the US was about double that of those living in China. However, this gap widened markedly as the benefits of new waves of technological adoption brought added benefits to people in the US, while the Chinese continued to rely on artisanal methods that predated the industrial revolution. By 1975, purchasing power in the US was thirty times that of China.[12]

From the second half of the nineteenth century until the First World War, fast-growing trade and migration led to a convergence of living standards between Europe and North America. During this period, more than 50 million Europeans migrated to the 'new world', and though Europeans had previously enjoyed much higher incomes than communities in North America, the migrants took their institutions, technologies and capital with them, spurring development in Canada and the US. The result was that wage differences between Europeans and people living in the resource-rich settlements of America fell from 100 per cent to 25 per cent.[13]

Since the middle of the nineteenth century inequalities between many, but not all, countries have come down, partly as a result of mass migration. Yet as the introduction to this chapter makes clear, inequality has been increasing within countries. This is because the benefits of international commerce and industrialisation have mostly been captured by the wealthiest households.[14] But it was not always this way. Creeping inequality within countries was significantly reduced from the 1930s to 1970s when governments introduced higher taxes on the wealthy. Increased government revenues also contributed to the strengthening of social welfare and investments in public health and education. Put simply, public policies significantly reduced inequality and led to an unparalleled flourishing of human welfare.[15]

These declines in inequality turned out to be short-lived. From the 1980s onwards the pendulum swung back against progressive taxation and state intervention. Margaret Thatcher, the Prime Minister of the UK from 1979 to 1990, and Ronald Reagan, President of the US from 1981 to 1989, led a rolling back of the state, broke the power of trade unions and initiated a race to the bottom in taxes that has continued to this day.[16] The result is widening inequality *within* the UK, US and other countries that adopted the 'neo-liberal' model, a topic discussed in the chapter on geopolitics. At the same time, the acceleration of globalisation has provided an unprecedented opportunity for many developing countries, not least those in Asia, helping to reduce income disparities *between* countries.[17]

The big picture today

The charts demonstrate how the distribution of income among the global population has changed between 1800 and 2019. In 1800, most people lived in dire poverty. Even in comparatively wealthy Europe (depicted in yellow) only the very well-off could afford decent housing and balanced diets, living to the left of the line that marks extreme poverty. By 1975, however, incomes had improved dramatically, particularly for people living in rich countries where average incomes were ten times those in poor countries.[18] The camel hump accounts for most of the income in rich countries and the neck of the camel captures most of the rest of the world. Finally, the difference between the distributions in 1975 and 2019 is remarkable and reveals the extent to which inequality declined over just four decades, due mostly to progress in Asia.

While inequality *within* countries is growing, the charts show that the gap *between* countries is narrowing. This is because developing countries are growing much faster than rich countries, allowing a

Global Income distribution in 1800, 1975 and 2019

Horizontally the figures go from 20 cents up to $200 a day in income, with anything to the left of $1.90 (in 2011 constant prices) defined by the World Bank as extreme poverty. Asia is red, Europe green, Africa blue and the Americas yellow with the height depicting the number of people. In 1800, 87 per cent of people were extremely poor and lived in Asia and Europe. By 2019 barely 10 per cent of the world lived in extreme poverty, mostly in Africa.

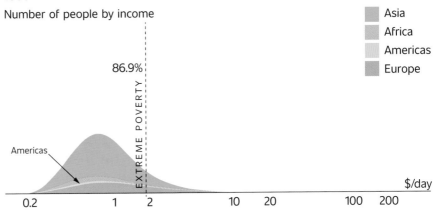

1800
Number of people by income

Asia
Africa
Americas
Europe

86.9%

EXTREME POVERTY

Americas

0.2   1   2   10   20   100   200 $/day

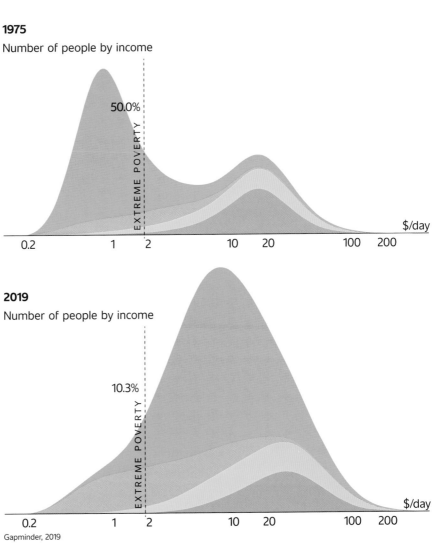

1975
Number of people by income

50.0%

EXTREME POVERTY

0.2   1   2   10   20   100   200 $/day

2019
Number of people by income

10.3%

EXTREME POVERTY

0.2   1   2   10   20   100   200 $/day

Gapminder, 2019

sizeable minority of poor countries to catch up with incomes previously experienced only in much richer countries.[19] Within wealthy countries inequality is widening owing to the combination of lower taxes and less redistribution, with country after country following the Reagan and Thatcher measures to roll back the welfare state and reduce the power of organised labour. The lowering of trade barriers and development of containerisation, together with the integration of China and East Asian economies into global supply chains, has exercised downward pressure on workers, particularly those performing routine, semi-skilled jobs. Meanwhile, owners of businesses, shareholders and skilled professionals, especially in the tech sectors, have benefited from lower taxes and more profitable global businesses. As the French economist Thomas Philippon has shown, this has also been associated with a rise in lobbying that further serves to entrench existing privileges and reduce redistribution, especially in the US.[20]

It turns out that a small number of countries are having an outsized effect on global inequality. China, India and Indonesia together account for more than 3 billion people. These countries are also responsible for most of the global achievements in reducing extreme poverty, with China achieving far more than any other nation in history.[21] Although Chinese growth slowed to historic lows by 2020, for much of the past forty years it was around 10 per cent a year.[22] This has led to a doubling of average incomes about every ten years, resulting in more than 800 million people being lifted out of desperate poverty.[23] While East Asia accounted for around half of the 2 billion people living in extreme poverty in 1988, today less than 9 per cent of the world's 800 million extremely poor people live there.[24]

Notwithstanding these astounding achievements, large pockets of deprivation remain. In 1988, about a third of the world lived in extreme poverty, mainly in India and China.[25] Since then, despite the world's population increasing by about 2.5 billion, about 1.2 billion people have escaped this bottom threshold. By 2019, barely 10 per cent of the world's population still live in extreme poverty, defined arbitrarily by the World Bank as having less than $1.90 a day.[26] Yet, more than 400 million Africans are still exceedingly poor, with that continent accounting for over half of the poorest people in the world, and the 2020 COVID-19 pandemic has compounded the depth of this problem.[27] In other regions, war and conflict have been associated with desperate poverty, including in Yemen, Afghanistan and Myanmar, and there is still much grinding poverty in South Asia and in Latin America and the Caribbean, not least in Haiti, which is one of the poorest countries in the world. But, with other regions making more rapid progress in reducing poverty, Africa is expected to account for nine out of ten of the poorest people in 2030.[28]

Inequality past and present

The graphic representations of inequality over time were created by Gapminder and are a heroic attempt to overcome the absence of long term comparable data to show the global distribution of income for all of humanity. The horizontal axis depicts income in dollars a day, starting at 20 cents up to $200. The different colours depict the continents: Asia is red, Europe yellow, Africa blue and the Americas are green. The height of the humps depicts the number of people, with the extreme poverty line of $1.90 (in 2011 constant prices) indicated vertically. In 1800, 87 per cent of people were extremely poor and mostly lived in Asia and Europe. By 1975, the distribution resembled a camel with two humps with about half the global population living in extreme poverty, mostly concentrated in Asia and with an increasing share in Africa, but no more in Europe. The improvement by 2019 is striking, with barely 10 per cent of the global population in extreme poverty but this now mainly concentrated in Africa.

The forward march of poverty reduction is far from guaranteed, especially in the wake of the COVID-19 pandemic. For one, the poor are still concentrated – more than 80 per cent – in rural areas. Balanced urbanisation and investments in education and health should continue eliminating extreme deprivation. But tackling the extreme poverty still endured by more than 700 million people requires that corruption, conflict and crime also be addressed, issues discussed in detail in the chapter on violence.[29] Conflict slows and reverses development and traps entire societies in extreme poverty, which has grown in war-torn areas such as Syria, Yemen, Somalia and South Sudan.[30] Corruption is less visible but is deeply corrosive and if unchecked can spread like a cancer through societies, crippling governments, distorting the private sector and increasing the costs for ordinary citizens of accessing whatever services continue to function.

Facundo Alvaredo and his co-authors have reconstructed a figure to show how the income of different income groups around the world grew between 1980 and 2016.[31] The chart has three distinctive features: a hump that includes the world's poorest, made up mainly of people living in developing countries; a valley that contains the working- and middle-class of the developed world and also the upper-class in developing countries; and a trunk made up of the global elite. They found that the richest 1 per cent captured more than twice the amount of growth than the poorest 50 per cent of people in the world. This was described originally by Chris Lackner and Branko Milanovic as the 'elephant curve', with the trunk of the elephant depicting the extent to which the top percentile has benefited. With the growth of the top 1 per cent, this has now morphed into a shape which more resembles the Loch Ness monster.

The global lower middle-class experienced significant improvements, not least in China, with this evident in the rising back of the Loch Ness monster. The bottom 10 per cent of extremely poor people experienced very little growth, and are stuck in a cycle of poverty and violence, as is depicted in the monster's sagging tail.[32] The financial crisis of 2008 – discussed in the globalisation and geopolitics chapters – led to job losses and stagnating wages from which working people have yet to recover, despite higher levels of employment. Meanwhile, the increasing automation of routine jobs and development of platforms that

Mex

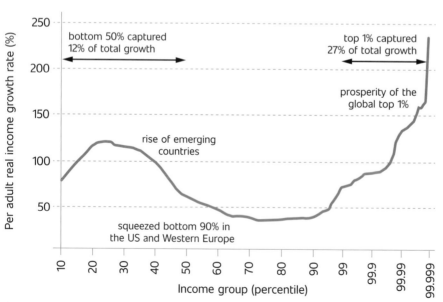

Loch Ness monster of global inequality
The figure shows how the incomes changed over the period from 1980 to 2016, with the long neck of the monster showing that the top one per cent of the population captured over 27 per cent of all the growth while the bottom half received only 12 per cent of the benefit of growth.

Alvaredo, Facundo, Lucas Chancel, Thomas Piketty, Emmanuel Saez, and Gabriel Zucman, 2018

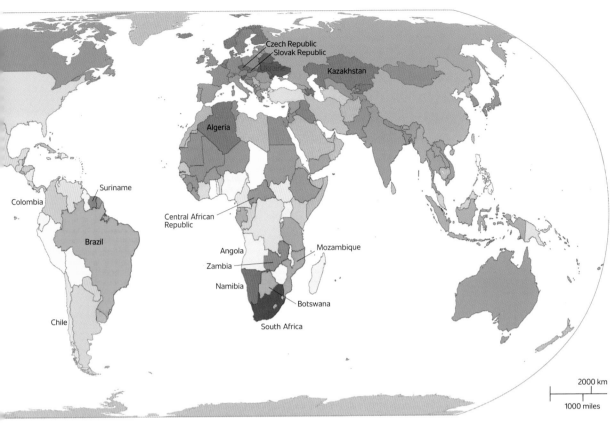

Gini coefficient

Red implies high levels of inequality and blue low: South Africa is the worst, followed by a number of its neighbours and Brazil. Asia is the most equal developing region and Europe the most equal, globally.

Gini coefficient of inequality
World Bank, Development Research Group, 2018

21   Most equal   Least equal   66   No data

facilitate gig work is creating highly unpredictable living standards and increasing insecurity. These factors help to explain the disillusionment and frustration that the poor share with the middle-class.[33]

The problem is at home

The map highlights the extent of income inequality within countries, using the Gini measure. The most economically unequal countries are shown in red and the most equal in blue. The latest data indicate that South Africa is the world's most unequal country. The country's racist and systemised segregation over centuries entrenched privileges for those classified as white, who accounted for barely 10 per cent of the population. Meanwhile, the 90 per cent defined as black were confined to barely 10 per cent of the land and were excluded from professions, universities and from living in the 'white' cities. In 1994 President Mandela became the first freely elected leader and although he vowed to create a non-racial democratic country, the terrible economic legacy of apartheid has yet to be overcome.[34]

In neighbouring Botswana and Namibia, the concentration of extraordinary wealth derived from diamond mines explains the high inequality, while in nearby Angola and the Central African Republic the deeply corrupt elite have managed to amass vast wealth at the cost of investments in the health, education, and infrastructure needed by the mass of citizens. Likewise, Brazil is also one of the world's most unequal countries, with more than 15 million people living in extreme poverty, and Colombia, Mexico and even Chile stand out for their

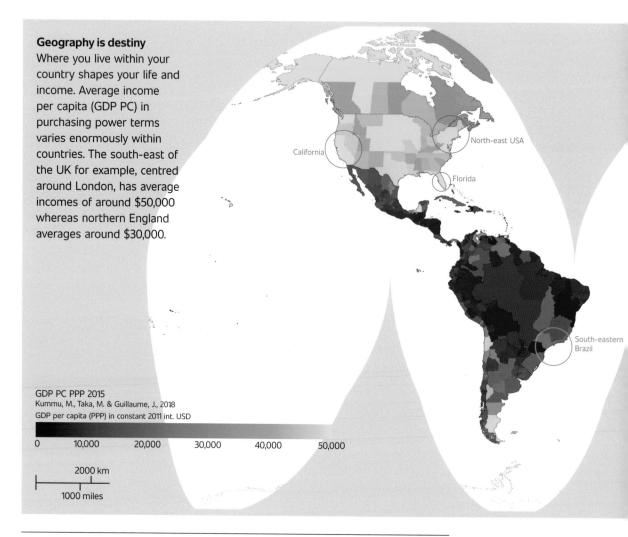

Geography is destiny
Where you live within your country shapes your life and income. Average income per capita (GDP PC) in purchasing power terms varies enormously within countries. The south-east of the UK for example, centred around London, has average incomes of around $50,000 whereas northern England averages around $30,000.

GDP PC PPP 2015
Kummu, M., Taka, M. & Guillaume, J., 2018
GDP per capita (PPP) in constant 2011 int. USD

0   10,000   20,000   30,000   40,000   50,000

2000 km
1000 miles

failures to overcome inequality, as evident in the protests that swept through these countries in 2019. By way of contrast, the Scandinavian countries, together with several formerly socialist states, notably Ukraine, Kazakhstan, the Czech and Slovak republics and Algeria, register low inequality.

Inequality is profoundly shaped by geography, and one's neighbourhood is a good predictor of your health, education, income and career prospects. Where people live has always exerted a strong influence over their destiny. Within countries, moving from down-and-out villages to dynamic cities offering jobs and opportunities has historically offered a route to a better life. This is becoming more difficult. Soaring housing prices have made some cities unaffordable and increasingly congested, and costly transportation has undermined

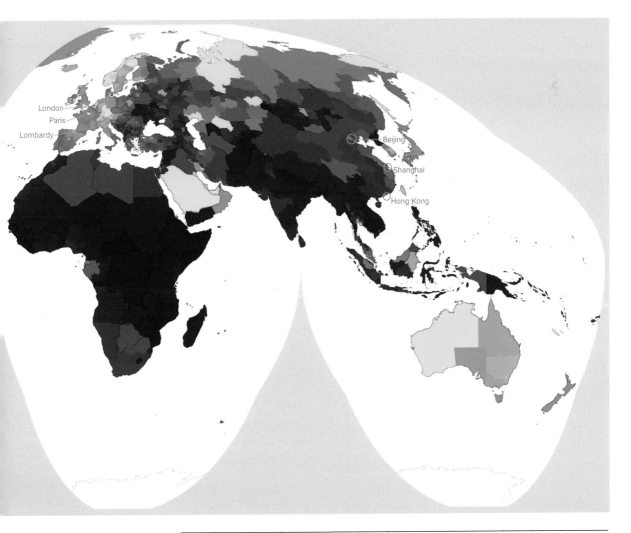

the benefits of commuting long distances. The result is growing geographical divides within countries and cities, with inequality defined by where people live.[35] The map considers the spatial distribution of incomes at the national and subnational scale, in income per capita for 2015, adjusted to take account of different costs of living, and inflation, to create constant 2011 values. The extent to which wealth in China is concentrated on the eastern seaboard, as it is in Australia, and the extent to which south-eastern Brazil is wealthier than the north-western and north-eastern areas is apparent, as is the concentration of wealth in the north of Italy.

Extreme wealth

The top 1 per cent of earners have done exceedingly well out of globalisation. A tiny minority of individuals have captured a vastly disproportionate share of the benefits.[36] At the most extreme, just before the COVID-19 pandemic, just three people – Jeff Bezos, Bill Gates and Warren Buffett – owned as much wealth as the 160 million poorest people in the US.[37] Meanwhile, the top 1 per cent – which in the US is reserved for those with an income above $750,000 per year – has doubled its share of national income over the past five decades.[38] Their incomes increased seven times faster than the incomes of the bottom 20 per cent. The top 1 per cent on average have about forty times more income than the bottom 90 per cent.[39] And, the uber-wealthy – the top 0.1 per cent, with incomes of more than $3 million a year – now earn 188 times more than the bottom 90 per cent.[40] The figure opposite was produced by the economist Thomas Piketty and his colleagues and reveals the extent to which income growth in the US is concentrated in a tiny fraction of the population.[41]

Similar trends are apparent in other wealthy countries, including the UK. There, just five rich families are worth more than the poorest 13 million people, who constitute 20 per cent of the population.[42] The 5,000 or so UK residents (of which 90 per cent are men) who constitute the top 0.01 per cent of the country's total wealth each make an average of £2.2 million ($2.7 million) a year. The extent to which they are living off their property and other investments is reflected in the fact that 'unearned' returns from investments rather than paid employment accounted for 40 per cent of their income.[43] Whereas

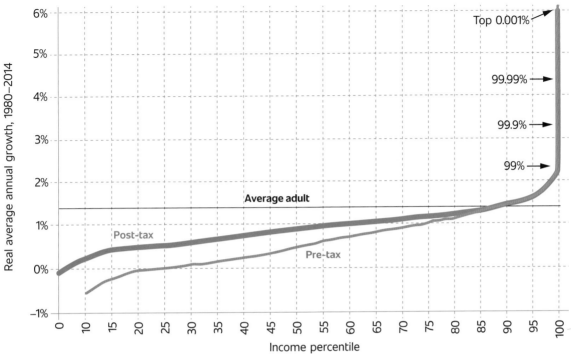

Real average annual growth, 1980–2014

Income percentile

Thomas Piketty, Emmanuel Saez, Gabriel Zucman, 2018

Soaring inequality: average annual real income growth by percentile in US, 1980–2014

The soaring incomes of a tiny elite, the top 0.001 per cent (1 in 100,000 people), saw their incomes go up by 6 per cent annually, while the bottom 20 per cent saw their pre-tax income contract, adjusted for inflation, over the period 1980–2014.

working people in the UK have seen a stagnation of their incomes in recent decades, which worsened following the financial crisis of 2008, the tiny elite – 1 in 10,000 British citizens – saw their incomes increase by 277 per cent over the past twenty years.

The fortunes of the world's richest 500 people grew 25 per cent in 2019.[44] Globally, the richest 2,153 people are estimated to have more wealth than the 4.6 billion people who constitute the bottom 60 per cent of the world's population.[45] The combined wealth of these billionaires in 2019 was $8.7 trillion, which is equivalent to the total income of the poorest 150 countries.[46]And the 22 richest men have more wealth than all the women in Africa. In April 2020, while COVID-19 was threatening the livelihoods of over half the world's workers and over 260 million more people were threatened with starvation, ten billionaires saw their combined fortunes rise by $126 billion.[47]

The failure of economic growth to translate to higher wages is reflected in the figure. It illustrates the stagnation of real wages in the US over the past fifty years, despite sustained economic growth. Clearly, economic growth has benefited the rich, but has not provided a much-needed engine for improvement in the incomes of

350
300
250
200
150
100

1972   1980   1990   2000   2010   2019

Rebased, Q4 1972 = 100 ——— Real weekly earnings ——— Real GDP

Refinitiv, 2019

ordinary wage earners. Meanwhile, in the UK, the income of the top 0.01 per cent almost tripled over the twenty-year period 1995–2015.[48] Most of these people are concentrated in London.[49] This also helps explain why so many lower- and middle-class citizens in the US and UK have grown so frustrated with what they see as 'out of touch' elites. No surprise, then, that many of them have called for a disruption of the status quo, voting in populist leaders promising to do precisely that.

Entrenched wealth

Many inequalities are sticky and deeply rooted, which is why they are so difficult to overcome. The biggest predictor of our income and wealth has surprisingly little to do with our personal decisions. Instead, it is almost entirely dependent on the status of our parents and the neighbourhoods in which we grew up.[50] This figure is named after F. Scott Fitzgerald's famous novel. Jay 'The Great' Gatsby was a bootlegger who escaped his lower-class background to become a high-society flyer.[51] The figure depicts the likelihood of remaining in the same social class as one's parents. The range of different levels of inequality and very different potential to attain social mobility is starkly illustrated. The measure of inequality used is the Gini coefficient and of immobility is Intergenerational Earnings Elasticity which is an economic representation of the persistence of incomes between parents and their children, showing the relationship between the earnings in one generation and the next. It reveals the ability, or inability, of generations from a family to move up or down over time.

Denmark has low inequality and it's relatively easy to enjoy upward mobility there. By contrast, Brazil, Chile, China and the US register very high levels of inequality and very low potential to escape one's background. The best places to be born in terms of both low inequality and the potential to move up in the world are in Australia, Canada, New Zealand and Europe.

The figure shows that the US suffers from not only high levels of Gini inequality, but also the fact that it is relatively difficult for individuals to exert upward social mobility. The fable that anyone can succeed if they just pull up their bootstraps is less true in the US than it is in many other countries. Denmark, as the figure shows, is not only exceptionally equal, but it is also a place where people can overcome disadvantages of birth more easily than elsewhere. Scandinavian countries have done a much better job overcoming inequality than, say, Brazil or Chile, where most people's destinies are determined by their parents' status and postal code.

The great escape: Gatsby relationship between inequality and social mobility
Denmark has low inequality and it's relatively easy to enjoy upward mobility. By contrast, Brazil, Chile, China and the US have very high levels of inequality and very low potential to escape one's background.
The best places to be born in terms of both low inequality and the potential to move up in the world are Australia, Canada and Europe.

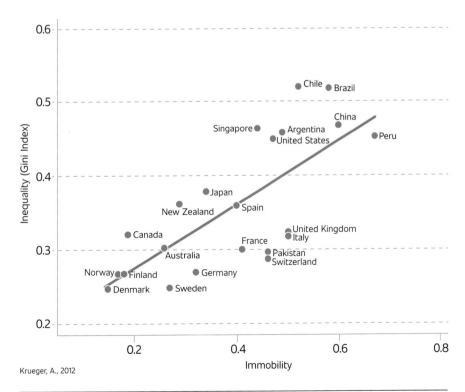

Krueger, A., 2012

Beating inequality

To overcome inequality, we need to decide which inequalities matter and why. Everyone is unique and we all have abilities to do different things. Some of us may be good at sport or drama and others at maths or music. Beyond these latent abilities, virtually all our life chances are shaped by where we are born, our parents' incomes and education, our schooling and other factors that are far beyond our control. Reducing inequality of opportunity by, for example, ensuring poorer young minority males from deprived neighbourhoods have access to decent schools, and focusing on giving everyone an equal chance to succeed, is vital. But to defeat inequality, it is also important to focus on outcomes, as deeply entrenched discrimination by race, gender, religion or other factors means that even if individuals have the same opportunities, they may still find they are unable to succeed in the way that more privileged individuals would and suffer very different end results.

To sustainably reduce inequalities, everyone should have the same opportunities. But even ensuring that everyone has an equal starting point is not enough.[52] As one of the grand doyens of the study of inequality, Tony Atkinson, makes clear, the race of life is unfair. Some people are running with one hand tied behind their back and may trip over obstacles placed in their path, while others are sprinting ahead unencumbered. Where winners take all (including prizes and the opportunities to compete) and the losers get nothing, inequality is entrenched.[53] As differences in wealth begin to widen, the opportunities for the next generation become increasingly constrained because children of successful parents have better preparation, training and nutrition. This reinforces and perpetuates a cycle of inequality.

Education is a powerful tool for overcoming inequality. As we show later in this book, levels of schooling differ vastly between countries and within them. Children of richer parents are five times more likely to receive pre-school education than the children of poor parents.[54] Inequality in primary education is carried through to later years. In the UK, 80 per cent of Cambridge and Oxford University students are from the top two social classes.[55] The inequalities of opportunity are even greater. While less than 1 per cent of the adult population in the UK are graduates of Oxford or Cambridge, just these two of the country's 100-plus universities have produced more than half of the prime ministers, senior judges and high-ranking civil servants.[56]

Creating a level playing field for opportunities is not just fairer on equity grounds, but also essential if the full talents of every individual are to be given the chance to flourish. This not only benefits specific people, but also entire societies. The risks of not doing so are also dangerously evident. Several studies underline the link between widening inequality and a growing range of social and economic challenges, including stagnating economic growth, increased crime, ill-health and depression.[57] Rising inequality is also strongly associated with rising populism and economic protectionism, as we discuss elsewhere in this book.[58]

There are strong ethical reasons for reducing inequality. Amartya Sen and Tony Atkinson both draw on *A Theory of Justice* by the pioneering philosopher John Rawls to show that there are intrinsic reasons, based on fairness and justice, to be concerned about inequality.[59] For Sen, inequality is above all about inequality of capabilities.[60] His focus on the opportunities available to people to lead fulfilling lives highlights the central place of education, as well as gender and human rights, in shaping equality. In his book *Development as Freedom*, Sen underlines why inequalities need to be overcome so that everyone can lead fulfilling lives that are both personally satisfying but also support the common good. Sen's writings led to the creation of the influential idea of 'human development', which goes beyond narrow measures of income and economic growth to striving for human flourishing in all its dimensions.[61]

Gender inequalities and unequal power relations can stunt human development. When large swathes of society are not treated equally, countries will operate below their potential.[62] Despite significant advances in the rights of women and girls, gender discrimination has yet to be eliminated in any country. Women everywhere still get lower pay for the same job and are under-represented in powerful positions. The extent of their unequal treatment varies. In most poorer countries, women's opportunities for gainful employment are limited to domestic work and subsistence farming, and they are often excluded from land ownership. Even in the wealthiest countries, including several oil-rich states, women are confined to home-based production or are forced to work in the informal economy where their incomes are low and working conditions are dismal. Around the world, women typically

Gender discrimination is a global problem

The World Economic Forum 2020 gender index ranks 153 countries on the basis of various dimensions of gender inequality. The best performers, are Iceland, Norway, Finland, and Sweden, followed by Nicaragua, New Zealand, Ireland, Spain and Rwanda, depicted in green. The worst, in red, include Pakistan, Iraq, Syria, Democratic Republic of Congo (DRC), Iran, Saudi Arabia and Morocco.

Nicaragua

Global Gender Gap Index
World Economic Forum, 2020

Inequality   Equality No data

endure the triple burden of relatively poorly paid employment, unpaid domestic work, including cooking and cleaning the home, and reproduction and childcare, which is also unpaid.[63]

The World Economic Forum (WEF) 2020 report on the global gender gap shows the extent of the challenge in overcoming gender discrimination. The index used identifies four dimensions of progress to gender parity: economic participation and opportunity, educational attainment, health, and political empowerment. The highly ranked countries, depicted in green, are not necessarily devoid of inequality. Even in these top performers gender inequality is predicted by the

WEF to persist for the next fifty years whereas in the worst countries it is projected to endure for well over 100 years.

In virtually every country for which meaningful data exists, women earn about a quarter less than men for the same job.[64] In the 1960s, women earned about 60 per cent of what men earned, but by 2000 this had improved to around 75 per cent.[65] Since then the pay gap in some nations, including wealthy ones such as Italy, has widened again.[66] Systemic discrimination against women is now receiving some, albeit still insufficient, attention. Other inequalities remain hidden. These include the daily difficulties experienced by elderly and disabled

Overcoming energy inequality: the world's largest concentrated solar plant
An inspiring example of how to overcome the energy deficit comes from Morocco, which is home to the world's largest concentrated solar plant at the Noor-Ouarzazate complex spanning 3,500 football pitches and producing enough electricity to power a city twice the size of Marrakesh.

1 km
1 mile

Google / Google Earth Engine, USGS, NASA, ESA, 2019

people, the discrimination against people on the grounds of sexual preferences or non-binary sex, inequalities experienced on the basis of race, caste or religion, and in numerous other areas. Discrimination experienced by migrants and refugees, and inequality arising due to unequal access to technology, are treated separately in other chapters.

Overcoming inequality

Our exploration of lights at night is a reminder of how far away we are from achieving full equality. New York state consumes more energy than the whole of Sub-Saharan Africa. But there are grounds for hope. Africa's limited dependence on fossil fuels, the declining costs of solar technology, and the fact that the continent is well-endowed with sunlight gives it an opportunity to leapfrog more advanced economies. Overcoming current deficits in electrification will require massive investment, with global support, in renewable energy infrastructure.

Governments have a central role to play in overcoming inequality. Inequality reduction is about much more than raising income and increasing economic growth, though this is still key. Ensuring equal access to education, health, energy, the internet and other services, as well as guaranteeing minimum standards, is equally critical. The fact that there are such different levels of inequality in different countries, and that in some it has been reduced while in others inequality has grown, is in large part due to government policies. These demonstrate that inequality is not a given and that smart policies can make a major difference.

The two panels reveal how deliberate changes in national policy from 1900 altered the share of total income going to the top 1 per cent. The contrast between English-speaking countries like the UK and US when compared to continental Europe and Japan is glaringly evident. In the latter countries, inequality has not risen since the Second World War. Meanwhile, in the English-speaking world, it has increased significantly. The reasons for this are principally to do with the different policy responses of governments, with European countries and Japan more determined to tax and spend to reduce inequalities.

Taxes and transfers, in the form of social security, housing, child, disability and other benefits, can all significantly help overcome inequality. Pre-tax inequality is almost as high in France as in the UK, with Gini coefficients of about 0.45, or 45 per cent, and even higher in Ireland (which has a Gini of 50 per cent, which means that without

Policy makes a difference: the share of income going to the top 1 per cent since 1900. The two panels reveal how deliberate changes in national policy from 1900 altered the share of total income going to the top 1 per cent. On the left are English-speaking countries like the UK and US where inequality has risen sharply, while on the right continental European countries and Japan where increases in inequality have been contained.

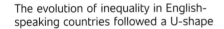

The evolution of inequality in English-speaking countries followed a U-shape

The evolution of inequality in continental Europe and Japan followed an L-shape

OWID / World Wealth and Income Database, 2018

redistribution it would have the most unequal income distribution of the thirty-four richest countries).[67] However, in Ireland and France taxation and redistribution have reduced the inequality index to 30 per cent and in the UK to 35 per cent. Meanwhile, in the US, the government has been more reluctant to use taxation and spending to overcome inequality, and as a result it is the most unequal of the rich countries.[68]

Brazil, where Robert lives, suffers from severe inequality, and Ian's homeland, South Africa, is in even worse shape. Concerted efforts on the part of past Brazilian governments succeeded in partially reducing this, even if levels have crept back up in recent years. The *Bolsa Família* programme provided parents with a monthly cash payment (of around $35) in exchange for sending their children to school and attending regular health check-ups.[69] The cash given to mothers was typically used to buy food, school supplies and clothes. This pioneering 'conditional cash transfer' initiative reached more than 50 million low-income Brazilians at its peak, a quarter of the national population, and is credited in reducing extreme poverty by half.[70]

Significantly, *Bolsa Família* reached people who really needed it and previously had not benefited from social welfare. When it was fully operational, an estimated 94 per cent of the funds reached the poorest 40 per cent of the population.[71] *Bolsa Família*, along with housing subsidies and an increase in the minimum wage, contributed to the marked reduction in inequality in Brazil. Originally trialled in Brazil in 1996, in Mexico (under the name *Progresa*) in 1997, and in Chile in 2002, the success of these and similar variants of the conditional cash transfer model have been adopted by more than twenty countries, including Indonesia, South Africa, Turkey and Morocco.[72] The concept has even been applied in New York with 'Opportunity NYC'. These and other experiments are more relevant than ever as the world adjusts to the effects of COVID-19.

Even so, good policies require good government and staying power. In Brazil, changes in government, austerity measures and the dramatic reversal of social policies have severely undermined inequality reduction efforts. Effective policies also benefit from a dose of good luck such as the commodity boom that propelled Brazil's economy during the 2000s. More recently, however, the combination of policy negligence and a sluggish economy have resulted in over 4 million people falling back below the poverty line. Making matters worse, the Gini index has crept back up to 53 per cent, which is among the worst in the world.[73] Meanwhile, in South Africa the Gini coefficient is a staggering 63 per cent.[74]

The good news is that the achievements of Brazil and other countries like France, Denmark, Bolivia, Thailand, Cambodia and South Korea demonstrate that inequalities can be overcome.[75] Reducing inequality is essential for many reasons. In 2015, Christine Lagarde, then head of the IMF, remarked how 'reducing excessive inequality is not just morally or politically correct but is good economics'.[76] The underlying reasons are straightforward: if only a few people gain and distort the rules in their favour through lobbying, corruption and avoiding paying their fair share of taxes, economic potential suffers and social cohesion dissolves.[77] Rising inequality is widely associated with rising anger with the urban elites and authorities.[78]

The rise of populism and nationalism is one of the most obvious reminders of how inequality frays the fabric of our societies.[79] Simmering inequality is also strongly connected to the Brexit vote in the UK, the election of Donald Trump in the US and the rise of populist and reactionary parties across Europe.[80] They also help explain the elections of Presidents Zuma (who was President of South Africa from 2009 to 2018) and Bolsonaro (who was voted in as President of Brazil in October 2018). The tragedy is that the policies implemented by these populist leaders benefit the few not the many, thereby deepening and entrenching inequality. The more hopeful news is that, with some exceptions, the incompetence of these populist leaders means that their terms could be short-lived.

Reducing inequality cannot be achieved through empty slogans. The so-called American Dream promises that if people work hard enough, they can succeed no matter how poor they are. This is an optimistic fantasy. One's parents' wealth is a far better predictor of future success than intelligence, education or one's willingness to slog away at work.[81] While we all celebrate the extraordinary stories of those who beat the odds, these are truly exceptional experiences. To overcome inequality, we need to overcome the root causes, not rely on the poor and vulnerable beating the absurdly adverse odds stacked up against them. COVID-19 has dramatically increased poverty and inequality.[82] Inequality is not a remote or abstract threat. It is real and dangerous. It must be reduced if we are to prioritise the well-being of people and our planet.

The world is more connected than ever before, with underwater
fibre-optic cables, rail networks and pipelines.
Parag Khanna and Jeff Blossom. Harvard World Map, 2017

Geopolitics

The 75-year-old liberal order is ending

We are shifting from unipolarity to multipolarity

Rising China–US tension is the dominant flashpoint

Populism and nationalism are weakening democracy

Global co-operation is more important than ever

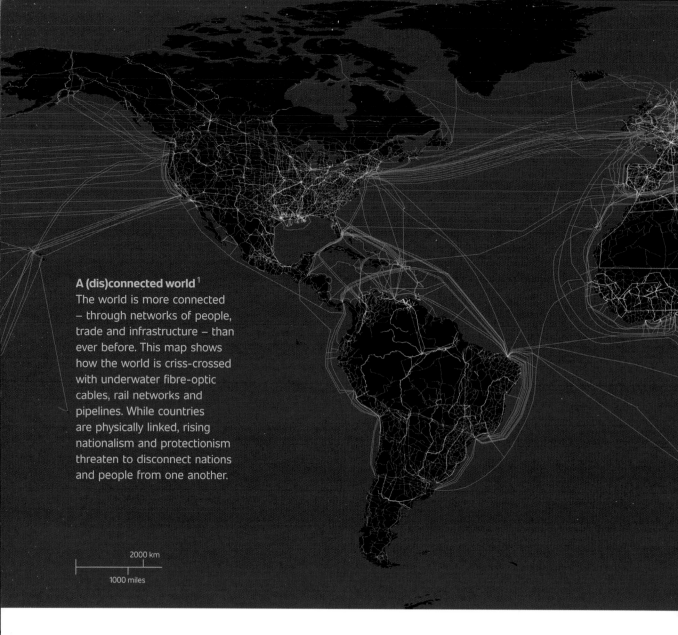

A (dis)connected world[1]
The world is more connected
– through networks of people,
trade and infrastructure – than
ever before. This map shows
how the world is criss-crossed
with underwater fibre-optic
cables, rail networks and
pipelines. While countries
are physically linked, rising
nationalism and protectionism
threaten to disconnect nations
and people from one another.

2000 km

1000 miles

Introduction

Even before the COVID-19 pandemic got underway in early 2020, the world was slipping into a turbulent geopolitical recession. The outbreak accentuated and accelerated trends that had already begun. The international system forged after the Second World War is giving way to something new. Many of the alliances and institutions that underpinned the global order for the past 75 years are rapidly coming unstuck. International relations are transitioning from a US-led uni-polar order back to a multipolar order.[2] What explains these wrenching

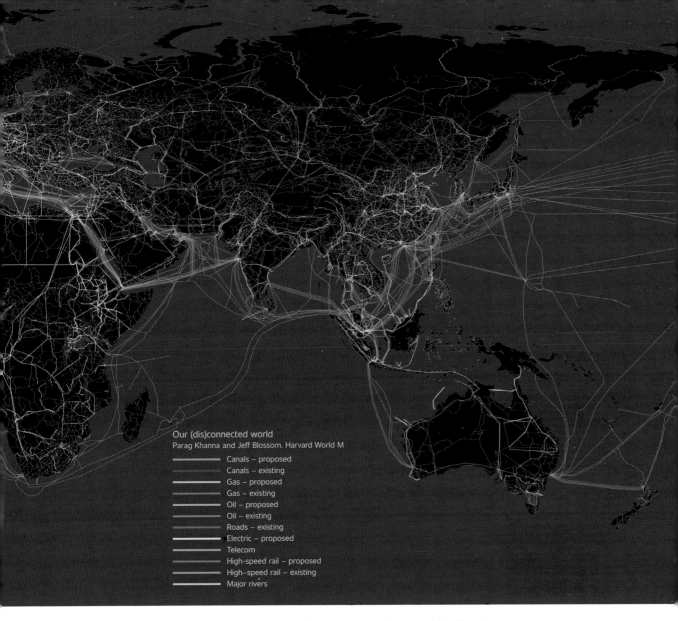

Our (dis)connected world
Parag Khanna and Jeff Blossom. Harvard World M

Canals – proposed
Canals – existing
Gas – proposed
Gas – existing
Oil – proposed
Oil – existing
Roads – existing
Electric – proposed
Telecom
High-speed rail – proposed
High–speed rail – existing
Major rivers

shifts? The principal factor is that the world's dominant superpower – the US – faces a powerful new rival in China. Another is that Western Europe, still a towering force, is experiencing internal disarray. Meanwhile, countries like Germany, India, Indonesia, Iran, Russia, Saudi Arabia and Turkey are flexing their muscles. As the tepid global response to the COVID-19 crisis amply showed, international co-operation is in short supply. No one is certain whether the emerging new order (or orders) will give rise to greater stability or more volatility. There is also considerable disagreement about what precisely is driving these changes. As we shall see, explanations range from the 'easternisation' of the global economy to military overstretch, wage stagnation, deepening inequality and debilitating political polarisation. With the 2020 pandemic triggering a massive economic shock, the willingness and ability to engage in collective action will be tested in ways unseen for over 75 years.

Dark clouds are settling over the western hemisphere. Many of the core pillars of the so-called global liberal order – common security agreements, open markets and the enduring commitment to democracy – are under assault, including from its chief architect, the US.[3] What were thought of as uncontested principles of this order – free and fair elections, the protection of human rights, pooled sovereignty and a free and independent press – are being called into question by leaders of mature democracies.[4] The failure to manage financial markets and the unravelling of social safety nets since the 1980s are also partly to blame. So is the hyper-concentration of wealth that has helped fuel resurgent populism, reactionary nationalism, rising protectionism and dangerous trade wars.[5] It is not at all certain that the current order will survive the multipolar era.[6] If it cannot be salvaged, what will replace it? The answer to this question may not be found in the West, but instead in the East.

At least before the outbreak of COVID-19, sunnier skies had broken out over large parts of the eastern hemisphere. This is because most Asians are looking forwards, not backwards, and they have good reasons to be optimistic. By at least one account, the Asian century has arrived, though COVID-19 may yet spoil the party.[7] Most of the region's economies are booming, having shrugged off the 1997 financial crisis and survived the 2008 collapse that almost decimated the global economy.[8] Asia is currently home to half of the world's middle-class and generates over 50 per cent of the world's economic output.[9] South and Southeast Asian countries are set to be the fastest-growing economies in the coming decade.[10] With the exception of India and the Philippines, the region's politicians have resisted the virulent strain of populism sweeping across Europe and the Americas.[11] One reason for this is that Asian leaders enjoy second mover advantage and will do all they can to avoid the mistakes made in the West.[12] Well before the most recent infectious disease outbreak that began in China, Asian governments had started resurrecting patterns of trade and cultural exchange that had thrived before the interlude of European colonialism and US hegemony.[13] Leading the pack is a resurgent China, which is building monumental maritime and terrestrial trading routes unlike any the world has ever seen.

Twenty-first century geopolitics look unstable from virtually every angle. Despite the best efforts of countries like Canada, Germany and France to keep it alive, the faith in the liberal multilateralism that characterised the 1990s and 2000s is evaporating. Old certainties and stable alliances based on shared principles and values are under scrutiny. The world is witnessing a return of great-power competition that is eroding the basis of post-Cold War co-operation. The most successful countries, companies and organisations will be those that make smart bets on future trends, discard outdated assumptions and find ways to work in constellations of dynamic strategic partnerships based on shared interests. In this chapter we draw on the maps to show how the global balance of power is changing and how the tipping of economic gravity eastwards is fundamentally transforming international relations. With geopolitics in disarray and the global economy in tatters, the future is more unpredictable than ever.

Geopolitical earthquakes or tremors

We live in uncertain times. The world is more connected than ever before, but it is also wracked by divisions. Why, at a time of un-precedented global interdependence, is international co-operation so fraught? Virtually every global leader would concede that pandemics and climate change are existential threats, yet they still struggle to forge collective action. A big part of the problem is that short-termism dominates the calculations of many elected leaders. Another challenge is the influence of powerful vested interests that are busily preserving the status quo. As a result, cynicism in the dividends of multilateralism is growing. The journalist George Monbiot believes we lack a shared story or narrative to motivate collaboration.[14] Making matters worse, growing numbers of people are resentful of traditional elites, especially the political class. The 2008 financial crisis intensified the loss of trust in public authorities and experts. The COVID-19 pandemic may deepen this antipathy further still. Not surprisingly, progressive politicians are finding it harder and harder to shore up the domestic support required to foster stronger multilateral ties.

There are structural explanations for why multilateral co-operation is getting harder. Specifically, many of the international norms, rules and decision-making bodies originally designed to enable co-operation are adapting too slowly to a fast-changing world.[15] Twentieth century organisations like the United Nations and the World Trade Organization are overburdened, underfunded and increasingly ignored. This is to be expected. After all, big shifts in the nature and distribution of geopolitical power necessarily require updating global institutions. A classic example of this is the United Nations Security Council that only includes five permanent members – the United States, China, Russia, France and the UK – yet excludes huge players like Germany, India, Brazil and Japan. It also lacks any representation from Africa. As a result, the Security Council is hopelessly paralysed. It took the members more than one hundred days to meet to discuss the COVID-19 pandemic and, even then, they could not agree on a joint statement. A more hopeful example is the G20 – a group of the world's most powerful national leaders, foreign ministers and central bank governors. The G20 was launched in 1999 (in the wake of the 1997 Asian financial crisis) to promote global financial stability. It has clout – its nineteen countries and the European Union account for 90 per cent of global world product, or GWP. It not only helped minimise the fallout from the 2008 financial crisis,[16] it is also being called on to calm global tensions in the present era.[17]

Multi, bipolar and unipolar systems[18]

Not surprisingly, newer forms of multilateralism are emerging that reflect an increasingly multipolar world. Alongside traditional Western alliances such as the G7 and the North Atlantic Treaty Organization, or NATO, is the Brazil, Russia, India, China and South Africa, or BRICS, coalition. Other networks include the Shanghai Cooperation Organization developed by China and Russia, and the Forum on China–Africa Cooperation that assembles more than 44 countries and 17 international and regional bodies. Rivalling the World Bank is the Chinese-backed and BRICS-led New Development Bank and Contingent Reserve Arrangement as well as the Asian Infrastructure Investment Bank, or AIIB. The AIIB is based in Beijing and has more than 57 shareholding countries. According to some political scientists, these and other networks do not presage the end of multilateralism but rather a new beginning. They are indicative of a new post-Western order pivoting around Asia. Given the way many Asian governments responded quickly to the COVID-19 crisis (in contrast to wealthy countries in the West), it is likely that Asia will continue to flourish.

There is no shortage of geopolitical threats in search of solutions. Alongside pandemics, global warming, nuclear war and AI are matters

Multipolar system

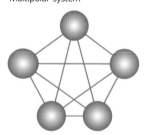

Bipolar system

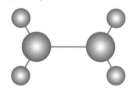

Unipolar system (hegemony)

Poles of influence
Credit Suisse Research Institute, 2017

4 highest   3   2   1 lowest

2000 km
1000 miles

Russia (2)
UK (3)
Eurozone (4)
Turkey (1)
China (4)
Japan (3)
Saudi Arabia (1)
India (3)
USA (4)
Brazil (1)
Indonesia (2)
South Africa (1)

Poles of influence [19]

The world is shifting from a US–European hegemony to more distributed and regional centres of power. This shift has economic and political dimensions. The dominant poles are the US, Europe and a China-centric Asia. Legacy powers such as Japan and the UK are losing steam, while other emerging markets such as Brazil, India, Russia and South Africa have yet to reach their potential.

of inter-state tensions. One of the most important is whether China's meteoric economic rise will generate manageable tremors or a catastrophic earthquake? International relations scholars have been puzzling over such questions for years. They often resort to theoretical models to help them make better predictions. Some of them distinguish between a unipolar, bipolar and multipolar balance of power. As the figure shows, unipolarity occurs when there is a single country (or group of countries) that exerts commanding political, economic and cultural influence. Bipolarity arises when two countries (or groups of nations) are the dominant powers. Multipolarity exists when power is distributed more or less evenly among three or more countries, groups of countries or regions. Understanding the particular state we're in (and where we're going) can help explain how countries interpret everything from trade disputes to questions of war and peace.

Some global orders are more stable than others. In a unipolar world, for example, minor spats may be forgotten after a few years. But in bipolar and multipolar contexts, small disputes may be connected to a wider strategic game triggering political, economic and even military responses.[20] So what world are we in right now? We appear to be exiting a short-lived unipolar era led by the US since 1989 and shifting to a multipolar world dominated by the US, the European Union and China, together with regional powers jostling for influence.[21] The map describes several identifiable poles of influence based on five criteria – economic output, ability to project hard power, soft power, the quality of governance and identifiable cultural distinctiveness.[22] Traditional powers such as the US, the European area, Japan and the UK still score highly, while other big players such as China, India, Russia and groups of developed countries are rapidly gaining ground.

No one has a clue how this new multipolar scenario will pan out. It is conceivable that there will be no single dominant player but instead a handful of regional powers, some with semi-imperial tendencies.[23] Political and economic influence is likely to be diffused across regional groupings of states and increasingly assertive non-state networks. Well before the COVID-19 crisis, multinational companies and groups of megacities were already exhibiting more global power. The political scientist Ian Bremmer describes the emerging scenario as a 'G-Zero world', one where the Security Council is irrelevant, the G7 obsolete and the G20 disabled by competing interests.[24] There are many people, the present authors included, who are fearful that fraying multilateral co-operation will undermine global co-operation at a critical juncture.[25] Others welcome a post-global liberal order as the first genuinely international expression of multipolarity in modern history.

In theory at least, multipolar systems can be stabilising. The more powerful states there are in a given system, the reasoning goes, the more webs of alliances and checks and balances on their use of force.[26] Yet as devastating wars during the nineteenth and twentieth centuries have shown, when multipolar orders crumble, they do so in spectacular fashion. Unipolarity can also contribute to a peaceful balance of power.[27] But one problem is that unipolar systems tend to be short-lived and prone to instability.[28] Although bipolar systems enjoy certain advantages such as a balance of power,[29] they can become unstable when the dominant country fears the rise of a rival. In fact, the most unstable phase of all is the transition from one of these states to another, which is precisely where we find ourselves today.

One of the earliest descriptions of a major power transition was recorded by the Greek historian and general, Thucydides.[30] He described how the rise of Athens contributed to the outbreak of the Peloponnesian War with the more powerful Sparta in 400 BC. The

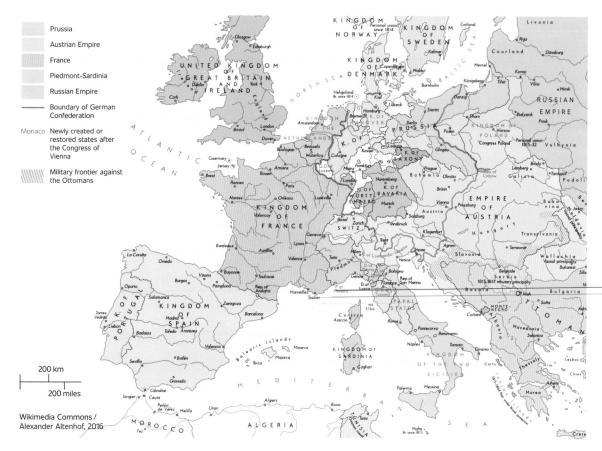

Concert of Europe, 1815–1914 [34]

The Concert of Europe was a dispute resolution system organised by the region's powers to maintain and consolidate their authority and maintain stability. The first phase was between 1815 and the 1840s and the second from the 1880s to 1914. It is widely considered to have contributed to a brief stable multipolar era.

Athenians' surging military and economic prowess terrified the Spartans. This basic insight – the idea that a hegemon might retaliate against a rising power – is often referred to as 'the Thucydides trap'.[31] Although Thucydides drew on a sample size of one, he was on to something: according to researchers at Harvard University, twelve of the last sixteen times a rising power threatened to displace the ruling one, the outcome was war.[32] Thucydides' insights are not lost on either Chinese or US policymakers today. In 2017, for example, China's Xinhua news agency took out a full-page advertisement in *The New York Times* urging both Chinese and US presidents to keep their countries from falling into the Thucydides trap and sparking a violent collision.[33]

Large parts of the world have lurched uneasily between multipolar and bipolar systems in recent centuries. Neither condition has correlated particularly well with peace and security.[35] As explained in the chapter on violence, human history is characterised by far more years of warfare than tranquillity. But there are notable exceptions, including the so-called Concert of Europe. The map highlights one of the first recorded periods of 'global' peace (1815–1847) between the five great powers of Europe at the time, namely Austria, Britain, France, Prussia and Russia. The German chancellor Otto von Bismarck helped usher in another period of stability (1871–1914) after brokering arrangements with Austria, Britain, Italy, France and Russia that held until the catastrophic outbreak of the First World War. The League of Nations in 1920 – the first intergovernmental organisation purpose-built to guarantee world peace – was another attempt to pacify the multipolar system. But the League was unable to restrain aggression by Axis powers in the 1930s and eventually collapsed. The Second World War set the stage for a new international architecture – the United Nations and the Bretton Woods system – to prevent violent conflict and the economic nationalism that gave rise to it.[36]

One of the most tense periods in history occurred during the Cold War. As the map shows, between 1947 and 1991 the US and the USSR divided the world into two competing blocs. Very generally, capitalist states across the western hemisphere fell in line with the US while communist and socialist countries lined up with the Soviets.[38] A host of other countries in Africa and Asia remained non-aligned, or oscillated between the US and Soviets. The two rivals waged many bloody proxy wars[39] yet still maintained a delicate balance of power backed by the threat of mutual annihilation. Nuclear deterrence was one of the chief factors keeping a cold war from becoming a hot one. Notwithstanding several close calls, the two superpowers avoided nuclear Armageddon. Bogged down by economic stagnation, exhausted by war in Afghanistan, and confronted with growing pressure from the US, the new Soviet leader Mikhail Gorbachev introduced a series of liberalising reforms known as *perestroika* (reorganisation) and *glasnost* (openness) in the 1980s. These processes unleashed waves of protest and nationalist movements and, surprising everyone, the demise of the USSR.

A short unipolar affair

The world instantly changed in 1989. Practically overnight the US became the sole superpower and we entered a unipolar era.[41] During the early 1990s, analysts compared the military and economic might

Cold War: 1947–1991 [37]
The Cold War divided the world into two blocs between 1947 and 1991. Led by the US, capitalist states more or less formed one bloc while the Soviet Union included dozens of communist and socialist countries in its bloc. There were also several countries that were non-aligned or switched allegiances, including in Africa and Asia.

Estimated nuclear warhead stockpile: 2019 [40]
There are around 14,500 known nuclear warheads still held by countries around the world. These are distributed between nine countries.

Estimated nuclear weapon inventories by country
Federation of American Scientists, 2019
● total weapon stockpile (not including decommissioned)

2000 km
1000 miles

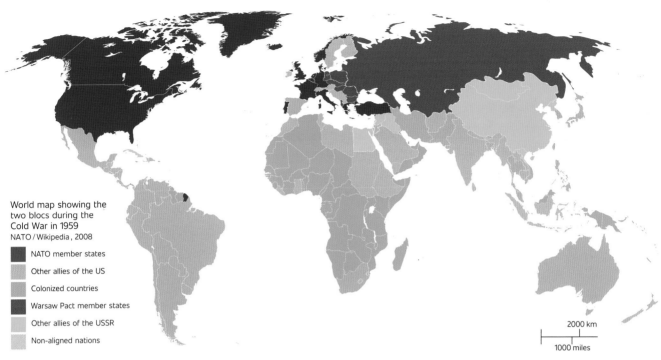

World map showing the
two blocs during the
Cold War in 1959
NATO / Wikipedia, 2008

NATO member states

Other allies of the US

Colonized countries

Warsaw Pact member states

Other allies of the USSR

Non-aligned nations

2000 km

1000 miles

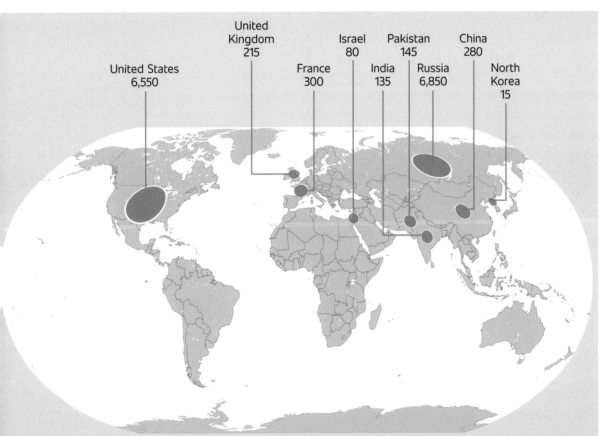

United States
6,550

United
Kingdom
215

France
300

Israel
80

India
135

Pakistan
145

Russia
6,850

China
280

North
Korea
15

of the US to the Persian, Roman, Mongol and Spanish empires of past centuries.[42] A former French foreign minister memorably described the US as the world's first 'hyper-power'. The US presided over the world's largest economy, accounted for more than half of all global military spending, commanded the largest blue-water navy in history, and amassed a daunting nuclear arsenal.[43] As the map shows, virtually all the remaining 14,500 nuclear warheads in the world are divided between the US and Russia, with smaller arsenals held by China, France, India, Israel, North Korea, Pakistan, and the UK.[44] That is down from an estimated 70,300 warheads in 1986.[45] The decision by US and Russian negotiators to reduce their stocks was not motivated by altruism or solidarity. Instead, it was a result of changing strategic calculations and a rash of treaties, alongside the creation of a new generation of hyper-sonic weaponry obviating the need for large and costly stockpiles.

US military bases abroad
David Vine, American University
Digital Archive, 2020

Bases

Lily pads (fewer than 200 personnel)

Naval fleet

Fleeting, unipolarity had its benefits. Compared to the nineteenth and twentieth centuries, the past three decades were atypically peaceful. Unopposed by its Cold War foe, the US expanded its cultural, economic and military assets unimpeded. The world witnessed its military might during the 1990s following interventions in Iraq and the former Yugoslavia. The US's vast security footprint grew dramatically in the wake of terrorist attacks in September 2001 and subsequent wars in Afghanistan, Iraq, Libya and Syria. As the map shows, the US supports more than 800 bases and 200,000 active troops deployed to 177 countries and territories. It is also likely that this military presence will contract in the aftermath of the COVID-19 pandemic. This unprecedented unipolar moment – what is known colloquially as the 'short peace' – was already coming to an end before the 2020 coronavirus outbreak. As international relations scholars well know, unipolarity is hard to maintain due to diminishing returns, rising costs, diffusion of power and the ceaseless counterbalancing of rivals.[46]

One of the reasons the short peace is ending is because the US is overstretched. The price tag for the so-called war on terror is in the trillions of dollars.[48] The annual costs of keeping the country's bases open is estimated to be at least $100 billion.[49] The US has been embroiled in one armed conflict or another for most of the past three decades, a source of concern to many voters.[50] It is not just war-fighting that is undermining US dominance. The 2008 financial crisis

The US's global military footprint [47]

The US dramatically expanded its global military footprint following the end of the Cold War. As of 2015 it supports more than 800 bases and 200,000 active troops deployed to 177 countries and territories.

also dented its influence and legitimacy. The crash signalled the end of a cycle of unbridled market capitalism that intensified inequality, dismantled organised labour and hurt the middle-classes of higher-income societies. The sluggish response of the US to the COVID-19 pandemic and its government's inability to contain the negative effects on the country's economy will be felt for years to come. The decline of US dominance is welcomed by rivals such as China and Russia that resent the centralisation of power in the US and Western Europe. [51] For decades, they have called for greater multipolarity. They are not alone. In the United Nations, for example, Brazil, Germany, India and South Africa have long demanded fairer trade rules and more representation in international institutions. [52]

The waning of US and Western European dominance coincides with the rise of China. After three decades of breathtaking economic growth, China is now the world's largest economy by purchasing power parity (PPP) and second largest by nominal GDP. [53] Despite facing intense pressure from the US, and a major economic contraction due to COVID-19, the country is busily rewiring global trade in its favour. China is not only an economic juggernaut. Its military

China's Belt and Road Initiative [57]
The Belt and Road Initiative links more than 100 countries across Asia, Africa, Europe and the South Pacific. The belt recreates the old Silk Road, while the road includes a series of maritime routes. Announced in 2013, the project is estimated to cost well over $1 trillion.

China's Belt and Road Initiative
Source anonymous

—— Maritime Road
● Sea ports
—— Overland Belt
● Railroad stops
—— Railroads

2000 km
1000 miles

expenditure jumped by more than 520 per cent between 2005 and 2018 (accounting for around 14 per cent of global military spending compared to over 36 per cent by the US).[54] In addition to being the planet's largest exporter of products, second largest importer of goods, and the fastest-growing consumer market, it vies with Japan[55] as America's largest foreign creditor.[56] China is playing a long game to consolidate its influence in its neighbourhood and beyond. One way it is doing this is through the so-called One Belt, One Road (or Belt and Road) Initiative. The Belt and Road is intended to expand foreign markets for Chinese goods and services, while also bolstering political influence and military ties.

Announced in 2013 by China's President Xi Jinping, the Belt and Road Initiative links at least 70 countries across Asia, Africa, Europe and the Americas. As the map shows, it consists of a lattice of pipelines, roads, railways, ports and even new cities designed to consolidate economic ties (and interdependence) with Beijing.[59] The Belt and

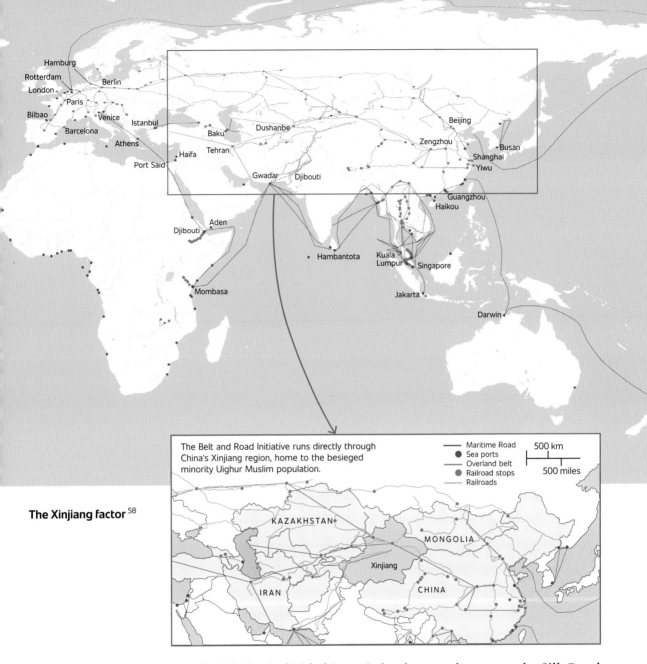

The Xinjiang factor [58]

The Belt and Road Initiative runs directly through China's Xinjiang region, home to the besieged minority Uighur Muslim population.

— Maritime Road
● Sea ports
— Overland belt
● Railroad stops
— Railroads

Road Initiative is divided into six land routes, known as the Silk Road Economic Belt, and one maritime route, called the Maritime Silk Road. Costing at least $1 trillion,[60] the Chinese initiative is possibly the largest and most ambitious development programme in history. To put it in perspective, it is at least four times larger than the US-led Marshall Plan to reconstruct post-war Europe between 1948 and 1951.[61] Taken together, it brings the equivalent of a third of all global GDP into its orbit.[62]

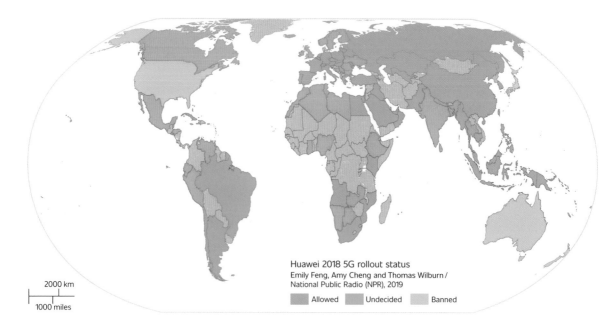

Huawei 2018 5G rollout status
Emily Feng, Amy Cheng and Thomas Wilburn /
National Public Radio (NPR), 2019

Allowed   Undecided   Banned

2000 km

1000 miles

Although China's Belt and Road Initiative is providing desperately needed credit and loans for building infrastructure in China and abroad, it is also triggering a domestic backlash. Within China, for example, a network of railroads and pipelines run through Xinjiang, home to the Uighur Muslim population.[63] As the map shows, Chinese authorities are converting Xinjiang into a 'core region' of national economic development. Yet as we show in the violence chapter, these modernising ambitions have coincided with the creation of a vast array of surveillance systems and detention centres to 're-educate' the locals.

While the Belt and Road Initiative is providing a powerful stimulus, it is also generating concerns abroad. In Cambodia, the rapid influx of billions of dollars in investment has cemented China's influence, including on sensitive issues of defence.[64] While many governments welcome the much needed investments, a number of Asian and African governments have complained about the poor quality of Chinese-built infrastructure, the displacement of local workers by Chinese labourers, and the lack of environmental protections.[65] Some recipients of Chinese aid worry that they are being issued loans they will never be able to repay and fear being caught up in 'debt trap diplomacy'.[66] Despite China's assurances of win-win co-operation and the queue of requests for more investments, not everyone is convinced. The Sri Lankan government was forced to agree to a ninety-nine-year debt-for-equity swap to operate a port. Malaysia, for its part, suspended

China's digital reach [68]
A global race between China and the US is underway to install 5G wireless networks. More than forty countries already have some version of 5G in place, with China's Huawei aggressively pursuing contracts around the world. Australia, Japan, New Zealand and the US have banned the company, claiming that it provides Chinese authorities with access to sensitive data.

China-backed deals valued at $23 billion, arguing that they were based on unequal trade agreements. India has also boycotted Belt and Road Initiative meetings,[67] anxious about how the initiative will dilute its own influence in the region.[69]

China is reshaping the global order and ensuring it has a more dominant role within it.[70] COVID-19 has dramatically reduced Chinese growth but, because it has had an even worse impact elsewhere, China remains on track to become the largest economy by 2030, generating a quarter of global GDP (depending on how you count).[71] In fact, depending on how it recovers from COVID-19, India could well be the second biggest economy in the world by 2030, followed by the US.[72] But future growth is not guaranteed. Even before COVID-19 struck, China's GDP had reached a twenty-seven-year low in 2019 in the wake of trade wars with the US.[73] In 2020 it is expected to sink to a forty-four-year low as COVID-19 cripples the economy. China, like many other wealthy countries, also faces a challenge with ageing: the total population will peak in 2030 and then start to decline.[74]

All of this partly explains why the Chinese government is dramatically expanding its trade linkages and ramping-up investment in a range of technology sectors. Some of China's bigger bets may pay off handsomely. Investment in AI is accelerating. The country is a green energy powerhouse, accounting for 40 per cent of all investment in clean energy technologies like solar and wind. Chinese firms have also moved rapidly to provide 5G,[75] having launched commercial services and pilots in dozens of countries despite US bans on high-profile Chinese providers like Huawei. The Chinese model, despite its limitations, is irresistible to many poor and middle-income countries intent on leapfrogging to the future. China is also turning the COVID-19 pandemic to its advantage, marketing its aggressive and technology-enabled response as an example for other countries to follow.

Returning to a multipolar world

Like it or not, the future is pluripolar.[76] There are many reasons to be uneasy, and it is not at all clear if multipolarity will result in more or less stability. The experiences of the Congress of Vienna and the Concert of Europe offer modest grounds for optimism.[77] But the uncomfortable truth is that today's world is far more crowded and complex than in the past.[78] The big powers (and some smaller ones) are bristling with nuclear, biological, chemical and cyber weapons and civilisation-ending miscalculations cannot be ruled out.[79] Forging global co-operation in a world with so many competing interests among countries and companies is difficult. It is even more complex when one considers that nation states are far from the only, and possibly not even the most important, players. As the late sociologist Benjamin Barber observed in the mid-1990s, big corporations and cities may be more pivotal players in shaping the direction of global affairs than national governments.[80] Part of the reason for this is the acceleration of new technologies. New technologies that facilitate communication and exchange generate important efficiencies, but they are also making governing more complex. As we shall see later in this book, they not only foster transnational alliances made up of governments, companies, philanthropists and non-governmental organisations, but also organised crime groups, networks of violent extremists and black hat hacker collectives.

Swelling rivalries between the big powers are generating geopolitical tremors. Political and trade relations between China and the US are at a low point. In 2018, for example, the US declared a new era of 'long-term strategic competition' and characterised China as a 'revisionist power' bent on creating 'a world consistent with their authoritarian models' [sic]. In 2019, the US military warned that China's dominance of 5G networks would result in overwhelming military advantage and the weaponisation of cities.[81] Not only did the US blacklist Huawei that same year,[82] it implemented global safeguarding tariffs for the first time in almost two decades. The US's shift from welcoming China's rise to strategically containing it is new. It is also fraught with risk. Without globally agreed guardrails, the simmering trade war could stumble into a military confrontation, dragging the rest of the world along and taking down the global trading system with it.

Another possible trigger for geopolitical tension is over the control of natural resources. A race is underway to secure key mineral deposits and control global supply chains in order to dominate the new economy. Governments and businesses don't just have their eye on untapped oil and gas or mineral reserves in the Arctic, but also rare earth elements such as indium, molybdenum and neodymium that are fundamental for building everything from computer chips

World mine production of rare earth elements and reserves in metric tons, 2018
United States Geological Survey (USGS), January 2020

	Production	Reserves
Australia	21,000	3,300,00
Brazil	1,000	22,000,000
Burundi	600	—
Canada	—	830,000
China	132,000	44,000,000
Greenland	—	1,500,000
India	3,000	6,900,000
Madagascar	2,000	—
Myanmar	22,000	—
Russia	2,700	12,000,000
South Africa	—	790,000
Tanzania	—	890,000
Thailand	1,800	—
USA	26,000	1,400,000
Vietnam	900	22,000,000

Rare earth elements production and reserves [83]

Demand for rare earth elements is growing as the world undergoes a major energy transition. China is by far the world's biggest supplier, though other countries have major supplies (Brazil and Vietnam) or are medium producers (Australia and the US). The dark blue circles refer to the known production in millions of tons while the light blue refers to reserves.

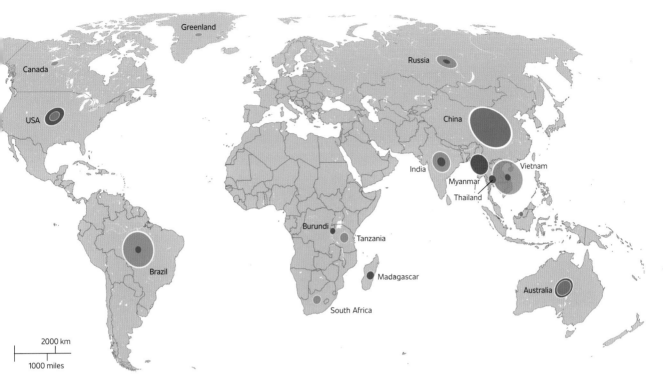

Canada

USA

Greenland

Russia

China

India

Vietnam

Myanmar

Thailand

Brazil

Burundi

Tanzania

Madagascar

South Africa

Australia

2000 km

1000 miles

Bolivia's Salar de Uyuni lithium mine

The salt flats are more than 10,000 km², making them the largest in the world. They are also 12,000 metres above sea level, making them the highest mines on the planet. Today, Bolivia is believed to hold up to 15 per cent of the global lithium supply.

200 metres

200 yards

NASA Earth Observatory, 2019

and batteries for electric cars to mobile phones.[84] As the map shows, just a handful of countries control the world's most sought-after resources.[85] For example, China controls the majority of global production.[86] Meanwhile, half of the world's known cobalt reserves are buried in the Democratic Republic of Congo and up to 15 per cent of the world's lithium supply is in Bolivia.[87]

In fact, the South American country has the second largest reserve of lithium in the world, mixed into mud that lies beneath massive white salt flats.[88] As the satellite image reveals, one of the largest deposits in the world is in the Salar de Uyuni mine, which is about 12,000 feet above sea level.

Another factor threatening global stability is the decline of the so-called global liberal order.[89] The basic foundations of the 75-year-old order consist of an array of overlapping political, economic, military and related agreements and alliances. At its centre are the United Nations, the International Monetary Fund and the World Bank (all founded in 1945), the General Agreement on Tariffs and Trade that later became the World Trade Organization (in 1995) and NATO (formed in 1949).[90] One of the gravest risks to the order is from its core members, including the US and some European countries. The reluctance of the US to take the lead is generating a leadership vacuum. This unwilling-ness was on display throughout the early period of the COVID-19 pandemic. The pre-occupation of Europe with its many political and economic divisions –

Declining freedom around the world[93] According to Freedom House, democracy has been in retreat for more than a decade. The group reports declining freedoms in long-standing democracies such as the US as well as in more authoritarian countries such as China and Russia.

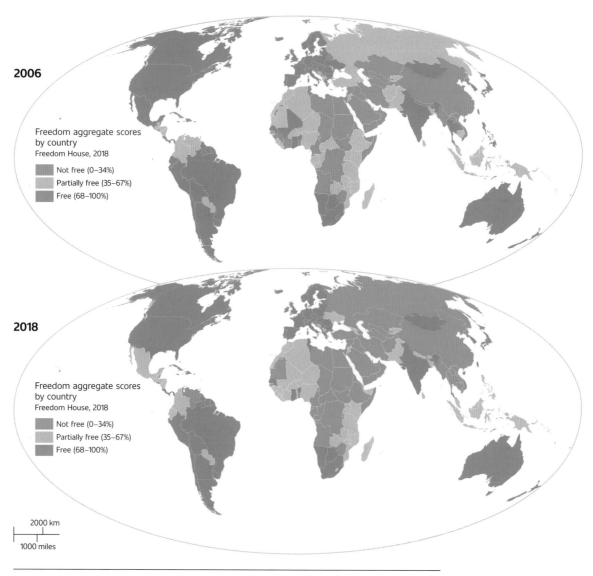

2006

Freedom aggregate scores by country
Freedom House, 2018

Not free (0–34%)
Partially free (35–67%)
Free (68–100%)

2018

Freedom aggregate scores by country
Freedom House, 2018

Not free (0–34%)
Partially free (35–67%)
Free (68–100%)

2000 km
1000 miles

not least the decision by the UK to leave the European Union – means that it is in no position to step up. While many US and European pundits have their doubts, China may soon be a more reliable provider of global public goods than either the US or Europe.[91]

The decline of the global liberal order started long before the emergence of either Donald Trump or the Brexit vote in 2016, much less fears of Russian meddling in Western elections or the COVID-19 pandemic. In the early 1970s, former US president Nixon decoupled the dollar from the gold standard, dealing a serious blow to the Bretton Woods system. Soon after, several Arab nations, angry about US support for Israel during the Yom Kippur War, stockpiled oil and the price of crude quadrupled. Food prices rose and the US fell into recession. Faced with 'stagflation' – a combination of inflation, recession and unemployment – the US raised interest rates and began dismantling regulations and constraints on capital flows. By the 1980s, what became known as the 'neo-liberal doctrine' was in full swing, with calls from its most ardent supporters to do away with capital controls, to balance budgets and to limit taxes and cut social spending. This formula was later exported to low- and medium-developed countries. With the curtailing of restraints on financiers and investors and the imposition of harsh austerity measures, the rich got richer and income inequality increased. The effects were felt around the world. In the US, for example, today's real average wage has the equivalent purchasing power it did forty years ago.[92] As of 2019, an astonishing 75 per cent of US residents live paycheck to paycheck.

A polarising world

One of the reasons why the global liberal order is flagging is because domestic support for it is declining. Indeed, people's support for democracy is waning around the world.[94] Support is not just declining in newer democracies, but in mature ones as well. A study of 154 countries and over 3,500 surveys between 1995 and 2020 determined that global dissatisfaction with democracy reached record highs in 2019. Trust in elected leaders also reached record lows and societal polarisation is off the charts. These dynamics are exacerbated by a combination of rising inequality and a pervasive sense of political hopelessness.[95] Many people simply feel that their elected governments

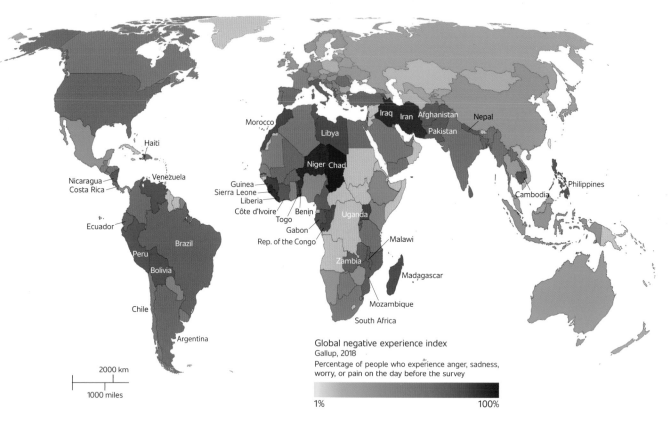

Global negative experience index
Gallup, 2018
Percentage of people who experience anger, sadness,
worry, or pain on the day before the survey

1%   100%

are not working for them. The spread of protests and unrest from
Cairo and Hong Kong to Barcelona and La Paz is a symptom of rising
frustration. National politics is increasingly charged and extreme
which has in turn fuelled the rise of far-right parties and populist
governments. Around the world, the political discourse is increasingly
moulded by identity politics defined along ethnic, racial, religious and
gender lines.[96] These tendencies could be dramatically accelerated
depending on how governments respond in the COVID-19 era.

   As the chapter on culture shows, social media platforms and
opinion-based media outlets are amplifying these divides. Anyone who
spends a few minutes using Twitter, Facebook, YouTube or WhatsApp
knows exactly what this looks like. The map illustrates how sentiments
associated with negativity have risen in more than 145 countries over
the past half-decade.[97] Almost 40 per cent of respondents to a Gallup
survey claimed to have experienced considerable worry or stress in the
past year and a significant proportion also experienced physical pain
and anger. Even before the immense grief and stress felt by many as a
result of the COVID-19 pandemic in 2020, at least one in five people
felt high levels of sadness and anger.[98] These feelings are especially
pronounced in parts of Africa, the Middle East, and Latin and North

Rising anger around the world

There are signs of
growing negativity
in some parts of the
world. Gallup produces
a Negative Experience
Index that captures
rising worry, stress
and anger. Negativity
is especially acute in
parts of North, Central
and South America,
Sub-Saharan Africa,
the Middle East, and
pockets of Southern
Europe and South Asia.

America. Writing before COVID-19 struck, the essayist and author Pankaj Mishra described our era as an 'age of anger' that is fuelled by, and fuelling, divisiveness.[99]

What explains rising anxiety, frustration and anger around the world? A big part of the story is the sense of declining social and economic status. Rising inequalities corrode social cohesion – the glue that binds societies together – and people's feelings of individual autonomy and control. Another factor might be mental health, depression and anxiety disorders that, as the health chapter reveals, have rocketed upwards since the 1990s, including among young people. One more piece of the puzzle is the hardening of group identities. When people feel threatened and vulnerable, they often turn inwards and become more nativist and tribal. This makes it harder to bridge political, cultural and economic divides between groups. As the migration chapter explains, many of those who feel left behind are directing their anger towards elites and minorities. This provides fertile ground for political opportunists to sweep in and rally supporters with promises to 'drain the swamp' and, if necessary, deny and deport outsiders.

Conscious of their sinking popularity, unable to communicate their message, and fearful of a backlash, growing numbers of politicians are going on the defensive. Support for multilateral co-operation (what some critics refer to as 'globalism') is waning because many of its supporters wish to avoid the inevitable anti-elite backlash. When commitment to the multilateral system fades, some states may be tempted to sidestep it entirely and move unilaterally. What this means is that co-operation on everything from pandemic response and climate change to the regulation of killer robots suffers.[100] It also means that the risks of misunderstanding spiral upward, making accidents and dangerous escalation more likely.

Democracy on the ropes – not yet out for the count

Democracies everywhere are taking a beating.[101] A cottage industry of democratic pessimism has emerged with grim talk of democratic 'undertow',[102] 'rollback',[103] 'recession',[104] and 'depression'.[105] There are fears that democracies are becoming 'partial',[106] 'low-intensity',[107] 'empty',[108] and 'illiberal'.[109] Students of democracy worry that they are

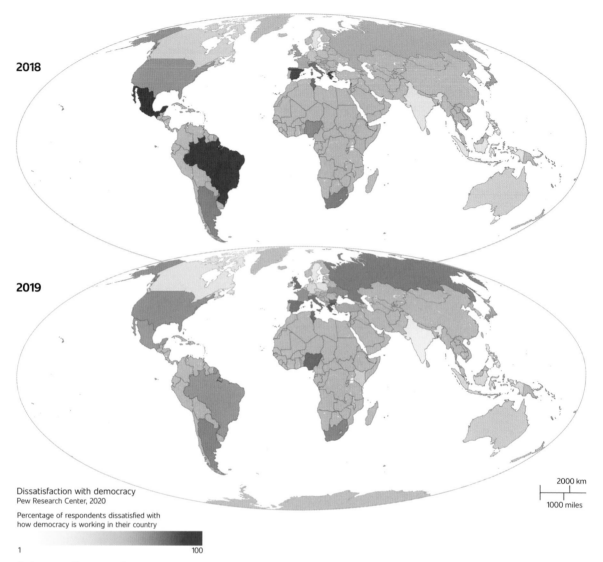

2018

2019

Dissatisfaction with democracy
Pew Research Center, 2020

Percentage of respondents dissatisfied with
how democracy is working in their country

1   100

2000 km
1000 miles

A democratic recession

The spread of new democracies is a recent achievement.[110] But the future of democracy is not assured. This map highlights dissatisfaction with democracy in 34 countries between 2018 and 2019. It shows more people are dissatisfied than satisfied with how their democracies are functioning. Discontent is more common among people with lower incomes or who believe the economy is doing poorly.

being hollowed out: elections still take place but civil liberties and checks on power are flouted. Consider the collapse of the 2003–2005 colour revolutions in Georgia, Ukraine and Kyrgyzstan or the violent push-back following the 2010–2011 Arab Spring protests in Egypt, Libya, Syria, Tunisia and elsewhere in the Middle East and North Africa. And with a significant minority of US citizens no longer in agreement about their role in the world, fewer and fewer societies are looking to it for assistance or inspiration.

Nationalist, anti-immigrant and populist parties have all gained strength in many parts of the world. Creeping authoritarianism is on display in Europe among recent democratic converts such as the so-called Visegrád Group: the Czech Republic, Hungary, Poland and, until recently, Slovakia.[111] Iron men are also emerging including Recep Tayyip Erdoğan in Turkey,[112] Matteo Salvini in Italy, and Vladimir Putin in Russia.[113] Brazilians also elected a former army captain, Jair Bolsonaro, who openly praises the country's dictatorship, thrives on disunity and advocates extreme police violence.[114] Meanwhile in the Philippines, president Rodrigo Duterte launched a ruthless crackdown on crime and steadily dismantled checks on executive authority.[115] And India's Narendra Modi has adopted authoritarian tactics to stifle dissent, including regularly shutting-down the internet. Watchdog groups that monitor the health of democracies are unnerved by how the world is becoming more illiberal and less free.[116] As the map shows, more and more countries are registering democratic deficits, according to the Economist Intelligence Unit.

A spate of popular books has added to the perception that democracies are on the rocks. In *How Democracies Die*, political scientists Steven Levitsky and Daniel Ziblatt argue that democracies typically end with a whimper and not a bang.[117] While demagogues such as Trump in the US or Orbán in Hungary can speed things up by undermining checks and balances, the real threats to democracy, in their view, emerge from within. Another social scientist, Yascha Mounk, warns that liberal democracies can succumb to 'undemocratic liberalism' and 'illiberal democracy'.[118] The former protects basic rights but delegates real power to supranational bodies like the European Commission – a frequent target of populist and extremist parties on the left and right. A growing number of political parties and democratically elected leaders are advocating for the restriction of minority rights and relaxation of the constraints on executive power.

When disaggregated by levels of pluralism, political participation and respect for civil liberties, several democracies register unmistakable signs of backsliding.[119] According to the Economist Intelligence Unit's *Democracy Index*, just twenty-two countries (13 per cent of the total), most of them in Western Europe, can be described as 'full democracies', as opposed to 'flawed democracies', 'hybrid regimes', or 'authoritarian regimes'.[120] Of the more than 160 countries in the index, eighty-nine registered signs of deterioration. There are growing concerns over the spread of autocratisation, which by one estimate now affects one-third of the world's population.[121] Despite the spread of multiparty elections and the rule of law over the past seventy-five years, these concepts risk losing meaning given growing pressures on media autonomy and freedom of expression alongside rising political exclusion in many parts of the globe.

A big problem seems to be that many people are dissatisfied with democracy.[122] As we discussed earlier, disaffection appears to be strongly linked to economic hardship and anger that the political elite are corrupt and out of touch.[123] These views are especially pronounced among youngsters. For example, the percentage of people living in the US and Western Europe who say it is 'essential' to live in a democracy drops precipitously from older to younger populations. The same goes for views on protecting civil rights, ensuring free elections and interest in democratic politics more generally.[124] Just under a third of Australians, Britons and US citizens born in the 1980s think democracy is essential as compared to three-quarters of those born in the 1930s.

Millennial scepticism with democracy and openness to illiberal alternatives is deeply personal. Many young people are frustrated with student debt, prolonged unemployment, limited social mobility and a feeling that the system is rigged and the establishment is dysfunctional. They are keen to see measures that ensure their economic security, even if they come at the expense of certain democratic principles.[125] In settings where there is already low affiliation with political parties there is less popular support for representative democracy.[126] In deeply polarised societies, where one side can literally win it all, elections are serving to deepen divisions and entrenched positions. While democracy is still the preferred model of governance around the world,[127] it is hardly the only model.

A democratic rebound

Although democracies are down, they are not out.[129] Despite dissatisfaction with how democracy is working, support for democratic ideals is still comparatively robust and procedural engagement is near record highs. It is worth recalling that 96 out of 167 countries

Three waves of democracy, 1800–2017 [128]

Political scientists, beginning with Samuel Huntington, often describe there being three waves of democracy since the nineteenth century. The first was a slow wave in the late 1800s. The second wave occurred after the Second World War and the third began in the mid-1970s and extended dramatically after 1989. This chart highlights the number of countries that are considered autocracies, anocracies and democracies since the 1800s.

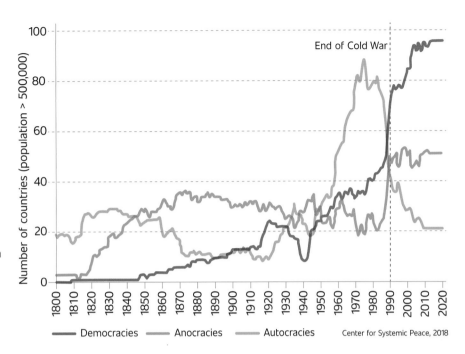

Three waves of democracy, 1800–2017. Center for Systemic Peace, 2018

Democracies — Anocracies — Autocracies

with populations of at least 500,000 people are democracies. Just twenty-one were autocracies (systems where a single person or party exerts absolute power) and forty-six were anocracies (systems featuring both democracy and autocracy).[130] Despite the present day cynicism and dismay, there has been an unmistakable upward trend in democratisation over the past half-century. Many people living in democracies still believe that representative forms of government are the best options available – even if these views co-exist with growing support for illiberal and undemocratic approaches such as rule by experts, strong leaders and the military.[131]

It turns out that an individual's support for democracy is strongly correlated with their economic situation. People who say their economies are in bad shape or that they cannot improve their standard of living are more inclined to be dissatisfied with democracy than those who say the economy is doing well.[133] This may help explain why in Europe, most Dutch people and Swedes believe democracy is working for them, while large majorities of Greeks, Italians and Spaniards say it is not. Meanwhile in Asia, with more people living under democratic governments than anywhere else on the planet, support for democracy is still on the rise. Another key factor is the extent to which people are affiliated (or not) with political parties. In countries where people are more unaffiliated, popular support for representative democracy also

1977

Political regimes over time
Marshall M., et al. – Polity IV Project, 2018

- Colony/non-sovereign
- Autocracy
- Anocracy
- Democracy

2018

2000 km
1000 miles

suffers.[134] This does not mean that recent upticks in the number of autocracies should be ignored, though it does suggest that the eulogies for democracy's demise may be premature.[135] Even so, democracies everywhere are facing one of their greatest challenges yet owing to the hugely negative economic effects of the COVID-19 pandemic and the perception that autocracies have been better able to manage and respond to the spread of the pandemic.

We should be cautious in our interpretation of polls showing declining support for democracy.[136] For one, it is hard to discern people's appetite for democracy in countries ruled by authoritarian regimes, where respondents are uneasy about disclosing their personal

The forward march of democracy[132]
Democracies have spread dramatically since the 1970s. Today, there are more democratic countries than autocratic or mixed ones. Notice the dramatic shifts in Latin America, Africa and Southeast Asia, in particular.

views. Findings from the Center for Systemic Peace's Polity Project[137] contends that the great third wave of democratisation, far from receding, may eventually give way to a fourth wave.[138] While there are admittedly different ways of measuring regime type, the share of the world's total population living in democracies (of varying types) is close to two-thirds. By comparison, a mere 1 per cent of all people lived in democratic countries in the early 1800s.

Shrinking trade unions and diminished church attendance

Renewing democracies will require rethinking and reinvigorating political parties. That is because in many places, political parties are struggling to recruit and retain members, get voters to turn out for elections, and sustain voter loyalty between elections. Since 1989 researchers have documented dwindling party membership, declines in voter engagement, and shaky party stability. Dissatisfaction with parties is evident everywhere, including in the heart of Western Europe.[139] In Sweden, a populist and anti-establishment party, the Sweden Democrats, won a fifth of the vote and was the third largest

Trends in union membership – 2000 and 2018 [140]

Trade union membership as a percentage of total employees (selected countries).

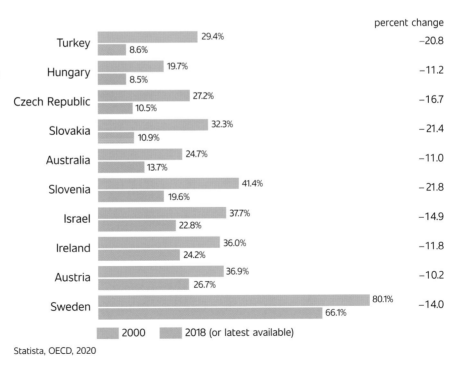

	2000	2018 (or latest available)	percent change
Turkey	29.4%	8.6%	−20.8
Hungary	19.7%	8.5%	−11.2
Czech Republic	27.2%	10.5%	−16.7
Slovakia	32.3%	10.9%	−21.4
Australia	24.7%	13.7%	−11.0
Slovenia	41.4%	19.6%	−21.8
Israel	37.7%	22.8%	−14.9
Ireland	36.0%	24.2%	−11.8
Austria	36.9%	26.7%	−10.2
Sweden	80.1%	66.1%	−14.0

Statista, OECD, 2020

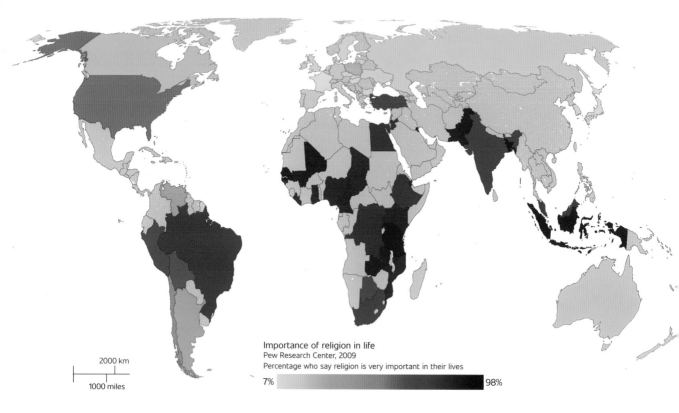

Importance of religion in life
Pew Research Center, 2009
Percentage who say religion is very important in their lives

7%   98%

2000 km

1000 miles

party in the country's 2018 elections. In Germany, traditional centre-left and centre-right parties declined dramatically as voters turned to the far-right Alternative for Germany in 2017 elections. Meanwhile, in Brazil, the virtually unknown Social Liberal Party was captured by an ultra-right populist in 2018, before being abandoned by the winner one year later. The moderate centre ground is ceding to the extremes.

One of the problems is that political parties in many mature democracies are in decline. These trends started decades ago. In Europe, party membership declined from 15 per cent of the adult population in the 1960s to less than 5 per cent by the end of the 2000s.[141] This was accompanied by steep declines in voter turn-out. Over the past half century voter turn-out in legislative elections in countries with competitive elections also fell from around 71 to 65 per cent. But in Europe, it dropped from 83 to 65 per cent. According to some analysts, these declines are due in part to the withering of trade unions and declining church attendance.[142] Deindustrialisation and anti-labour efforts have severely diminished labour union membership in many Western countries since the 1970s. The chart on the previous page

Religious importance to daily life
Declining membership of Catholic and Protestant churches has affected centre-right Christian democratic parties.

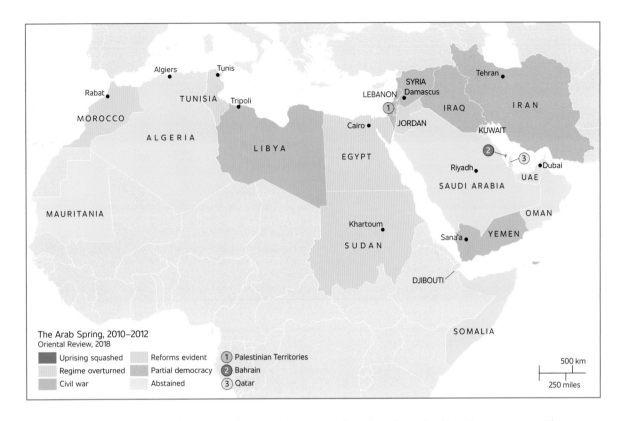

The Arab Spring, 2010–2012
Oriental Review, 2018

- Uprising squashed
- Regime overturned
- Civil war
- Reforms evident
- Partial democracy
- Abstained
- (1) Palestinian Territories
- (2) Bahrain
- (3) Qatar

500 km
250 miles

Arab Spring [145]

The Arab Spring involved a wave of pro-democracy and anti-government protests, uprisings and armed revolts that spread across several North African and Middle Eastern countries between 2010 and 2012.

illustrates how union membership has declined across a wide range of advanced economies. Electoral support for centre-left parties has also tumbled: witness the collapse of the French socialist party in 2017. Meanwhile, declining membership of Catholic and Protestant churches has also affected centre-right Christian democratic parties. This speaks to a wider crisis of class and religion as cornerstones of group identity.

The decline of political parties is associated with social unrest. Without parties, protesters struggle to convert grievances into policy proposals and measurable change through the ballot box. Take the case of the Arab Spring between 2010 and 2012. Most of the protests there lacked strong membership-based organisations and were rapidly extinguished. The hope that these movements could put an end to corruption, increase political participation and bring about economic inclusivity was quashed. On the contrary, they produced few leaders, limited credible programmes for action and limited new ideas. [143] As the map shows, some of them led to even sharper crack-downs on democratic freedoms and civil liberties, and even outright civil war. Tunisia was the only country to consolidate democracy through the moderating effects of its national labour union. Or consider the Occupy Wall Street movement that reached almost 90 countries between 2011 and 2012.[144] While decentralised online and street protests rallied millions to the cause, they did not result in tangible legislative victories and soon faded. Yet another example is the Yellow Vest movement that started in France in 2018 and has spread to two dozen countries.

In the long run, the waning of political parties is corrosive for democracy. This is because weak parties undermine accountability, not least by weakening the organisational capacities of citizens to hold governments to account. As we are seeing around the world, populists circumvent traditional parties by appealing directly to voters through social media and encrypted channels. Their outsider status – and their lack of experience – gives them a 'pass' to do so. While populist and authoritarian regimes often collapse under the weight of ineffective performance, they can still do a great deal of damage on the way down.

Can international co-operation be restored?

The big geopolitical shifts are hard to miss. The entire international system is being restructured as China expands its influence, the US recedes from its leadership role, and Europe struggles to cope with division. The COVID-19 pandemic simply accelerated this trend. Different facets of the global economy are adapting differently to the end of the global liberal order. Other systemic changes may generate even greater consequences for global affairs. New technologies are already changing the face of energy production and the future of work. Climate change is currently multiplying tensions over resources. There is no doubt that the COVID-19 pandemic – and the way governments did or did not respond – will shape global affairs for decades to come.

The first step to navigating the future is to recognise that these structural disruptions are occurring simultaneously. The emerging global order, then, is more likely to be self-organising, decentralised and regionalised rather than exclusively mediated by a small number of powerful nation states. To be fair, nations have always sparred about whether and how to co-operate. There has always been resistance to actions that might interfere in their sovereign affairs. But the world's biggest threats today are precisely the ones that no single government, or business consortium, or coalition of philanthropists can address on their own. If the emerging global order cannot find ways to co-operate and empower creative multi-stakeholder coalitions, we may not survive this century.

So what is to be done? Clearly, great compromises and sacrifices are required. At a minimum, global institutions will need to structurally adapt to the vastly changed political, economic, health-related and demographic realities. For example, the Security Council must expand to reflect a changed world, even if this seems diplomatically improbable. New organisational structures like the G20 will also need to be created to help states, companies and civil societies to establish the minimum rules for co-operation so they can more effectively navigate the accelerated pace and dynamics of change. Regional institutions including inter-governmental and investment entities will have to assume greater importance. After all, the day-to-day business of global affairs is regional, and has increasingly less to do with the core centres of power. But time is not on our side. Although a more pluralist world order may be desirable, there are very few realistic proposals, much less agreement, on how to achieve this.

Without global leadership, there is a real possibility that an ad hoc and patchwork international order emerges by default. We risk moving from a rules-based order to a transactional, deal-based framework instead. Markets hate uncertainty and volatility. How will businesses manage unclear regulations, safeguard their reputations and settle disputes when they arise? The risks of disastrous miscalculation are real. So too are the dangers of norm-breaking unilateralism such as the annexation of Crimea by Russia in 2014, the repeated use of chemical weapons in Syria or cyber attacks from advanced actor groups backed by nation states. In a divided world, who sets the red lines, and what happens when they are regularly flouted?

Around the world, nation states are hedging their bets. On the one hand, support for the United Nations is holding, albeit precariously. At the same time, new political, economic and security arrangements are emerging, some of them sidestepping legacy structures built for a previous century. Struggling to secure unanimity for global trade deals, there are risks that some countries may abandon them altogether. There are good reasons to diversify global institutions to regulate security, improve governance and facilitate fairer trade rules. Such institutional

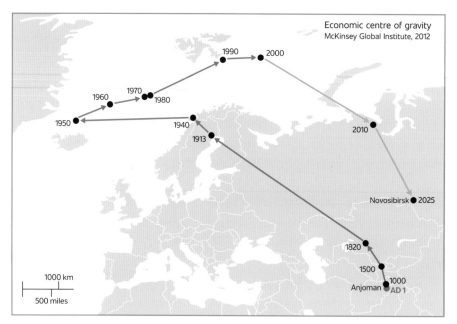

Economic centre of gravity
McKinsey Global Institute, 2012

The eastern drift of the centre of economic gravity[146]
The economic centre of gravity is estimated by weighting locations by GDP in three dimensions and then projecting them to the nearest point on the earth's surface. We can see that the centre shifts first west between 1000 and 1950 and then dramatically eastwards through to 2025.

evolution reflects a more plural world and the shift of gravity away from the West towards Asia, as the map shows. But without the 'engines' of shared norms, ethics and interests, there are real dangers ahead. The proliferation of new bodies with competing values could just as easily reinforce (rather than reduce) competing spheres of influence. When combined with reactionary nationalism and economic protectionism, they could easily fuel wider geopolitical turbulence.

It is hard to be optimistic that global co-operation will improve anytime soon. Stopping future pandemics and comprehensively addressing the issue of climate change are arguably our highest collective priorities. Despite growing consensus that the world has reached a climate change 'tipping point', implementation of the 2015 Paris Agreement is lagging. Before the considerable drop in greenhouse gas emissions precipitated by the COVID-19 pandemic, carbon emissions had risen to the highest point in 3 million years.[147] As the chapter on climate shows, the world is facing a potentially civilisation-ending crisis and the world's most powerful country has decided to withdraw, a perfect illustration of the mortal perils of a transaction-based nation-centric world.[148] With nation states struggling to take collective action on pandemics and climate change, it is not clear how they will deal with other issues on the near horizon. How can they possibly muster common cause to regulate hypersonic weapons, biotechnology or AI?

The promise of democracy

Strong, inclusive and ethical leadership is needed more than ever. Yet given the extreme flux of politics, there is a deficit of capable, ethical and effective politicians. This is not to say there are not powerful leaders out there. Authoritarians like Xi Jinping and Vladimir Putin offer clear direction and tend to follow through on their promises. These tendencies are likely to sharpen, especially as the remaining European leaders with any global stature exit the global stage. Notwithstanding the crop of young socially progressive newcomers in Canada, the US, parts of Europe and New Zealand, there are still comparatively few convincing personalities who can articulate a compelling alternative to the populists.

In the face of all these existential threats, it is important not to forget the promise of democracy. It comes down to the Churchillian idea that despite its many flaws, democracy is preferable to the alternatives. It allows people to replace their representatives without resorting to bloodshed. In well-governed democracies people can complain, publish, organise, protest, strike, threaten to secede and even move their money elsewhere without being detained, tortured or worse. On the contrary, governments often respond to their pleas. Take the case of Extinction Rebellion, which has as its very premise the idea of non-violent civil disobedience. This does not mean democracies cannot and should not be updated and improved. The most mature liberal democracies are works in progress that need constant grooming and improvement.

Yet for a democracy to flourish, citizens (especially younger ones) must be convinced that it is a better alternative to theocracy, the divine right of kings, colonial paternalism or authoritarian rule. Over the past few centuries, people around the world came to recognise that it is, and the idea of liberal democracy became contagious. Despite their limitations, democracies have proved remarkably effective at curbing the more sinister instincts of governments. Such profound changes are a reminder of why it is so important to fight for free and fair elections, the rights of minorities, freedom of the press and the rule of law. While many democracies have faced a crisis of confidence in recent years, their victories – and their continuing superiority to the alternatives – are grounds for some hope.

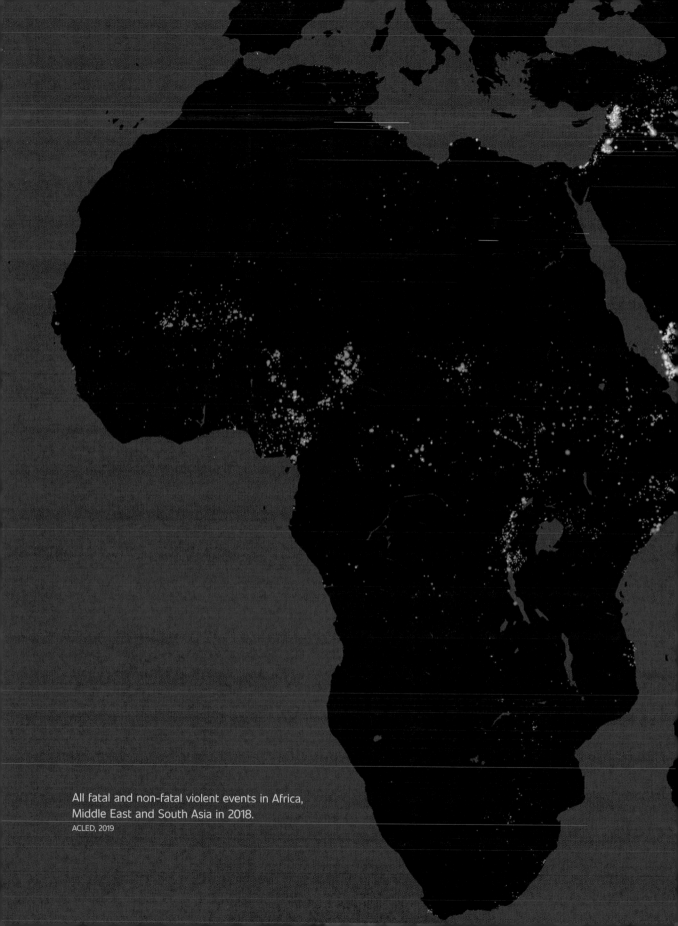

All fatal and non-fatal violent events in Africa,
Middle East and South Asia in 2018.
ACLED, 2019

Violence

The world is less violent but more disorderly

Crime and repression kill more people than war

Armed conflicts are harder than ever to resolve

State repression and criminal violence are increasing

New military technologies are difficult to regulate

Global co-operation to reduce violence is essential

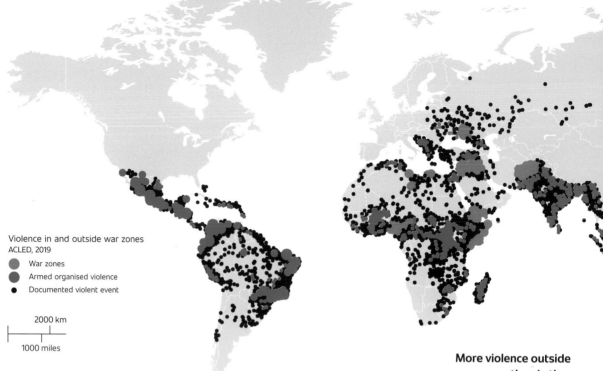

Violence in and outside war zones
ACLED, 2019

- ● War zones
- ● Armed organised violence
- ● Documented violent event

2000 km

1000 miles

Introduction

Humans are natural born killers. Estimates of the number of people who have died in wars varies from a few hundred million to more than one billion.[1] This map highlights the distribution of casualties of violence in war zones as well as outside them in 2019. Casualties in war-torn countries – marked in red – stand out. In Afghanistan, over 32,000 civilians have been killed and more than 60,000 wounded over the past decade.[2] Another half a million people are believed to have died violently in Syria since the civil war began there in 2011.[3] And in Yemen, at least 100,000 civilians have reportedly been killed since the outbreak of armed conflict in 2015.[4] All of these statistics are best guesses. It is hard to know the 'true' cost because vital registration systems frequently fall apart, armies and armed groups fix the numbers, and academics and activists who study death tolls vehemently disagree over estimates. Try as we might to count the dead, truth is one of the first casualties of war.

The map reveals how armed conflicts are just one part of the story when it comes to measuring the global burden of violence. While it may come as a surprise, every year far more people are killed outside conflict zones than in them. Violent extremism, organised crime and

More violence outside war zones than in them – 2019
The vast majority of organised violence occurs outside the world's war zones. This map highlights the distribution of fatal and non-fatal events arising from warfare (red dots) and other forms of armed organised violence (blue dots) in Latin America and the Caribbean, Africa, the Middle East, Eastern Europe and Central, South and Southeast Asia. It is a reminder of how non-conflict violence is much more widely distributed than often assumed.

state repression are responsible for hundreds of thousands of deaths annually. More people were killed by gangs, militia and police in countries like Brazil, Colombia, Mexico, the Philippines and South Africa last year than in virtually every war zone combined. Other forms of organised violence are so poorly documented that they are practically invisible, not least when they target migrants, minorities, women and children. While we cannot say precisely how many people die each year from violence, our best guess is that around half of the world's population has been touched by some form of violence in the last ten years.

The world certainly *feels* more dangerous and volatile. One of the reasons for this is that our television, computer and mobile phone screens are saturated with images of bloodshed. Today, most of the weapons doing the killing and maiming are relatively low-tech, especially handguns, rifles and landmines. Tomorrow's technologies – whether they are hypersonic gliding missiles, lasers, biological agents or self-organising swarming drones and nanobots – could be even more destructive.[5] But is the world really more violent today than in the past? While it may seem counterintuitive, fewer people are violently killed today as a proportion of the entire population than at any time in history.[6] The *rate* at which people are dying in international and civil wars is a fraction of the rate in previous centuries. And it's not just war-related deaths that are falling. Terrorist-related killings are also dropping in most countries. Homicidal violence has also sharply declined in virtually every part of the world.

Recent improvements in safety do not imply that the future is going to be stable. Indeed, rising unemployment and food insecurity in the wake of the COVID-19 pandemic significantly increased the risk of social unrest. Rather, they are a reminder that we might be doing some things right to make the world more secure. One of the most important lessons of this chapter is that many forms of intentional violence have declined over the past half-century. Another message is that most forms of violence are highly concentrated in specific countries, cities and neighbourhoods. We also find that seemingly disparate forms of violence – whether it is perpetrated by warlords, gangsters or police – are often a function of similar types of risk factors, including inequality and impunity. While the history of violence is dark and disturbing, these common-sense observations provide hope that it can be further prevented and disrupted.

The past was anything but peaceful

War is one of humanity's oldest pastimes. Peter Brecke's catalogue of more than 3,700 armed conflicts over the past 600 years is reproduced here.[8] The red dots indicate individually recorded events – the larger the dot, the more people are estimated to have been killed. The red line represents the estimated conflict fatality rate – the number of victims per 100,000 people. Very generally, most recorded armed conflicts killed between one and ten people per 100,000.[9] In some shorter wars, rates jumped as high as 200 killings per 100,000. To put these numbers in perspective – the average death rates from car accidents and non-communicable diseases are around 17 per 100,000[10] and 536 per 100,000 respectively.[11] As the image shows, the average conflict fatality rate has oscillated over time but began declining over the past century.

Archaeologists who study such things are fairly sure that most of human existence has been dominated by constant raids, massacres and atrocities.[12] These ancient 'primitive wars' were probably an order of magnitude more lethal than modern warfare.[13] One reason for this is that many pre-modern societies lacked basic moral codes to limit the slaughter of innocent civilians. Another is the absence

Documenting conflict deaths from the fifteenth to the twenty-first centuries[7]
There have been thousands of wars throughout history. This charts more than 3,700 of them and tracks each approximate death rate per 100,000. The red line is the moving average (over 15-year increments) and the blue line represents annual trends since 1900.

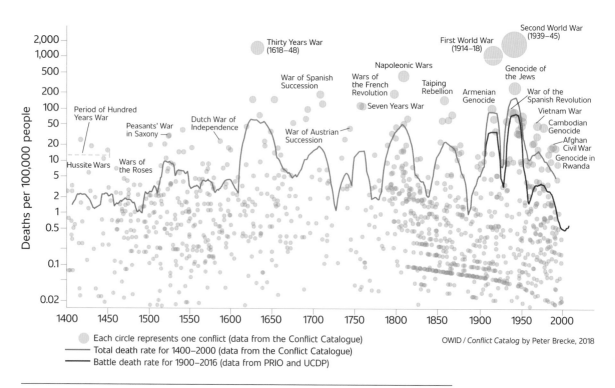

Each circle represents one conflict (data from the Conflict Catalogue)
Total death rate for 1400–2000 (data from the Conflict Catalogue)
Battle death rate for 1900–2016 (data from PRIO and UCDP)

OWID / *Conflict Catalog* by Peter Brecke, 2018

of medical care and antibiotics to treat the wounded. While there is abundant evidence of cooperation, humans have spent far more time perpetrating violence than refraining from it. Our ancient past was less Jean-Jacques Rousseau's 'peaceful man in nature' than Thomas Hobbes' 'war of all against all'. So why did we spend so much time fighting one another? Much of the motivation boils down to scarcity. Put simply, when times were good – which was almost never – people put away their clubs, spears and arrows. When times were rough, they reached for their weapons. [14]

In his book *Constant Battles: Why We Fight*, Harvard University archaeologist Steven LeBlanc explains how chronic shortfalls in food, water and land triggered violent raids.[15] LeBlanc uncovered fossilised evidence – preserved pollen, plants and human remains – that demonstrated the ways droughts, floods and storms contributed to food deprivation, migration and, more often than not, violence. Primitive wars started declining in number and intensity only after humans started becoming less nomadic and more sedentary. When nomads decided to settle – they sometimes produced and stockpiled more food and formed complex systems of governance – their tolerance of feuding started to wane. Over time, the spread of mercenary and professional armies, the expansion of international borders, and the growth of trade and shared value sets reduced zero-sum thinking – though not everywhere as we shall soon see.[16]

Homo sapiens have perfected the technology of killing over the past few thousand years. For most of history, we relied on a combination of blunt and bladed objects to murder and maim our enemies. All this changed with the invention of gunpowder. Around AD 850, Chinese alchemists stumbled across a game-changing substance while experimenting with sulphur and saltpetre. In one of the greatest ironies of recorded history, they accidentally invented gunpowder while trying to concoct life-extending elixirs and potions.[17] The implications for war-fighting were profound. Without a doubt, guns and ammunition have slain more people than any other weapons system. Yet while the absolute number of people fatally injured by firearms and bombs has increased over the past few centuries, the killing rate has actually stayed relatively stable (so the increases in mortality are due to a steadily growing population and not changes in human proclivity for violence).

Although twentieth-century wars generated unprecedented slaughter, they were not necessarily the most lethal in per capita terms. The

War of Austrian Succession (1740–1748) generated a similar mortality rate as the Vietnam War (1955–1975), averaging roughly fifty violent deaths per 100,000 people. Even so, twentieth century warfare was the most murderous if measured by absolute numbers of people killed. Three of the most violent wars in recorded history occurred in the past hundred years including the Second World War, in which as many as 85 million people were killed, the First World War, with as many as 22 million deaths, and the Russian Civil War (1917–1922), in which up to 9 million people lost their lives. But then something strange started happening in the 1950s. The intensity of all types of armed conflicts started to decline.

The long peace – more years of peace than war for the first time

The decline in war-related deaths since the middle of the twentieth century is one of humanity's great unheralded achievements. As the graphic shows, deaths from most kinds of war – civil conflicts, conflicts between states and imperial conquests – have fallen dramatically since the 1950s. The proxy wars of the 1960s, 1970s and 1980s also became progressively less violent. War-related deaths tumbled further still after the Cold War ended in 1989. This period is known colloquially as the 'long peace' and was enabled by norms and institutions (and a

Tracking the decline in war-related deaths since 1946[18] The overall number of soldiers and civilians killed in armed conflict has declined dramatically since the mid-1940s. Today, inter-state conflicts are comparatively rare events.

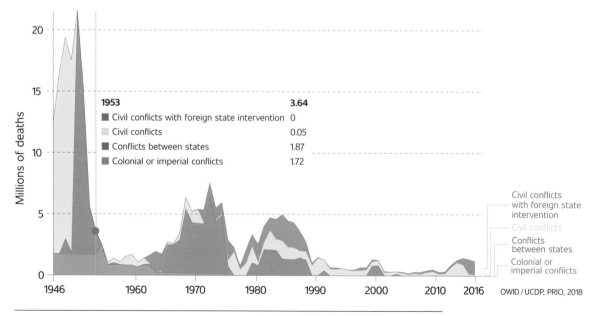

1953 — 3.64
- Civil conflicts with foreign state intervention — 0
- Civil conflicts — 0.05
- Conflicts between states — 1.87
- Colonial or imperial conflicts — 1.72

Civil conflicts with foreign state intervention
Civil conflicts
Conflicts between states
Colonial or imperial conflicts

OWID / UCDP, PRIO, 2018

Fatal events by type
of violence
UCDP, 2018

State-based violence

Non-state violence

One-sided violence

Each point represents
the location of an event

2000 km

1000 miles

Mapping multiple types of conflict fatalities – 2018[19]

This map captures fatal events arising from state-based, non-state and one-sided violence in 2018. Notice the concentrations of fatalities in the Sahel, Central and East Africa, the Middle East, and central, South and Southeast Asia. Surprisingly levels of non-state violence are also apparent in Mexico, Central America and South America.

dose of good luck) that helped the US and Soviet Union keep their rivalry – and their nuclear arsenals – in check. The United Nations was regularly called on to dispatch political and peacekeeping missions to keep wars from reigniting. Meanwhile, wealthier countries ramped up their development aid to poorer and fragile countries as detailed in the globalisation chapter. These and other activities not only helped prevent and reduce the incidence of wars between states, but also helped to stamp out some civil ones as well.

The end of the short peace and the rise of crime wars

A golden age of stability ensued between 1990 and 2010 known as the 'short peace'. The short peace was tragically short-lived. An indicator of this has been the rising incidence of armed conflicts around the world over the past decade. There are twice as many armed conflicts today – around fifty at the last count – than just two decades ago.[20] And while their death tolls are still comparatively low by historical standards, these conflicts are nasty affairs. They are typically waged by badly trained armies and non-state armed groups equipped with a fearsome arsenal of modern weaponry. The circles on this map highlight the distribution of conflict-related fatalities in 2018. The violence footprint is most apparent in parts of Colombia and Mexico (which is now often classified as a conflict zone), across Sub-Saharan Africa and into the Middle East, Central, South and Southeast Asia. Notwithstanding the vast geographical dispersion of conflict fatalities, only a handful of countries – especially Afghanistan, Nigeria, Somalia, Syria and Yemen – generate the vast majority of conflict deaths worldwide.

Today's armed conflicts are devilishly hard to bring to an end. A big reason for this comes down to the sheer number of armed groups involved in fighting them. In war-torn countries like Libya and Syria there may be dozens, even hundreds, of rebel and militia entities vying for control of key cities, oil reserves and ports. In the Central African Republic,[21] the Democratic Republic of the Congo[22] and Mali,[23] jobless young men are easily recruited into mercenary armies and terrorist organisations.[24] The more armed groups there are on the battlefield, the harder it is to negotiate a lasting peace. Yet another reason these simmering conflicts are hard to stop is that they are often waged over high-value resources. Whether groups are fighting over cocaine trafficking routes in Colombia or coltan mines in the Democratic Republic of the Congo, the insatiable global appetite for commodities fuels violence over their control and distribution.

Contemporary wars can be bewildering to outside observers. It is not uncommon to find drug cartels, mafia groups, criminal gangs and extremist organisations competing and colluding with conventional armies, organised rebels and private security companies.[25] Consider the Taliban in Afghanistan which is a rebel group that has fought for political and territorial control, but also doubles as a wildly successful drug cartel that dominates the global heroin market (valued at hundreds of millions of dollars a year), including smuggling routes to Europe.[26] After almost two decades of trying, the US quietly ended its campaign to destroy the Taliban's drug labs and eradicate its poppy fields as it hadn't made a dent on local production.[27] On the contrary, the production and availability of heroin has actually increased since 2001.

The confluence of conflict, criminal and extremist violence makes it hard for United Nations peacekeepers to 'keep the peace'.[28] There have been at least seventy United Nations-mandated peacekeeping missions since the 1950s.[29] Most of these operations were severely under-funded. While routinely accused of being too timid, the blue helmets are usually operating with one arm tied behind their back.

This is because peacekeepers are constrained by conservative mandates set by the Security Council and fears that they might interfere with the sovereignty of host countries. These kinds of restraints made sense when peacekeeping duties were restricted to monitoring ceasefires and enforcing peace agreements between clearly defined soldiers and rebels. They are much less convincing when there is no peace to keep and when organised violence is intimately connected to the interests of local elites and globalised war economies.[30]

Climate conflicts – rising temperatures are triggering violence

Rising temperatures, prolonged droughts, massive flooding and soil degradation are all violence multipliers. While the relationships are not always linear or clear-cut, climate change is triggering more frequent crop and livestock failures, increasing food prices, greater hunger and the outbreak of social unrest and organised violence. A textbook example is the African Sahel, where livestock herders and subsistence farmers are increasingly fighting over diminishing arable

Water stress and conflict in the Lake Chad Basin of the Sahel[31] Climate change is a conflict multiplier. This map highlights the dramatic fluctuations of water availability in the Lake Chad Basin, a key resource in the Sahel and one of the largest waters sources in Africa. Lake Chad is estimated by some researchers to have shrunk from 10,000 square miles to 580 square miles over the past half-century, dramatically increasing food insecurity and contributing to tensions over diminishing food supplies. The map also highlights how water sources have increased in some areas, sometimes resulting in floods.

Water change, 1984–2018
JRC, Google / Google Earth Engine, USGS, NASA, ESA, ACLED, 2020

water decrease　　　　　　water increase
Approximate shoreline of Lake Chad 1963

land. For centuries, herders in northern parts of Mali, Burkina Faso, Niger and Nigeria have taken their cows and goats south to more fertile grazing areas during the dry season. For the most part, farmers welcomed them since their livestock fertilised depleted cropland with manure. Outbreaks of violence were typically averted with the help of respected chieftains who mediated between hotheads. All this started to change with the prolongation of dry seasons, shortening of wet seasons and rapidly diminishing water supplies. One especially hot spot is the Lake Chad Basin, Africa's largest water reservoir. The basin spans eight countries – about 8 per cent of the continent – and is relied upon by 30 million people. According to some researchers (not all agree), the lake has purportedly shrunk by 90 per cent since the 1960s owing to rising temperatures, soaring population growth and intensive irrigation.[32] Not surprisingly, it is the site of escalating intercommunal violence.[33] Today it is one of the most water-stressed and volatile places on Earth.[34]

The Sahel is one of several canaries in the mine when it comes to climate conflicts. The situation there is about to get a whole lot worse. Temperatures in Sahelian countries may rise by between three and five degrees Celsius by 2050 in a part of the world where monthly temperatures are already sizzling.[35] About 80 per cent of the Sahel's potential farmland is already severely degraded because temperatures are rising so much faster than the global average.[36] Extreme weather events are now the norm and growing populations are straining already overstretched water and food supplies. With herders forced to arrive earlier and stay longer, disputes with farmers are mounting in intensity. And it's not just the Sahel that's in trouble. Similar trends are evident elsewhere in Africa,[37] as well as in Central America,[38] the Middle East[39] and South Asia. According to political scientists Joshua Busby and Nina von Uexkull, countries with a recent history of conflict, a large proportion of the population dependent on agriculture and a sizable number of citizens excluded from political power are most susceptible to climate conflicts.[40] As the map overleaf shows there may be as many as forty countries that tick all these boxes.[41]

War-torn Syria offers another reminder of how climate change exacerbates the risk of violent conflict. Although it is roughly the same size as Spain, Syria is mostly an uninhabitable desert. Syrians are crammed into densely packed areas that are one-thirteenth the size

Syria's drought–conflict connection (2000 and 2019) [42]

Syria has experienced severe water shortages for centuries. But one of the worst droughts in the country's history struck between 2006 and 2011 causing 75 per cent of Syrian farms to fail and 85 per cent of livestock to perish. The drought forced at least 1.5 million Syrians to migrate to cities like Damascus and Homs. Not long after, civil war broke out.

of Spain, about as large as Switzerland. Well before the outbreak of civil war in 2011, the country was ravaged by another kind of crisis. After suffering from a string of brutal desert storms between 2001 and 2005, Syrians experienced one of the worst ever droughts that lasted from 2006 to 2011.[43] Record-breaking heat and dramatically reduced rainfall destroyed the topsoil and most local food production. At least 800,000 farmers lost their jobs and 200,000 of them abandoned their land altogether.[44] More than 75 per cent of the country's crops failed and 85 per cent of its livestock died from starvation. Complicating matters, the government sold off the country's strategic grain reserves just before the 2006 drought and had to import wheat to keep its own population alive. As food prices climbed, over 1.5 million people fled to cities searching for non-existent jobs. Predictably, Syrians took to the streets to protest as social and economic conditions worsened across the country.

The spark that ignited Syria's civil war came in 2011. A group of school-age teenagers in the southern city of Deraa were arrested for painting political graffiti. When crowds demanded their release, they were violently crushed by government troops. Fearful that social unrest might spread, President Assad authorised a country-wide crackdown. What started out as a water and food crisis rapidly metastasised into a political and religious uprising fragmented across sectarian and ethnic

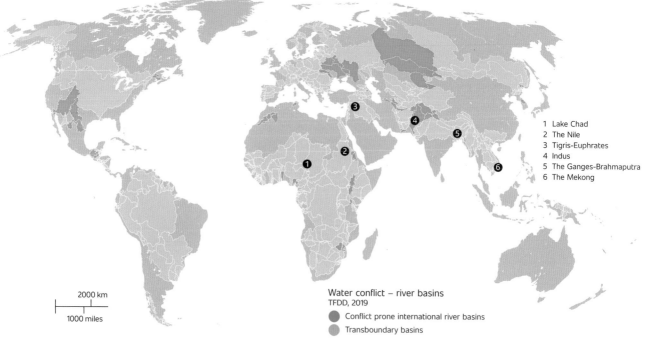

1 Lake Chad
2 The Nile
3 Tigris-Euphrates
4 Indus
5 The Ganges-Brahmaputra
6 The Mekong

2000 km

1000 miles

Water conflict – river basins
TFDD, 2019

● Conflict prone international river basins
● Transboundary basins

Flashpoints for water tensions [46]
With droughts intensifying around the world, the risks of violent disputes breaking out over the control of fresh water supplies are real. This map highlights the many separate trans-border river basins that are especially vulnerable to disputes, especially the Nile, Euphrates, Indus and Mekong. These flashpoints are based on a model that predicts future tensions on the basis of past conflict and co-operation. Historically there has been at least as much co-operation as conflict, but climate change may increase the risk of tension.

lines. The civil war that ensued is dizzyingly complex. To date, more than 500,000 Syrians have been killed in the fighting and half the country's population displaced.[45]

Climate conflicts will become more common as countries fight it out over water resources. Tensions can flare when water-stressed communities are hit by rapidly growing populations and an inability to manage shared common resources, as in the case of Lake Chad, described earlier.[47] Violent disputes can also flare-up over access to fresh water or as a result of the targeting of water systems outright.[48] Conflicts over water are hardly new: researchers have documented hundreds of them over the past few thousand years.[49] But owing to global warming, they are likely to grow more frequent. Open clashes over water occurred in at least forty-five countries in the last decade, especially in parched areas of the Middle East and North Africa like Algeria, Somalia and Sudan.[50] The map above highlights a number of potential water conflict hot spots in red, including in the Americas, Eastern Europe and Eurasia. Areas that are especially at risk – the

Ganges-Brahmaputra and Indus in South Asia, the Tigris-Euphrates in the Middle East and the Nile in Africa – are also where local residents are already struggling to access fresh water.[51]

The threat of violent conflict is heightened where there is diminishing rainfall and a high dependence on sharing water reserves between two or more countries.[52] Take the case of Ethiopia, where public authorities are building the monumental Grand Renaissance Dam. Although the dam could dramatically increase agricultural production in Ethiopia, it also has the potential to disrupt industrial and subsistence farming in neighbouring countries. At least eleven African states rely on the Nile for irrigation, and Ethiopia stands to control most of the water passing into neighbouring Egypt and Sudan.[53] If the water available to Egypt or Sudan declines for whatever reason, this would negatively affect their domestic food production. These countries will be left with two options: find a diplomatic solution or take military action.[54] These challenges are only likely to grow: the combination of rising temperatures and population growth could escalate the chance of so-called 'water wars' by 75 to 90 per cent in the next fifty to 100 years, according to some researchers.[55]

The new arms race – from small arms to high-tech swarms

One reason why warfare in places like the African Sahel and the Middle East is so deadly is the availability of high-powered, low-tech weaponry such as assault rifles, rocket-propelled grenades, mortars and landmines.[56] In most of today's conflicts, AK-47s, M16s and AR-15s are the real weapons of mass destruction for the simple reason that they are incredibly easy to use, highly durable and portable, cheap and widely available. Groups like the Small Arms Survey have documented at least 1,000 companies in 100 countries involved in manufacturing firearms and ammunition and their parts and accessories.[57] Close to 1 billion small arms and light weapons are already in circulation and tens of millions more are added to the global stockpile every year.[58] Many of them are illegal, having slipped into black markets after a huge surplus became available from downsizing militaries after the Cold War ended. Despite their massive human cost, the total value of the authorised (legal) small arms market is (relatively) modest,

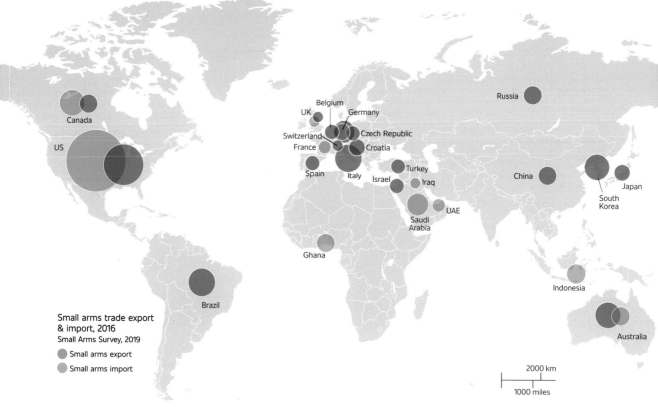

Small arms trade export & import, 2016
Small Arms Survey, 2019

● Small arms export
● Small arms import

2000 km
1000 miles

Tracing the flow of small arms, light weapons and ammunition – 2016 [59]

There are an estimated 1 billion small arms in circulation around the world, the vast majority of which are held by civilians. This map highlights the global reach of the trade in small arms, light weapons and ammunition, which is valued at close to $10 billion a year. The largest exporters of small arms, light weapons and ammunition are Austria, Belgium, Brazil, Germany, Italy, Russia, South Korea, Switzerland and the US. The largest importers are Australia, Canada, France, Germany, Saudi Arabia, the UK and the US.

The bubbles depict the financial value of small arms exports/imports.

Top Exporters ($ millions):

United States	1100
Italy	618
Brazil	600
Germany	498
Australia	480
South Korea	405
Croatia	225
Czech Republic	218
Turkey	192
Russia	182
Israel	165
Belgium	164
China	121
Spain	116
Canada	115
Japan	108
Switzerland	106
United Kingdom	101

Top Importers ($ millions):

United States	2510
Saudi Arabia	333
Indonesia	281
Canada	249
Germany	203
Ghana	197
Australia	174
France	137
United Arab Emirates	136
Iraq	121
United Kingdom	120

probably no more than $10 billion a year.[60] The illegal market may be closer to $1 billion a year. By way of comparison, Netflix was valued at $258 billion at the end of 2019.[61]

The world's arms industries literally make a killing from global instability. Global military spending on all types of weapons – including small arms, light weapons and ammunition – exceeded $1.8 trillion in 2018, the highest level since the end of the Cold War.[63] This amounts to 2.1 per cent of global GDP, or roughly $239 a year for every person on earth. As the pie-chart shows, the US is the world's most profligate exporter and accounts for over a third of the global trade, more than all Western European countries combined. Meanwhile, Saudi Arabia, India, Egypt, Australia and Algeria were ranked among the top importers between 2014–2018, accounting for over a third of all arms imports. That said, the largest and longest expansion in

Top conventional arms exporters (by country) – 2014–2018 [62]

The US share of global exports increased from 30 to 36 per cent between 2014–2018, making it the world's leading arms trader. The US shipped weapons to over 97 countries during this period. Russia, France, Germany and China rounded out the top five.

Top arms producers by sales – 2017

The top arms producers in the world are located in North America, Europe and Asia. The US and China are the major leaders, with others like France, Germany, UK, Israel and Russia also significant players. This chart, produced by the Stockholm Peace Research Institute, provides insight into the dimensions of the key countries and companies involved.

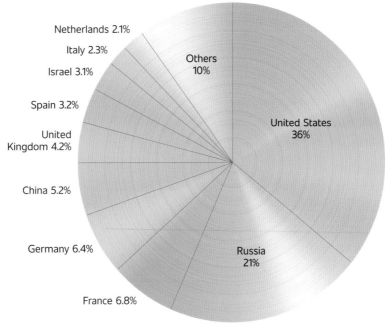

Stockholm International Peace Research Institute (SIPRI), 2019

military spending is occurring in places like China, India, Pakistan and South Korea. And in Europe, countries such as Bulgaria, Latvia, Poland and Ukraine scaled up military spending in 2018 owing to fears of Russian incursions. But it is the Middle Eastern governments like Saudi Arabia, Oman, Kuwait and Lebanon that are the biggest military spenders relative to the sizes of their economies.[64]

While rarely used, weapons of mass destruction – nuclear, chemical and biological – are among our most dire existential threats. The world's nuclear powers have dramatically reduced their arsenals of

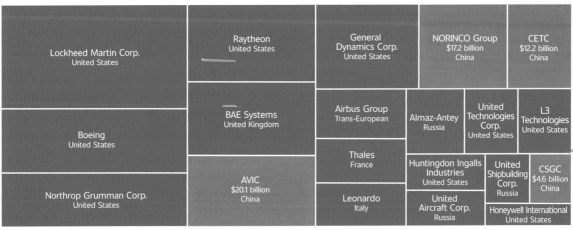

SIPRI, 2018

Countries committed to the non-proliferation treaty (NPT) [65]
A total of 191 states have joined the NPT, including the five countries that have declared possession of nuclear weapons. More countries have ratified the NPT than any other arms agreement. The map highlights recognised nuclear weapon state ratifiers (light blue), all other ratifiers (light green) and non-signatories – including countries known to have nuclear weapons such as India, Israel, and Pakistan (red). North Korea, which has nuclear weapons, withdrew from the NPT in 2003.

Non-proliferation treaty
Wikipedia / UN Office for Disarmament Affairs, 2016

- Recognised nuclear weapon state ratifiers
- Recognised nuclear weapon state acceders
- Other ratifiers
- Other acceders or succeeders
- Unrecognised state, abiding by acceders
- Withdrawn
- Non-signatory

1000 km

500 miles

nuclear warheads over the past three decades as the geopolitics chapter makes clear. But there are still thousands of nuclear weapons held by a small number of states. Although eight countries have declared that they possess them (US, Russia, UK, France, China, India, Pakistan and North Korea, though Israel has not), only five have signed on to the nuclear non-proliferation treaty (NPT).[66] Chemical weapons are also still a menace. The repeated use of prohibited weapons such as sarin gas, chlorine and sulphur mustard in Syria between 2012 and 2018 shocked the world.[67] A United Nations-led fact-finding mission confirmed at least twenty-eight incidents involving chemical weapons in Syria,[68] although some human rights groups believe the number could be as high as 336.[69] The repeated use of such weapons – in flagrant violation of the Geneva Protocol of 1925 and the 1997

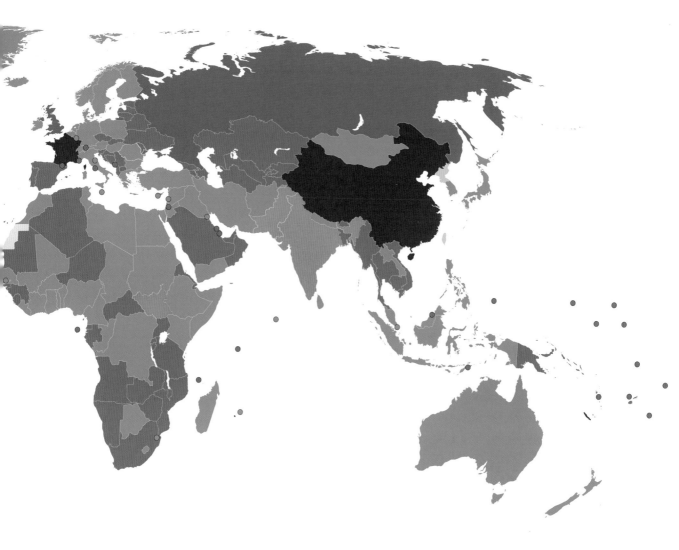

Chemical Weapons Convention – could embolden authoritarians, despots and extremists around the world.

Complicating matters even more, an AI arms race is also heating-up.[70] Automated surveillance drones are now commonplace[71] and autonomous weapons systems, or 'killer robots', are already being marketed by BAE Systems, Dassault, MiG and Raytheon.[72] Some military experts are convinced that unmanned naval, aerial and terrain vehicles, autonomous drone swarms,[73] and fire-and-forget missile systems will reduce human error and civilian casualties. Not everyone is convinced that these next generation arms will civilise war. Hundreds of engineers, roboticists and researchers including the entrepreneur Elon Musk and Google DeepMind co-founder Mustafa Suleyman are advocating for much stricter regulations over

States calling for ban on lethal autonomous weapons

Algeria	Guatemala
Argentina	Holy See
Austria	Iraq
Bolivia	Mexico
Brazil	Morocco
Chile	Namibia
China[1]	Nicaragua
Colombia	Pakistan
Costa Rica	Panama
Cuba	Peru
Djibouti	State of Palestine
Ecuador	Uganda
Egypt	Venezuela
El Salvador	Western Sahara
Ghana	Zimbabwe

Regulation of autonomous weapons
Campaign to Stop Killer Robots, 2020

- States calling for a ban on lethal autonomous weapons
- States opposed to a ban on lethal autonomous weapons
- States known to be researching and developing lethal autonomous weapons

Note 1: China says its intention is to ban fully autonomous weapons, but not their development or production.
Note 2: There are now 30 states that have called for a ban on lethal autonomous weapons.

autonomous weapons.[74] A small number of governments are also pushing for more restraints on their development and deployment. The map shows how more than twenty countries are calling for an outright ban on lethal autonomous weapons systems.[76] But several other nations, such as Australia, China, Israel, South Korea, Russia, the UK and the US, have not. These countries have blocked efforts to develop an international treaty, fearing they will lose out in the race to automise war.

The diminishing risk of terrorism

Terrorism is an old problem but a relatively new global priority. Arguably one of the first reported uses of terrorist tactics was by Sicarii Zealots in the first century. They acquired celebrity status for the quiet efficiency with which they dispatched their Roman overlords.[78] Long before the rise of al-Qaeda or ISIS, there was a secret order of Ismaili assassins called the Hashshashin ('users of hashish'), who carried out professional hits across Iran, Syria and Turkey between the tenth and twelfth centuries.[79] Among the earliest instances of European

States calling for more regulation of autonomous weapons[75]
Ethical, legal, operational and other concerns are rising over the prospect of removing human control in the use of force. At least 6 states are known to be developing weapons systems with decreasing levels of human control over the critical functions of selecting and engaging targets (marked with a red dot). However, many states now see a need for a new international treaty to retain meaningful human control over the use of force, including 30 states seeking to ban fully autonomous weapons (marked in green).

terrorism was the 'Gunpowder Plot', a scheme to blow-up the English Parliament in the early seventeenth century.[80] Other well-known European terrorist groups include the Fenian Brotherhood, the predecessor to the Irish Republican Army (IRA),[81] and the Narodnaya Volya ('People's Will') in Russia that was made up of revolutionary socialists who wanted to bring about an end to Tsarism.[82] All of these groups claimed at one time or another to be fighting for liberation from their oppressors. Those who were accused of oppressing them, predictably, referred to them as 'terrorists'.

Terrorism ranked low on the international agenda until the late twentieth century. For the most part, terrorist acts were treated as a domestic nuisance, largely dealt with by law enforcement. Before September 2001, the single deadliest act of terrorism in US history – the 1995 Oklahoma bombing[83] – was perpetrated by an American citizen intent on destroying the federal government. Meanwhile, residents of London, Madrid and Rome were rattled by separatist groups like the IRA, Basque Homeland and Liberty (ETA) and Italy's Red Brigade, among others. Most of the roughly 16,000 documented terrorist incidents in Europe over the past half-century occurred before the 1990s.[84] The al-Qaeda-led attacks on the US and the subsequent invasion of Afghanistan (2001) and Iraq (2003) changed everything.

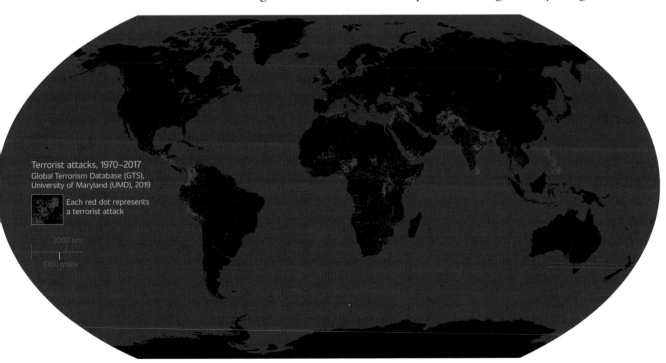

Terrorist attacks, 1970–2017
Global Terrorism Database (GTS),
University of Maryland (UMD), 2019

Each red dot represents a terrorist attack

2000 km
1000 miles

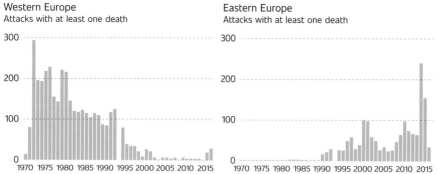

Western Europe
Attacks with at least one death

Eastern Europe
Attacks with at least one death

Cumulative reported terrorist events in Western and Eastern Europe – 1970–2015 [85]
Europeans have suffered from terrorist events for decades. This map reveals the geographic distribution of events between 1970 and 2015. A closer inspection of the data reveals that most events are clustered in very specific geographic regions of the UK and Spain, the Balkans, Ukraine and Russia. Interestingly, reported terrorist events declined spectacularly in Western Europe and increased equally dramatically in Eastern Europe from the 1990s onwards.

But are North American and European societies really bearing the brunt of terrorist violence? And is today's extremism more widespread than, say, in the 1970s or 1980s when it simmered in Northern Ireland and Spain's Basque Country? As the graphs show, despite a small uptick in incidents over the past decade, terrorist attacks have become comparatively rare events in Western Europe. By contrast, reported terrorist events have increased dramatically in Eastern Europe, especially in parts of eastern Ukraine, southern Russia, and the North Caucasus republics of Chechnya, Dagestan, Ingushetia and Kabardino-Balkaria. But therein lies a problem. The measurement of terrorist events depends on how terrorism is defined. Some definitions restrict terrorism to indiscriminate violence perpetrated by extremists against civilians to achieve political, religious or ideological change, while other definitions are expanded to include acts committed by police and soldiers. Not surprisingly, most governments are strongly opposed to the idea that states can perpetrate, much less be prosecuted for, terrorism. This helps explain why diplomats have repeatedly failed to agree on a unified definition of terrorism, much less an international convention.[86] Negotiators are still bogged down with the age-old problem that one person's terrorist is another person's freedom fighter.

The truth is that neither North America nor Europe are the regions most badly affected by terrorism. Today, less than 2 per cent of all reported terrorist attacks occur there.[87] The probability of dying from a terrorist attack in Europe is around 0.02 per 100,000, about the same as being killed by a lightning strike and significantly less risky than drowning in a bathtub.[88] By contrast, more than 90 per cent of all terrorist attacks and associated deaths over the past few decades occurred in a handful of Central Asian, Middle Eastern and African countries: Afghanistan, Iraq, Nigeria, Pakistan, Somalia, Syria and Yemen were among the worse affected.[89] The map illustrates the concentration of terrorist events in an arc extending from West and North Africa across the Middle East to South and Southeast Asia. Terrorism tends to flourish where there are ongoing armed conflicts, which helps explain why they are increasingly intertwined.

Terorist attacks 2018, concentration and intensity
GTD / UMD

High   Low
Intensity value is a combination of incident fatalities and injuries

1000 km

1000 miles

A small number of terrorist groups are responsible for most extremist violence. More than 200 organisations have been designated as terrorist by national governments.[90] Some of them operate transnationally while others are hyper-local. About half of all violent incidents classified as terrorism in recent years can be attributed to just four of them. According to the Global Terrorism Index, over 10,000 of the roughly 19,000 documented terrorist killings in 2018 were perpetrated by either ISIS, the Taliban, al-Shabaab or Boko Haram.[91] Today, among the greatest 'terrorist' threats facing North Americans and Western Europeans is probably not from groups espousing political Islam,[92] but instead from white nationalist and supremacist movements. The Southern Poverty Law Center has tracked more than 1,000 hate groups in the US, and law enforcement groups are adamant that domestic terrorists are far more worrying than foreign ones.[93] The map highlights the distribution of Nazi and white supremacist groups as well as anti-immigrant, anti-LGBT and anti-Muslim entities. More positively, terrorist-related deaths and injuries have recently declined. After a surge of terrorism in the wake of US-led

The spread of hate groups in the US – 2018[94]
The hate map, produced by the Southern Poverty Law Center, depicts the approximate locations of specific groups that vilify people because of their race, religion, ethnicity, sexual orientation or gender identity. They include a wide range of categories from neo-Nazi and white nationalist groups to black nationalist, anti-immigrant, anti-LGBT and anti-Muslim entities.

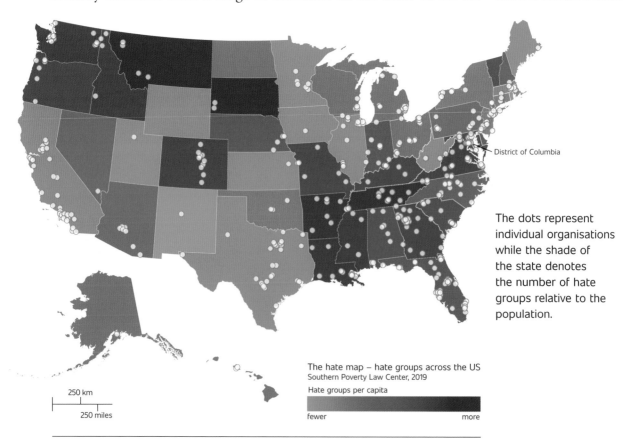

District of Columbia

The dots represent individual organisations while the shade of the state denotes the number of hate groups relative to the population.

250 km
250 miles

The hate map – hate groups across the US
Southern Poverty Law Center, 2019
Hate groups per capita

fewer   more

interventions in Afghanistan and Iraq, killings have dropped by over 40 per cent since 2014.[95] Whether this is a blip or a trend remains to be seen. Either way, it makes little sense to organise international and national security planning around such a tactic. Instead, the focus should be on preventing attacks that could generate widespread harm, particularly the use of dirty nuclear bombs, biological agents[96] and chemical weapons that have become thinkable again due to the Syrian war.[97] Deterring terrorism will not be achieved by simply throwing military assets at the problem. Instead, its disruption and reduction require investment in police intelligence, the hardening of possible targets, the prevention of radicalisation in prisons, and concerted efforts to minimise the underlying social and economic factors that give rise to it in the first place. One approach that patently does not work is ramping up state repression since it can lead to recruitment to terrorist causes.

Surging state repression

Despite the dramatic spread of democracy over the past half-century, billions of people in both democratic and non-democratic societies alike still face state repression when they try exercising their most basic human and civil rights. The stakes are highest in authoritarian countries, where citizens who dare to assemble or organise can face harassment, torture, prosecution, imprisonment or even death.[98] While abhorrent, it is worth recalling that such repression used to be the norm around the world. Despots and dictators have ruthlessly cajoled and controlled their populations for centuries. Such offences are still widespread. According to a 2019 Freedom House survey, at least fifty countries are classified as 'not free' where the excessive use of force by soldiers, police and paramilitaries against locals is commonplace.[99] There are also 'free' countries, including Brazil, the Philippines and South Africa, where thousands of citizens are 'lawfully' killed each year by state forces and their proxies.

Governments get away with murder when they can bend the rule of law in their favour. Even when a country's security forces are flagrantly abusing the rights of citizens, other nations may be reluctant to intervene. This is because national sovereignty and non-interference – two concepts discussed in the geopolitics chapter – routinely trump

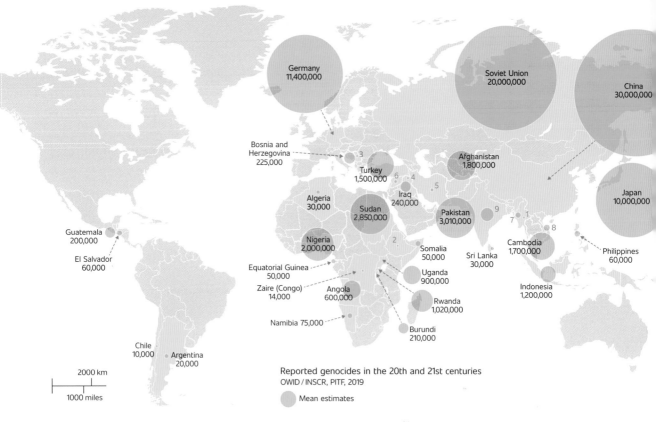

Documented genocides in the twentieth and twenty-first centuries [100]

Only recognised as an international crime in 1946 and codified in 1948, genocidal violence has existed for thousands of years. Genocides are acts undertaken with the express intent to destroy, in whole or part, a national, ethnical, racial or religious group. It includes efforts to kill, cause serious bodily and mental harm, impose measures to prevent births within certain groups, and forcibly transfer children from one group to another. The map highlights a sample of 'genocidal events', some of them contested, since 1900. It is not comprehensive and the estimated death tolls are widely disputed.

fundamental human rights. The idea that states should avoid meddling in each other's domestic affairs goes back a long way. Despite efforts to revisit and rewrite norms first established in 1648, many countries strongly resist efforts to water them down, a subject we will return to shortly.

Intended to restrain states from perpetual war, the concepts of sovereignty and non-interference also gave them latitude to repress people within their own borders. These concepts became, in effect, a licence to commit genocidal violence.[101] As the map shows, the most serious genocides of the past century – China's cultural revolution, the Russian gulags, the Holocaust, Cambodia, Rwanda, Srebrenica and Darfur – were perpetrated by government forces and paramilitaries, and not foreign soldiers, rebels or terrorists. The late political scientist Rudolph Rummel coined the term 'democide' to describe the estimated 260 million people killed by their own governments over the course of the twentieth century, several times more than all wars combined during the same period.[102] In just the former Soviet Union, China and Cambodia, between 85 and 110 million people died from executions, labour camps, ethnic cleansing and induced famines in the twentieth century alone.

A group of diplomats, lawyers, academics and activists started questioning the limits of national sovereignty in the early twenty-first century. They felt an urgency to do so after the Rwandan genocide and the mass atrocities perpetrated in the former Yugoslavia in the 1990s. Appalled by the inability of governments to prevent these atrocities, they reframed sovereignty as the 'responsibility' to protect all populations from mass violence. The so-called 'Responsibility to Protect' doctrine gathered steam in the United Nations and was officially endorsed in 2005. It specifies that when states are unable or unwilling to protect their people, then they forfeit their claims to sovereignty. The controversial idea was mobilised (with mixed effects) to deter an escalation of election violence in Kenya (2007–2008) and reduce a massive outbreak of violence in the Central African Republic (2013). But the concept came under fire after North American and Western European governments invoked it to justify military interventions in Libya (2011)[103] and Syria (2011).[104] China and Russia, in particular, accused the US and its allies of instrumentalising the doctrine not to protect civilians, but to pursue regime change instead.

Most instances of state repression fall short of genocide and mass atrocities. Take the case of extrajudicial and arbitrary killings committed by soldiers, police and paramilitaries pursuing a 'war' on crime and drugs.[105] There is rarely any systematic accounting of the number of citizens illegally killed or 'disappeared' due to actions by their governments. Reporting and investigations are often sporadic and inaccurate. In Brazil, for example, military and civil state police were involved in more than 6,100 killings in 2018 and even more in 2019.[106] In El Salvador, police were implicated in 400 of the country's roughly 4,000 murders in 2017.[107] Another 5,000 citizens have been officially killed by police and militia in the Philippines since 2016 – although human rights activists say the numbers could be four times higher.[108] And in the US, police are implicated in the deaths of at least 1,000

1 Burma 5,000
2 Ethiopia 10,000
3 Yugoslavia 10,000
4 Iraq (Yazidis) 10,000
5 Iran 20,000
6 Syria 30,000
7 Myanmar (Rohingya) 40,000
8 Vietnam 500,000
9 India 1,000,000

people a year,[109] more than virtually every OECD country combined. The incidence of police brutality in some of these places is so widespread that legal and human rights associations have called on the International Criminal Court in The Hague to investigate possible crimes against humanity.[110]

The incarceration explosion

Another indicator of state repression is mass incarceration. At least 11 million people are currently detained and held in penal institutions around the world.[111] At the top of the rankings is the US with at least 2.2 million people behind bars, followed by China, Brazil and Russia. Once locked into the criminal justice and penal system, it is exceedingly difficult for people to get out. Although some countries are decreasing their prison populations,[112] the overall number of individuals being put behind bars is rising by about 3.7 per cent a year.[113] One of the reasons prison populations are swelling

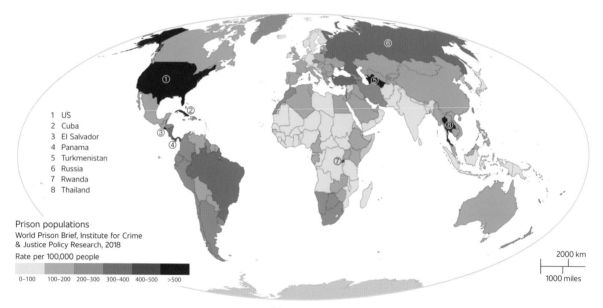

1 US
2 Cuba
3 El Salvador
4 Panama
5 Turkmenistan
6 Russia
7 Rwanda
8 Thailand

Prison populations
World Prison Brief, Institute for Crime & Justice Policy Research, 2018
Rate per 100,000 people

0–100   100–200   200–300   300–400   400–500   >500

2000 km
1000 miles

The prevalence of incarceration around the world [114]
Mass incarceration is commonly used not just to impose law and order, but also to stifle dissent and suppress opposition. The map highlights the rate of imprisonment (as opposed to absolute numbers of inmates). The US, China, Brazil, Russia and India have the highest total number of inmates. By comparison, Cuba, El Salvador, Panama, Turkmenistan, Russia, Rwanda and Thailand, as well as the US, register among the highest prevalence of inmates.

is because of the popularity of incarceration as a strategy to tackle minor offences and drug-related crime. Penal facilities are also big business and many of them are privatised.[115] The US alone spends more than $80 billion on maintaining its prisons every year – about $260 for every citizen.[116]

Pre-trial detention and sky-high recidivism ensure that prisons are swelling well beyond their intended capacities. In South America and Southeast Asia, for example, the prison population has risen by more than 175 per cent and 122 per cent respectively since 2000. Countries like Brazil, El Salvador, Panama, Cambodia, Egypt, Indonesia, Jordan, Nicaragua, Turkmenistan, Rwanda, the Philippines and Turkey have all registered sharp increases in incarceration in recent years. Although they are concealed from public view, prisons are typically filled with poorer, under-educated and minority populations, many of them languishing there for years even before receiving a sentence. Non-violent offenders are often crammed into prisons cheek by jowl with convicted murderers and sexual predators. Despite the efforts of many human rights and humanitarian groups to access prisons, monitor conditions and publicise maltreatment, overcrowding, illness and abuses are commonplace.[117]

In most parts of the world, prison conditions border on the medieval. Measures designed to curb violations and care for inmates are routinely flaunted.[118] In Brazil, for example, prison riots involving beheadings occur with alarming regularity.[119] In Mexico, approximately three-quarters of the country's inmates claim to have been tortured while in prison, suffering beatings, asphyxiation, electric shocks and sexual harassment.[120] In Syria, political prisoners are routinely executed and thousands have died while in custody after being subjected to prolonged solitary confinement, exposure to extreme temperatures, and restricted access to food and medical treatment.[121] Meanwhile, in Nigeria, thousands more have died while serving time in government detention centres or in the custody of soldiers and paramilitary groups.[122] And in North Korea, as many as 130,000 people are believed to be languishing in concentration camps that have been compared to those in Nazi Germany.[123]

China has also undertaken a campaign to detain large segments of the population and keep them under surveillance. According to human rights observers, at least 1.7 million people are locked up in Chinese prisons.[124] As many as 2 million additional Uighurs and other

Suspected internment camps in Xinjiang [125] China has reportedly detained more than two million Uighurs and other Muslim groups in its western Xingjian province. This map highlights 18 're-education camps' that have been verified from space using satellite imagery. Researchers believe there could be as many as 1,200 in total.

Map labels: RUSSIA — KAZAKHSTAN — MONGOLIA — KYRGYZSTAN — Urumqi — TAJIKISTAN — Xinjiang — Beijing — AFGHANISTAN — CHINA — PAKISTAN — INDIA — Confirmed or likely re-education camps are shown in red. The total number of camps is estimated to be as high as 1,200. — 500 km — 250 miles — ASPI, 2018

minority Muslim populations in China's Xinjiang province have also been forced into detention centres for 're-education' and 'de-extremification', or are serving longer sentences in prison camps.[126] The map highlights just eighteen confirmed 're-education centres', but foreign researchers believe there could be as many as 1,200 more.[127] Satellite images reveal a sample of these camps despite the denial of Chinese authorities.[128] The detention programme was set up to address what China's president refers to as the 'three evils' – terrorism, religious extremism and separatism.[129] Chinese authorities scaled up their so-called 'strike-hard' and 'people's war' campaigns following a string of bombings that kicked-off in the 1990s, including an attack in Kunming that left more than thirty-five people dead and 140 injured in 2014.[130] Since then, the Chinese government has dramatically ramped-up everything from social media monitoring and phone-tapping to facial and biometric recognition technologies.[131]

China's surveillance systems are intended to deter crime and contain political dissent. Today, at least 200 million CCTV cameras comb city streets, shops, bus stations and border crossings – the equivalent of one camera for every seven residents.[132] The Chinese government has also rolled out a social credit score system in order to track citizens' trustworthiness, including their bill payments, blog posts and daily purchases. A score is then used to determine the

United States of America

Trafficking of illegal commodities
UNODC, 2014

→ Heroin
⇢ Cocaine
┈┈▸ Firearms
••••▸ Smuggling of migrants
▬▬▸ Female trafficking victims (main sources)
— Counterfeit consumer goods
••••• Counterfeit medicines
Piracy off the Horn of Africa
→ Wildlife
---▸ Timber
→ Gold
─··─▸ Cassiterite

Mexico

Mapping the trafficking of illegal commodities around the world [136]

The global trade in illicit commodities affects every country in the world. The map highlights the suspected direction of a wide range of illegal commodities from the point of origin to their destination. Specifically, it shows how cocaine is trafficked from Colombia to the US and Europe while heroin flows from Afghanistan to Europe and Russia. Meanwhile, illegal wildlife is moving from Africa to Asia, and sexual trafficking is rife between Brazil and Russia to Europe.

citizen's access to basic services, credit ratings and school and university placements.[133] In Xinjiang, the authorities have deployed AI-powered surveillance to monitor the population's movements to banks, hospitals, shopping centres and parks.[134] Similar technologies were also used to improve contact tracing and enforce quarantines following the outbreak of COVID-19 in early 2020. The government is also exporting its surveillance expertise to other countries.[135]

The sinister expansion of organised crime – a $2.2 trillion-a-year business

Crime is an ancient profession. Throughout history organised criminals included pirates, bandits and highwaymen, often operating out of the reach of royal or state authorities. For thousands of years, criminal organisations – mafias, syndicates and gangs – have also provided protection, albeit for a price. The sociologist Diego Gambetta has described how one of the most formidable organised crime networks in history – Italy's Sicilian mafia – became successful by satisfying a market demand for what we now refer to as public safety.[137] Around the world – from China[138] and Japan,[139] to the US[140] and Russia[141] – organised crime

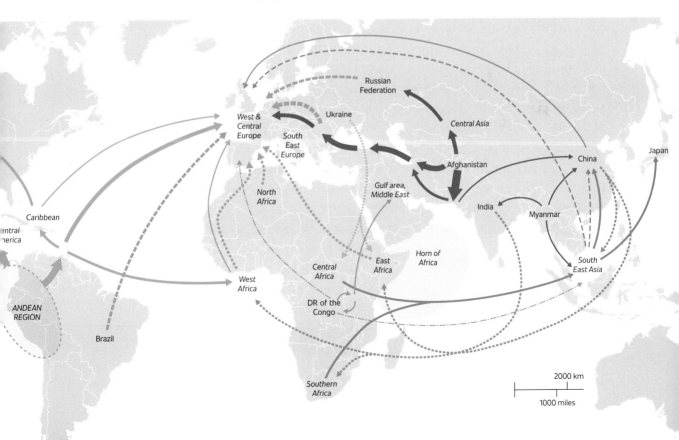

groups offer services and sell commodities, often deploying threats and acts of violence to extract financial returns. The map from several years ago describes how organised crime has become a truly global business with eye-watering profits. Cocaine produced in Colombia, Bolivia and Peru and sold in North America and Western Europe is part of a $650 billion-a-year illegal drugs industry. But narcotics are just one part of a sprawling ecosystem of organised crime. One study estimates that organised crime groups rake in as much as $2.2 trillion a year.[142]

Organised criminals do not operate in a vacuum – they regularly conspire and collude with elected politicians and civil servants. Their businesses are intimately intertwined with state institutions like the police, judicial system, and customs services. Organised crime groups usually operate as businesses that respond to market signals. This is something that the US government learned after it prohibited the sale and importation of alcohol between 1920 and 1933. Before Prohibition, gangs were 'thugs for hire', recruited by their political bosses to provide extra muscle, intimidate politicians and channel votes to preferred candidates. In return, politicians and law enforcement officers turned a blind eye to the involvement of gangsters in gambling and prostitution rings. After Prohibition, gangs took charge and set up protection rackets, paying off politicians to keep their illicit enterprises running. As the money poured in, mobsters recruited lawyers and accountants, created national and transnational partnerships, and set up legitimate businesses to launder their illegal profits.[143]

Organised crime flourishes when governments are unable or un-willing to control territory or deliver basic services.[144] But government authorities are also frequently on the payroll of mafia, cartels and gangs. This helps explain why some elected officials may politicise and undermine their own security services. In Rio de Janeiro, for example, heavily armed police are battling with gangs and militia not just to restore law and order, but also to dominate and control illegal markets for drugs, racketeering and illicit services. This map illustrates how big metropolitan areas can be overrun by a patchwork of criminal groups, alternately competing and colluding with state authorities. The areas marked in red, yellow and green are dominated by heavily armed drug trafficking factions. The blue zones are controlled by the fearsome militia – groups of current and former police officers, firemen and soldiers who are also heavily involved in everything from land grabbing to targeted assassinations. Left to fend for themselves and

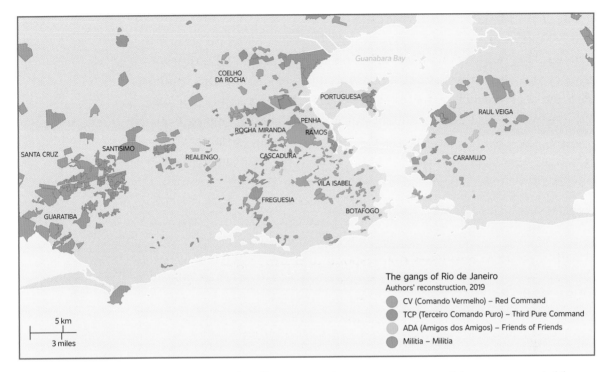

The gangs of Rio de Janeiro
Authors' reconstruction, 2019

- CV (Comando Vermelho) – Red Command
- TCP (Terceiro Comando Puro) – Third Pure Command
- ADA (Amigos dos Amigos) – Friends of Friends
- Militia – Militia

The gangs of Rio de Janeiro

Rio de Janeiro is criss-crossed with drug trafficking factions and militia. The red, yellow and green areas are those dominated by territorial drug-trafficking gangs. The blue areas are those held by militia. The high concentration of gangs and militia in a small area generates considerable violence.

with limited police protection, poor communities turn to neighbour-hood watch groups, vigilantes, and mafias, who offer security for a hefty price.[145]

Organised crime corrodes democracy. When people fear crime, or are themselves victims, their faith in democratic institutions declines.[146] They are also statistically more likely to support authoritarian solutions to restore order than democratic ones. Criminal groups are expert at flooding political systems with corruption, privilege, and perks for the well-connected. Colombians have considerable experience with this, and refer to it as 'narco-politics'. Throughout the 1990s and 2000s the survival of the country's elected politicians and local businessmen depended on keeping the drugs trade operating smoothly. Similar dynamics are now apparent in Mexico, where hundreds of municipalities are overseen by 'shadow governments' made up of criminalised political and bureaucratic systems of patronage and protection.[147] Many Brazilian cities and towns are also controlled by drug-trafficking factions, militia and corrupt public officials.[148] There, some of them are auctioning off the 'vote' of entire neighbourhoods to the highest bidder.[149]

Governments may also use the threat of organised crime as a pretext for militarising public security.[150] There are good reasons

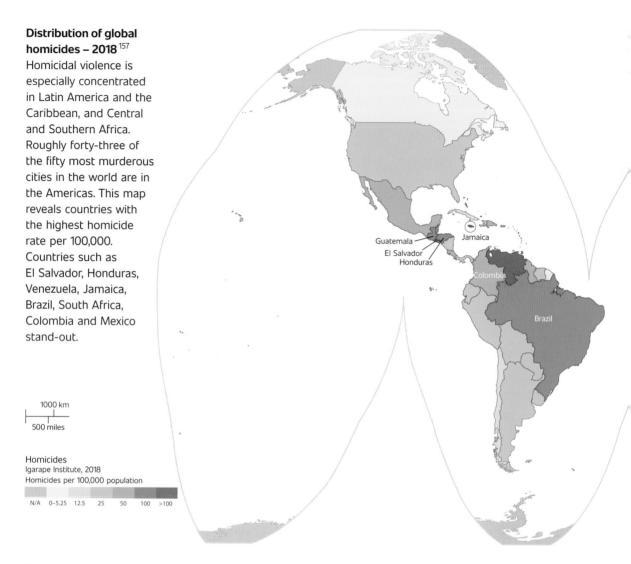

Distribution of global homicides – 2018 [157]

Homicidal violence is especially concentrated in Latin America and the Caribbean, and Central and Southern Africa. Roughly forty-three of the fifty most murderous cities in the world are in the Americas. This map reveals countries with the highest homicide rate per 100,000. Countries such as El Salvador, Honduras, Venezuela, Jamaica, Brazil, South Africa, Colombia and Mexico stand-out.

1000 km
500 miles

Homicides
Igarape Institute, 2018
Homicides per 100,000 population

N/A   0–5.25   12.5   25   50   100   >100

Guatemala
El Salvador
Honduras
Jamaica
Colombia
Brazil

for this: politicians who appear weak on crime typically suffer at the ballot box,[151] while a tough-on-crime posture usually helps them win elections.[152] Paradoxically, these so-called iron fist or *mano dura* policing strategies that result in mass arrests can actually supercharge criminal groups.[153] This is because incarcerated young men hone their criminal skillsets inside prisons.[154] Across Central and South America, for example, gangs such as the MS-13 and Primeiro Comando da Capital essentially run large parts of the criminal justice and prison systems – they are judges, jurors and executioners.[155] Making matters worse, there is ample evidence that repressive policing does not just fail to deter organised crime, it can actually increase criminal violence. Citizens, especially those living in neighbourhoods dominated by

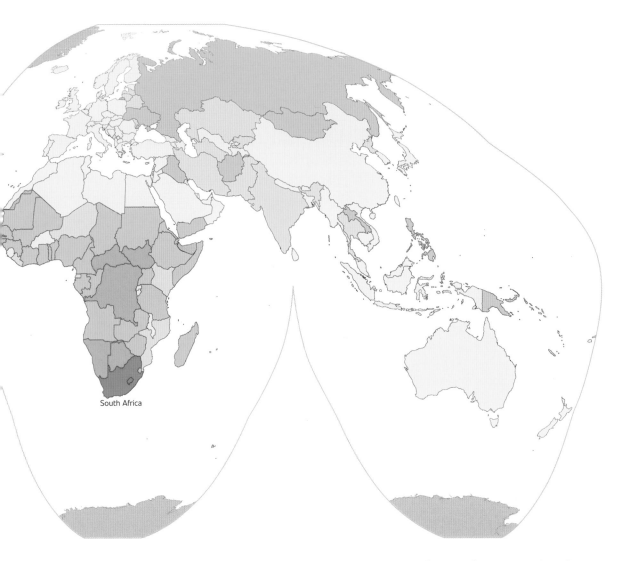

South Africa

gangs, are forced to choose sides. Some of them fear reprisal violence from gangsters, while others are worried about predatory, criminalised police officers.[156]

One part of the world where all these factors are converging is Latin America and the Caribbean. Since 2000, at least 2.5 million people from the region have been murdered. Every fifteen minutes, a Latin American or Caribbean citizen – usually a poor black adult male – is gunned down. In some countries, gun-related homicide is the number one cause of death for adolescent males, outpacing accidents, cancer, suicide and disease.[158] The map shows how Latin America and the Caribbean features seventeen of the twenty most homicidal countries on Earth. Brazil, Colombia, Mexico and Venezuela alone concentrate about a quarter of all homicides worldwide. On average, roughly 340 people are murdered in these four countries every day. By way of comparison, about 57 South Africans are murdered every day.

Sky-high crime in places like Brazil and South Africa are often the result of inequality, stubborn youth unemployment and breathtaking impunity. Both countries rank among the world's most unequal, and most crime goes unpunished. In Brazil, for example, less than 8 per cent of all reported homicides result in a conviction. Compare this to Japan, where 98 per cent of all murders lead to a prison sentence. Things aren't terrible everywhere. Countries like Argentina, Costa Rica, Ecuador, Peru, Uruguay and, in particular, Chile (with its homicide rate of 2.7 per 100,000) are much safer than El Salvador, Honduras, Trinidad and Tobago, or Venezuela. Even so, the average homicide rate of the region's least violent countries – around 6.5 per 100,000 – is still twice that of North America. The costs of all this criminal violence are enormous, about 3.5 per cent of total regional GDP, a tax of $300 per person.

Reining in the dogs of war

Notwithstanding these grim headlines, there are proven ways to effectively prevent and reduce organised violence.[159] Once the challenge is recognised and political leaders express commitment to do something about the issue, governments, businesses and civil society groups need to focus their energy on the specific places where violence concentrates. They must not focus just on policing, but also on comprehensive prevention programmes that tackle the determinants of violence. In the end, reversing trends in armed conflict, terrorism and organised crime will require making a serious investment treating both the symptoms and causes. It will also necessitate more cooperation to support peace, disrupt extremism, legalise and regulate drugs, and fight crime, not less.[160] This is difficult at a time when populist and nationalist govern-ments are denouncing 'globalism' and questioning the effectiveness of multilateralism.

Notwithstanding criticism, the United Nations has had some success in stemming and reversing violence.[161] The organisation rarely gets much credit since few people notice when wars or genocidal violence doesn't happen. Not only have the United Nations and other

UN peacekeeping missions
UN Peacekeeping, 2019
● Active mission HQ
▦ Active mission
▨ Completed mission

2000 km
1000 miles

Deploying the 'Blue Helmets' to prevent conflict – 1954–2019
Peacekeepers have been dispatched around the world for more than half a century. There have been more than seventy peacekeeping operations since 1954 involving millions of military personnel, police and other civilians from 120 countries. The map highlights the thirteen ongoing operations (in yellow) and the many completed missions (green). The headquarters of active missions is also noted in red.

organisations helped reduce the number of civil wars and associated death tolls, but it has done so at a reasonably low cost.[162] Whether we like it or not, the United Nations is still the only genuinely global mechanism the world has to keep international and civil wars from spiralling out of control. Yet the global appetite for using the United Nations to solve tricky problems is ebbing, and fissures between countries like China, Russia, and the US have crippled the Security Council. As the chapter on geopolitics makes clear, Security Council reform is essential. But now is not the time to abandon those parts of the organisation that actually work.

The changing nature of conflict, terrorism, crime and state repression also require new ways of promoting peace. New types of policy responses and public–private co-operation are urgently needed to target and shut down the sources of illicit revenue that fuel disorder and instability.[163] Today's insurgents, terrorists and mafia depend on global supply chains – from financial institutions and commodity traders to producers, suppliers, retailers and shipping companies – to survive. At the height of its influence in 2015, for example, ISIS generated more than $6 billion a year from selling oil, taxing hotels and hospitals, running car dealerships and trading gold – a value equivalent to the economic output of Montenegro.[164] These kinds of dependencies offer entry-points for intervention, including through greater transparency of financial transactions, operations to end money-laundering, restrictions on the use of the dark web of illicit online vendors, and the exertion of greater pressure on private firms and tax havens to clean up their acts.

In the long term, the most effective route to preventing conflict, terrorism and crime is by addressing the underlying structural factors that give rise to them in the first place. Inclusive economic growth and the reduction of social inequalities is critical since they are key determinants of violence. Targeted support for at-risk communities suffering from concentrated disadvantage and political exclusion is essential.[165] Creating pathways for more participatory decision-making in divided societies is crucial to keeping them from fragmenting. All this means engaging more directly with political and economic elites and the pacts they often make to sustain a violent status quo;[166] also involving a much wider group of people in forging solutions than is often the case, including faith and inter-faith based groups.[167] To be sure, societies that expand the political and economic participation of women and young people and encourage greater social mobility tend to register less violence than those that do not.[168] Research shows that reversing high rates of gender inequality and gender-based violence can decrease vulnerability to civil and interstate war.[169] All these remedies are as important in the COVID-19 era as they were in the period that preceded it.

Fighting state violence and crime

In the end, the prevention of many forms of violence comes down to the quality of the relationship between citizens and state institutions. Organised violence thrives in divided and fragmented societies – and especially when the state loses its legitimacy and credibility. When the social contract crumbles, violence is often inevitable. One of the greatest threats to peace, then, is the suppression of free speech and the right to assembly, the extinguishing of opposition voices, and the intentional polarisation of societies. The impetus for peaceful change must emerge from within and cannot be forced on societies by outsiders. In the long-run, it is citizens themselves who will need to compel their governments to be more responsive and responsible.

Supporting free and independent media to ensure that people know what their government and business elites are doing is essential. So is investment in local organisations and citizens' assemblies that can gather and disseminate information in ways that diminish partisan pitfalls and bring citizens together across divided and polarised communities. Targeting foreign aid towards building an independent and vibrant middle-class is key, rather than focusing narrowly on 'economic growth' targets alone. And where corrupt politicians are fuelling the violence they claim to be fighting, government donors and private investors should withhold aid and investment. The private sector also has a critical role to play. Financial hubs and offshore havens should be curbed so that criminals and politicians cannot launder ill-gotten gains.[170] Consumers and companies around the world can boycott criminal activities and ensure that their supply chains and products are not tainted by criminal activities or produced in repressive conditions.

While we've painted a bleak picture, the good news is that levels of violence seem to be at historic lows. The steady spread of the rule of law and gradual improvement of overall living conditions are at least partly responsible for this. The bad news is that violence is a chameleon and is assuming new forms that are harder than ever to eradicate. The intellectual effort and political capital invested in diminishing war in the twentieth century must now be deployed towards reducing organised crime and state repression in the twenty-first. We know what needs to be done. What is needed is the political wherewithal to get started. The pay-off for the one in six people worldwide affected by global violence – and the countries buckling under the influx of migrants fleeing bloodshed – will be well worth the effort.

Beijing and Tianjin at night from space
NASA, 2010

Demography

Population growth is slowing dramatically

World population will peak at around 11 billion by 2100

The world is ageing rapidly, only Africa remains young

Fertility is below replacement levels in most countries

Tokyo from space
The photograph taken by a NASA astronaut of Tokyo at night shows the population density of this megacity. With over 21 per cent of its population aged 65 and over, it has become what the UN terms a 'super-aged' society. With more people dying than being born, and migration into the city negligible, Tokyo's population has peaked, as it has for the whole of Japan.

5 km
3 miles

NASA, 2008

Introduction

Is the global population growing too quickly or too slowly? The answer is not straight-forward. What is clear is that the world is ageing rapidly. For the first time in history there are more people on Earth over the age of sixty-four than under the age of five.[2] By 2040, there will be 2 billion people over sixty, twice as many as there are right now.[3] It could well be that within the next fifty years our planet will have too few – not too many – young people.[4] For reasons we explain in this chapter, it is almost certainly the case that the global population will experience increasingly skewed gender distribution, with many more elderly women than men, and many more young boys than girls.[5]

The idea that the world is overpopulated has deep roots. It stretches back at least to the early Christian theologian Quintus Septimius Florens Tertullianus, or Tertullian, who, in the second century AD when the world population was under 200 million, wrote about population numbers being in excess of what the world could support. Perhaps the most widely known population pessimist was Thomas Malthus who in 1798 warned that excessive population growth was slowing progress. More recently, in 1968, Paul R. Ehrlich's *The Population Bomb* adapted this thinking to the modern era, arguing that rising population would lead to food shortages and condemn a growing share of the world population to famine. The latest thinker to worry that the world's problems boil down to too many people is Stephen Emmott who wrote *10 Billion*.[6] When people speak about overpopulation they are not usually referring to too many of 'our' people, but rather too many of 'them'. The others. But is overpopulation the problem it is imagined to be, especially in Asia and Africa?[7]

Half of humanity lives inside this circle

More people live inside the circle drawn on this map than outside it. The encircled Asian countries are home to more than 3.8 billion people.[1] The country with the most people is China, whose population has peaked at around 1.42 billion and is currently being overtaken by India's with 1.35 billion, which is followed by Indonesia with 274 million people.

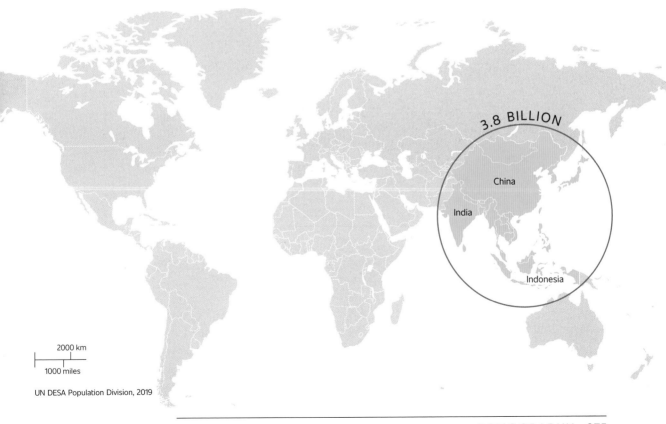

3.8 BILLION

China

India

Indonesia

2000 km
1000 miles

UN DESA Population Division, 2019

World population growth in history

By the end of the most recent ice age, some 20,000 years ago, there were about 1 million humans on earth, sparsely scattered across Africa, Asia and Europe.[8] By 5,000 BC, the world population had reached an estimated 5 million and every continent was settled.[9] It took almost 7,000 years for the global population to reach the 1 billion mark, in

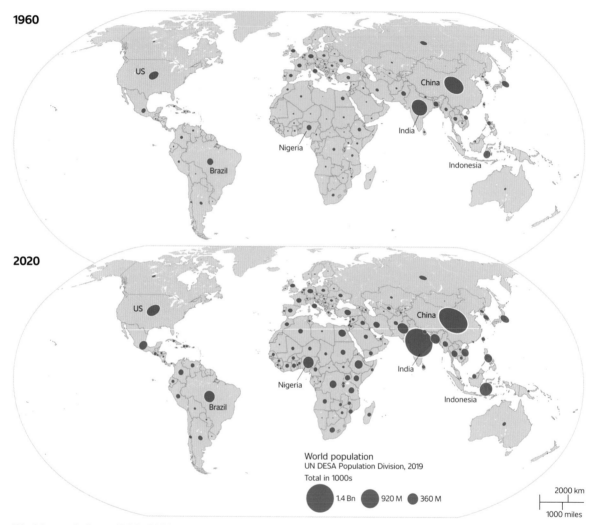

1960

2020

World population
UN DESA Population Division, 2019
Total in 1000s

1.4 Bn · 920 M · 360 M

2000 km
1000 miles

World population – 1960–2020
There are two factors that determine the size of the population: how many people are born, or fertility, and how long they live, known as life expectancy or mortality. The circles on these maps represent the number of people living in different countries. All regions have experienced population increases, but comparing the 1960 and 2020 maps, the growth of India's and China's populations is striking, as is that of African countries.

around AD 1800.[10] As the figure shows, the population doubled to 2 billion around 1925, and then doubled again to reach 4 billion in 1975, a period of just fifty years. By 2000 it had reached 6 billion, by 2020 7.8 billion and by 2050 it is forecast to be just shy of 10 billion, a threefold increase in just 100 years.[11] Then, population growth is likely to slow dramatically, taking another fifty or so years to reach a peak population of about 11 billion people, before beginning a long contraction. In a century's time, many demographers predict that the world population may actually sink back to current levels, below 8 billion.[12]

The biggest surprise is not how fast the global population has grown, but how fast fertility rates have fallen. More than half of the world's countries are now below replacement rate, which is the level of fertility required to keep the population the same from generation to generation.[14] The magic number is 2.1: a ratio slightly above 2 is required because not all girls survive to be women and not all women go on to give birth.[15] The marked global decline in fertility rates comes

Population growth around the world, 1700–2100

The long-term trends in population growth are evident from this figure which shows that growth rates have slowed dramatically since the 1970s, at the end of the baby boom, and will continue plummeting. By 2100, according to this projection, the world's population will have peaked at close to 11 billion, about ten times more than it was at the time of the industrial revolution, but then will decline steadily.[13]

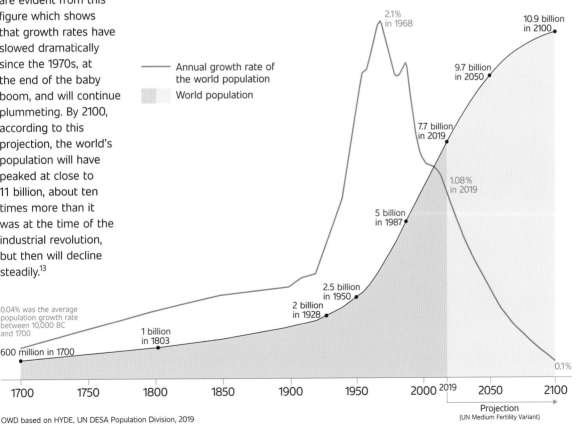

OWD based on HYDE, UN DESA Population Division, 2019

down to political, social and economic factors. Put simply, a growing share of women around the world are getting better educated, entering the workforce, becoming urbanised and increasingly deciding how many children they will have and when to have them. More and more women are choosing to delay and reduce the size of their families.

The decline in fertility can happen quickly. In a single generation, entire communities can see their fertility rates plummet in ways that cannot be attributed to cultural, religious and ethnic factors.[16] In Iran, Ireland and Italy, many mothers who are one of seven or more children elect to have just one child. As a result, in these countries which a generation ago were characterised by rapid population growth, fertility rates are now 1.6, 1.9 and 1.3 respectively, well below replacement levels. Yet these countries are markedly different in terms of their dominant cultural beliefs and none has undergone a metamorphosis in religious values. Instead, what they have in common is a sharp reduction in infant mortality, rapid improvements in income, higher levels of female education, rising levels of female employment, the development of social security to support the elderly, and the availability of contraception. Leadership from politicians, religious leaders and educators also makes a significant difference, not least if this helps girls envision a role beyond being a mother.[17]

As explained in the health chapter, improvements in child health and education are also key to reducing fertility. For one, the fewer children that die prematurely, the fewer babies mothers need to have to achieve their ideal family size. Educating young girls improves their chances of getting meaningful jobs and may lead them to delay pregnancy and have fewer children. Their growing awareness of contraception, along with its availability, can allow women to better plan how many children they have. Education also expands horizons and allows individuals to envisage alternative lifestyle choices, beyond raising children.[18] There are other positive externalities as well. Literacy and education not only raise incomes but also contributes to better nutrition, sanitation and access to health. In fact, the education level of a mother is the single most significant predictor of her child's life or death. That more than 100 million children still do not attend school and more than half of all African girls do not complete primary school is deeply worrying.[19] This tragic failure needs to be remedied for many reasons that extend far beyond questions of fertility.

Countries do not have to become rich to reduce population growth and many developing countries have lower population growth rates than far richer countries. Thailand has a lower fertility rate than the UK, and Vietnamese women are having fewer children than Swedish mothers. Countries as varied as Bangladesh, Chile, Egypt, Indonesia, Iran and Tunisia are all barely at replacement level and are likely to soon fall below.[20] India's fertility rate of 2.3 is half of what it was in 1980 and is expected to fall below replacement level within the next ten years, which given it will be the most populous country has a significant impact on global population projections.[21]

Spain, Taiwan and Hong Kong share the distinction of having among the lowest fertility rates in the world.[22] Very low fertility rates in Germany and Italy are not far from those in Barbados, Vietnam, Mauritius, Chile, Tunisia, the US and Myanmar.[23] In Singapore and South Korea, fertility rates are just above one.[24] The abolition of China's one-child policy is not anticipated to lead to increases in its current fertility rate of 1.6.[25] In Italy, the number of births has fallen to levels not seen at any time since the nation was created in 1861. This has made Italy's the second oldest population in the world, after Japan's, with almost a quarter of the population above sixty-five in both countries.[26] In the US there also has been a sharp fall in fertility over the past decade from 2.1 in 2007 to 1.7 in 2018, which is well below replacement level.[27]

The forty-five or so countries that still register high fertility rates – above four children per woman – are mostly in Sub-Saharan Africa.[28] An important question is whether these countries, especially the three with the largest populations and highest fertility rates (Niger, Nigeria and Tanzania), will follow the demographic transition that other countries have experienced in tandem with improvements in incomes, girls' education, the availability of contraception, and urbanisation.[29] In Ethiopia and Kenya, for example, fertility rates have halved in the past forty years, giving cause for optimism that African countries are following the sharp reductions in fertility experienced in other regions.[30] The general consensus is that in most African countries, fertility will fall to below or near replacement by 2050.[31] This implies that the the African population will increase from 1.3 billion now, to around 2 billion by 2050. Depending on how quickly fertility declines, the population could be more than 2.7 billion by the end of the century, at which point the fifty-four countries on the continent

1960

2000 km
1000 miles

Sweden
Estonia
Latvia
Czech Republic
Hungary
Switzerland
Japan

Fertility rate, total 1960 and 2017
UN DESA Population Division, 2019
Births per woman

1–2.1   2.2–4.9   5–9.0

2017

2000 km
1000 miles

Mali Niger
Chad
Burkina Faso
Nigeria
Somalia
DRC
Uganda
Burundi
Angola
Mozambique

Rapidly declining fertility

The maps highlight the rapidly declining fertility rates around the world between
1960 and 2017. In 1960, only a handful of countries – notably Sweden, Switzerland and
Japan, depicted in yellow – had fertility rates that were below replacement levels.
By 2017, we observe a very different world, with the majority of countries below
replacement, depicted again in yellow. Africa is the only continent where there still are
countries that have high fertility rates, and even within Africa, as a comparison of the
two maps shows, high rates are now the exception.

will have a combined population equivalent to the current combined population of China and India.[32] It is likely that Africa will probably be the only continent in which populations will increase in the second half of this century.[33]

Population pyramids provide graphical representations of the evolution of populations over time, by age and gender. Population pyramid is, however, a misnomer and dates back to a time when the young outnumbered the elderly and each successive step in age had fewer people than the previous step. Nigeria has a very large base owing to its high fertility rate of 5.5 children per woman, and a rapidly narrowing pyramid owing to low life expectancy of barely fifty-four years.[34] The result is that an astonishing 44 per cent of the population are younger than 15 years old and 54 per cent of the population are younger than 20.[35] This is in stark contrast with South Korea, which has one of the world's most rapidly ageing populations. There, the traditional population pyramid looks like a spinning top in which a relatively narrow base of young people supports the growing weight

From population pyramids to coffins

Population pyramids provide an easy way to represent the number of people in a country by age and gender. In the pyramids below, the vertical steps are divided by five years from birth to 100 years, and the width of each step indicates the percentage share of the population that each five-year group occupies. The very different demographic shapes of the profiles for Nigeria, South Korea and the US reflects how age structures can change dramatically.

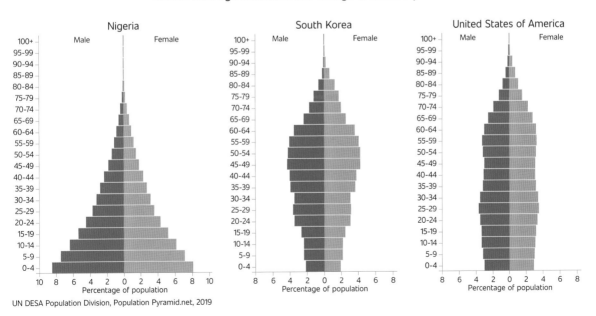

UN DESA Population Division, Population Pyramid.net, 2019

of an increasingly elderly population. Meanwhile, the US population pyramid is shaped ominously like a coffin standing up, with a gradual reduction in fertility being more than matched by increases in longevity.[36] As a result, the number of newborns is lower than it was in 1987, when the population was 80 million below the current total of around 330 million.[37]

Fertility declines don't necessarily translate into falling populations right away. The total size of populations depends on the balance between a smaller number of babies being born and people living longer. In addition to ageing there may be a lag of a couple of decades since girls born when fertility levels were higher also become mothers a few decades later. Even though most countries will register negative fertility rates by 2030, the world's population is not expected to peak before 2050, at which time Africa and Asia combined will comprise about three-quarters of the total.[38] By 2030, it is very likely that one of every three babies born in the world will be African.[39]

Getting older . . . all the time

A reader of this book can expect that in the next hour their life expectancy will increase on average by around ten minutes, and over the next year by over two months.[40] This raises some intriguing questions. If you could live beyond 100, would you want to? Even if one is physically able to lead a full life, Parkinson's, Alzheimer's and other causes of dementia might restrict mental abilities for those who are fortunate enough to live over the age of ninety.[41] In a few decades, these degenerative diseases of the brain will hopefully be conquered and, as we show in the health chapter, living well beyond 100 should become more pleasurable and less of a burden on society.

The graphic representation of the changing shape of societies over time shows that, while total births have almost stopped increasing, population will keep growing now and into the future, due to ageing. The blue shows the shape in 1950, light green the 2018 shape, and the yellow area shows projections until 2100. In the coming decades virtually all population growth will come through ageing rather than increased births, with the height rather than width at the base changing as the global population increases. From around 2018 there is almost no increase in the total global under-ten population, as high

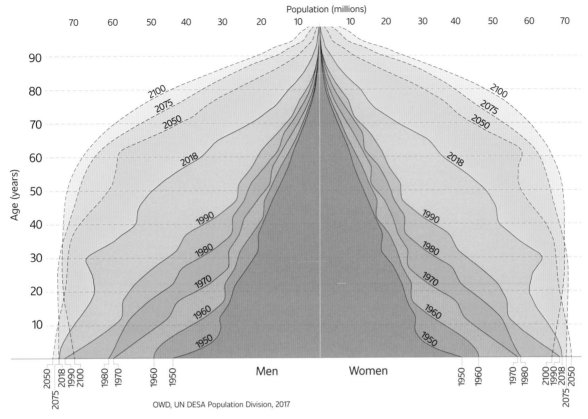

Population (millions)

70 60 50 40 30 20 10 10 20 30 40 50 60 70

Age (years)

2100
2075
2050
2018
1990
1980
1970
1960
1950

Men   Women

2050 2018 1990 2100 2075 1980 1970 1960 1950   1950 1960 1970 1980 2100 1990 2018 2050 2075

OWD, UN DESA Population Division, 2017

The changing shape of societies as they age and fertility declines: 1950 to 2100

Demographic change is represented by a series of tents constructed over each other, with the inner tent showing the global population distribution in 1950 and each outer layer showing how it is changing each decade, until we reach 2100.

birth rates in Africa are offset by sharply declining fertility elsewhere, with virtually all the global increase in population coming from ageing.

Disappearing workforces

In the wealthier OECD countries, the working-age population is projected to decline from around 800 million to 600 million over the coming thirty years, with Europe's workforce declining by a third.[42] As the number of elderly dependants rises relative to those who support them through taxes, drastic policy changes will be required. Many of them will be unpopular. They include the extension or even abolition of retirement ages, along with other reforms to pension systems including the ending of defined benefits, so pensions will be determined on the basis of contributions and performance of investments, rather than a fixed entitlement, which has typically been based on salaries at retirement.

A growing share of the elderly will work well into their seventies and eighties. The need to sustain personal incomes and support dependent family members will be the most compelling reasons to continue working. Already today, individuals of retirement age face the daunting prospect of dwindling savings and being dependent on increasingly indebted corporate and government pensions systems. The combination of rapidly ageing populations and lower returns on investment means that many existing pension systems are unsustainable, and this is going to deteriorate further as ageing and low interest rates undermine the values of pension funds while increasing their liabilities to the elderly who are living longer.[43]

The implication of longer working lives is that the elderly are likely to hold on to their jobs and their senior positions for longer, reducing the opportunities for advancement of young people. Not surprisingly, younger generations will become increasingly frustrated and disillusioned. In France, each elderly person currently depends on three working-age people.[44] In Japan, there are only two workers for every pensioner.[45] By 2050 it is forecast that half of all European countries will be in the same position that faces Japan today.[46] The stark reality is that to avoid overburdening future generations, the elderly will have to work longer and become more active and independent. In the short-term, these shifts may well be muted by COVID-19 which is especially dangerous for older people.

For decades, Japan has led the world in terms of rapid population ageing. Because fertility there collapsed, Japan's population is declining by around 430,000 people per year, equivalent to a mid-size city.[47] In Japan's northern prefectures of Aomori and Akita, the population is falling by more than 1 per cent each year, and villages are dying out, inhabited almost entirely by people over seventy.[48] Already its population of 126 million has declined by 2 million from its peak and it is estimated that within thirty years Japan's population will be contracting by almost a million people a year, equivalent to the population of Glasgow or Austin.[49] Without radical changes to immigration policy, Japan's population by the end of this century will probably fall to just 50 million.[50] As the number of people under sixty-four gets smaller and those sixty-five and above continues to grow at more than 2.5 per cent a year, the growing number of elderly will place a greater burden on families, communities and public finances.[51]

The Japanese response to this predicament has been to look for technological fixes rather than opening the door to migrants, as other countries have done. It is no accident that the first country to show the most dramatic signs of ageing also registers the highest rate of uptake of robots. Proactive policies encourage the wider use of machines, including not only attractive depreciation allowances for investment in robots, but also publicity and marketing campaigns that seek to increase acceptability of the use of robots and automated systems. Meanwhile, a number of European countries, including Italy, Bulgaria, Poland and Romania, are shrinking even faster than Japan, owing to a combination of exceptionally low levels of fertility and high rates of emigration.[52] In the period 2017–2019, the antipathy of the Italian government towards immigration meant that the Italian population contracted rapidly, with 150,000 more people leaving than arriving each year.[53]

The decline in fertility and increase in longevity is leading to rapid increases in median ages. As the figure shows, median ages are expected to double over the century to 2050, rising in Japan from around twenty-two years in 1950 to an anticipated fifty-five years in 2050. Far from this trend being confined to rich countries, in China the median age has increased from twenty in 1975 to thirty-seven today and by 2050 is expected to reach forty-seven. In Mexico it has risen from around seventeen in 1965 to over thirty in 2019 and by 2050 is expected to be over forty.

Doubling in median ages, 1950–2050
Median ages are expected to double over the century to 2050, rising in Japan from twenty-two years in 1950 to an anticipated fifty-five years in 2050, in China from just over twenty to forty-seven, in Italy from under thirty to over fifty and in the UK from thirty-four to forty-four.

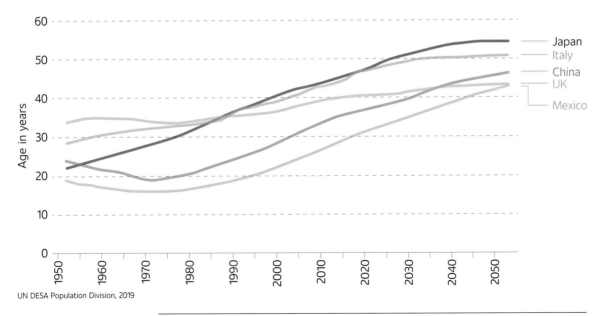

UN DESA Population Division, 2019

1970

2020

Sweden

Germany

France

Italy

Greece

Japan

Dependency ratio of elderly to working population, 1970 and 2020
World Bank, UN Population Division, 2020
Number of elderly dependents (age >64) to working-age population (age 15–64)

| <11 | 11–30 | >30 |

Dependency ratios: More elderly depending on fewer workers – comparing 1970 and 2020

In 1970 the world was young, with low numbers of elderly dependants per worker, depicted in yellow. Since 1970, dramatic increases in life expectancy and declines in fertility have increased dependency ratios as is shown in the blue shading in France, Italy, Greece, Germany, Sweden and Japan.

In 1970, as the map shows, the whole world was young. It could be divided between countries where the over sixty-fours represented a very small burden on the working age, with fewer than eleven retirees for every 100 of working age, depicted in yellow, and those with a higher range of 11 to 30 per cent (depicted in pink), with at least three workers per retiree. Since 1970, global average life expectancy has increased from 59 to 72 years.[54] At the same time global fertility levels have decreased markedly, from an average of 4.7 children per mother in 1970 to 2.4.[55] The combination of these factors has led to a rapid increase in dependency ratios, calculated by comparing the number of dependent children (defined as under fifteen) and dependent elderly (defined as over sixty-four) with the population aged 15 to 64. The European countries of France, Italy, Greece, Germany and Sweden, have dependency ratios exceeding 30 per cent, as does Japan. China, Chile and Brazil have also seen marked increases in their dependency ratios.

More elderly societies

Demography reveals stark social and economic trends. Around the world, there are more elderly women than men. This is because, on average, women don't smoke, drink or die violently as frequently as men which helps explain why they generally live longer. Meanwhile, in more patriarchal societies in which pre-natal sex testing can lead to the termination of pregnancies, more boys relative to girls are being born. Fewer young women than men adds to the downward trend in fertility, which is a worldwide phenomenon.[56]

When populations stop growing, average incomes rise with economic growth, as increases in national income are divided among the same fixed number of people. Because there is virtually no population growth in China and national economic growth rates are similar to India's, Chinese citizens are on average getting richer more quickly than Indians. Even though India's economic growth rate may begin to exceed that of China, the benefits are being shared among a growing number of people. On average, Chinese people are twenty-two times richer today than they were thirty-five years ago owing to the combination of exceptionally high economic growth rates and exceptionally low population growth.[57] Rapid economic growth in India, coupled with sharp reductions in fertility, would lead to marked improvements in average incomes in the coming decades. As noted above, India's fertility rate has more than halved from 5.5 in 1979 to 2.3 today and by 2030 is forecast to be below replacement level.[58]

Demography influences politics

Changes in demography exert a profound impact on national politics. For one, younger people tend to vote differently to older generations. As median ages rise, the cost of elderly care and pensions soar, placing an additional burden on public finances. This means that young workers can be expected to contribute more in taxes. As the elderly become more politically influential (by virtue of their growing numbers), their interests, rather than those of their more youthful cohorts, will increasingly shape the direction of political debate. Given that the world is ageing, we can expect fireworks ahead.

When the pensions and retirement systems currently operating in Europe and the US were developed in the early 1970s, average life expectancy after retirement at 65 was about 5 years and average returns on investments, adjusted for inflation, were 4 per cent.[59] Now average life expectancy at the lowered average retirement age of 60 is around 19 years for men and more than 20 years for women.[60] Meanwhile, inflation-adjusted (real) returns on investment have fallen to below 0.5 per cent.[61] This implies that people will have to save a hundred times more than they would have done forty years ago to sustain an equivalent standard of living. As a growing share of the population save more, they will spend and consume less. Instead, the savings will be devoted to health and home care and, if they can be afforded, on leisure and travel. Relatively less will be spent on purchasing consumer goods, from houses and cars to clothes and entertainment.

Older people will need to hang on to their money and assets for later life. This means that they will hand over less money to their children and be less willing to pay for school or college fees. They will also need to hold on to their houses for longer periods of time. If the elderly are lucky enough to own their own home, their children can expect to inherit the house or other assets when they are well into their seventies, after their parents eventually pass away in their nineties, or beyond. The implication of all this is a dramatic shift in wealth in favour of the elderly, at the expense of the young.

These demographic trends have dramatic implications for individual and family finances and also for national economies. The forces that will shape personal and government finances in wealthier countries are also at work in middle-income and poorer countries (other than Africa where, owing to the youth bulge, the economic challenges are very different). The need to support a growing number of elderly while workforces are declining increases the pressure on public finances, and will almost inevitably lead to higher taxes to fund health and care costs. China has in excess of 3 million fewer workers every year, and the challenges are particularly acute for it and other countries that are getting old before they have got rich.

Rising savings globally will exert further downward pressure on interest rates. Lower interest rates in turn mean that people have to save even more to earn returns on which they can live into their retirement. Regulations introduced after the 2008 financial crisis are generating perverse impacts that are compounding the problems facing pensioners. Governments have insisted that pensions and long-term insurers invest in 'low risk' assets, such as government bonds and other supposedly secure products. As more and more people invest in these assets, the interest rates are pushed down even further, and this reduces the returns even more and forces people to save more, causing a vicious circle of more savings pushing down returns and requiring even more savings.

The pensions crisis is a looming disaster and could well precipitate a future global financial meltdown.[62] This is both a private pensions crisis, as many of the biggest companies are effectively insolvent if their pension obligations are taken into account, and also a public pensions crisis, since governments are increasingly unable or unwilling to meet the expectations of retirees who are entitled to their promised pensions. The ordeal of Greek pensioners, who found their pensions halved following the financial crisis of 2008, is likely to become a more common experience.

Health-care costs are escalating dramatically as populations age. The costs of caring for the elderly in a growing number of countries exceeds the combined costs of care for all other age groups. The widening gap between mental and physical life expectancy will continue to place additional strain on government and personal finances. With physical life expectancy rising by at least two years a decade and no major breakthroughs with mental longevity expected in the next ten years, the prospect of hundreds of millions of people who are utterly dependent on subsidised care poses a looming threat for many countries, both rich and poor.[63]

What can be done: coping with 100 year life

A challenge for governments is how to cope with 100-year-life societies and ensure that the elderly can afford to live long, fulfilling lives. This is daunting, but it is also a welcome development, not least because it reflects the success of efforts to curtail population growth and to ensure that people live longer, two enduring objectives of human progress. Key to managing the demographic transition in ageing societies will be ensuring that enough people are working productively to support those who are dependent on them.[64] Immigration, which the US relied on until recently, and the design and deployment of new technologies, which has been the Japanese response, are both part of the solution and should be pursued.

What else will societies need to do to deal with their ageing populations? For one, they need to raise the education levels and productivity of workers, who must remain active and productive for longer. Increasing the participation of women in the workforce is also vital. Governments will have to extend retirement ages and expand part-time and flexible work arrangements both for parents and caregivers. There are ample opportunities to encourage able elderly people to work, to earn income and to contribute and gain fulfilment as mentors, volunteers and part-time workers. Finally, enhancing the mental and physical well-being of the elderly is essential, including through the development of active and supportive community care, physical activities, healthy diets and other public health interventions.

An aged society need not be a decrepit one. However, it is vital to guard against the growing political and economic weight of the elderly having an outsize influence on politics, as this can result in over-prioritising their immediate concerns and undermining investment in the future. As the elderly become more politically powerful, there is a risk that the youth will become more frustrated and angrier. As dependency ratios rise and the pressures on budgets to sustain the elderly increase, politicians and pundits in countries undergoing rapid ageing will be bemoaning the lack of young people and eventually even become more welcoming of immigrants. The likely result is that by the end of this century most countries in the world will have adapted to a smaller and older population, and fears of the dreaded 'population explosion' will be consigned to history, where they belong.

Refugee flows 2015. Each dot represents 17 separate refugees.
UNHCR, 2016

Migration

- Migrant share of population not rising

- Migrants are source of innovation and dynamism

- Migrants contribute more economically than they take

- Record numbers of refugees, mostly close to home

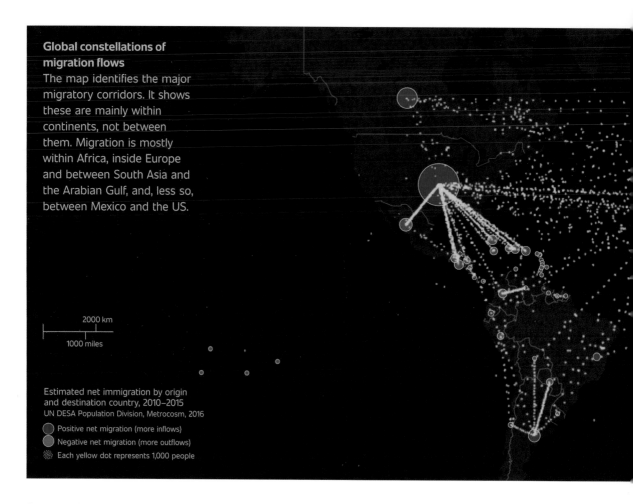

Global constellations of migration flows

The map identifies the major migratory corridors. It shows these are mainly within continents, not between them. Migration is mostly within Africa, inside Europe and between South Asia and the Arabian Gulf, and, less so, between Mexico and the US.

2000 km

1000 miles

Estimated net immigration by origin and destination country, 2010–2015
UN DESA Population Division, Metrocosm, 2016

Positive net migration (more inflows)
Negative net migration (more outflows)
Each yellow dot represents 1,000 people

Introduction

Before the COVID-19 pandemic temporarily closed borders, migration and refugees were a hot topic which generated much heat but very little light due to the disconnect between the politics and evidence. The map reveals how population movements within Africa, inside Europe and between South Asian and Arab nations have far outpaced migration between continents. The fastest growth in migration was within Asia and Africa and not to Western Europe or North America, as anti-migrant advocates shrilly proclaim.[1] Roughly 9 in 10 African migrants never leave the continent. The same goes for more than 80 per cent of East Asians and people from Latin America and the Caribbean, and over 60 per cent of Central and South Asians. Within Europe, with the exception of the temporary suspension of movement due to COVID-19, there are no restrictions on travel for Europeans

between the twenty-six countries of the Schengen area, and this forms a densely packed, star-shaped network of migratory flows, with strong links to other regions.

Migrants and refugees are at the top of the political agenda in many countries. Indeed, they have never been far from it. In this chapter we take the long view of migration, showing how it is as old as humankind. When seen from a historical perspective, the recent levels of migration are not particularly unusual. This is especially the case when one considers that more than a hundred new countries have been created over the past 100 years, so there are many more borders. These frontiers are much more monitored and controlled than ever before. Bombarded by sensationalist and xenophobic media headlines, many people now see migrants as a problem – as a threat to be walled out. This is a mistake. In many countries, although not everywhere, more migrants are needed.

We would do well to start this chapter by distinguishing between migrants and refugees since the two categories are frequently conflated. Migrants are born in one country and move voluntarily as either immigrants or emigrants for economic motives, family reunification purposes or to study. Refugees are forced to move due to threats to their lives. Countries have the right to refuse entry to migrants, but in the case of refugees there is a moral and ethical imperative to offer sanctuary. What is more, there are legally enshrined treaties obligating governments to accept and protect people whose lives are in mortal danger. In legalistic terms, a refugee is someone who has a well-founded fear of being persecuted because of their race, religion, nationality, social affiliation or political opinion.[2] If refugees are to be genuinely protected and cared-for, all of us must get better at burden-sharing, helping them integrate into new communities and addressing the underlying causes of their flight.[3]

Rapidly declining communication and transportation costs made possible by globalisation have reduced barriers to migration. Increasing access to information for migrants has widened their knowledge of the choices available and the pros and cons associated with different destinations. Nevertheless, the spread of international borders and expanding control of people crossing them – especially in a time of COVID-19 – means that migration for most people is more difficult than in the past. Although international agreements have been established to facilitate financial, trade and other flows, migration is still something of an orphan of globalisation and global governance.[4] The challenges facing refugees, the internally displaced and stateless people are doubly complex.

Migration has shaped our world and will fundamentally define our future. By taking risks, innovating and adapting, migrants and refugees frequently advance their own lives and the positive overall conditions of their host societies. It is no accident that the world's most dynamic cities and countries are often those with a relatively high share of migrants, while those that are falling behind tend to be closed and homogeneous. Take the case of Toronto, regularly ranked as one of the world's most liveable cities, where almost 50 per cent of the population is foreign-born.[5] Cities like London, New York and Sydney, with more than a third of their populations coming from abroad, make a virtue of their diversity, actively bringing new people and ideas together.

The routes of our ancestors
Humans started their long trek out of East Africa over 100,000 years ago. This map highlights the migration patterns of our earliest ancestors. Based on an analysis of genetic markers, it shows how humans moved first across Africa, then to the Middle East, Europe and Asia, then finally made it to the Americas. The numbers on the map describe roughly how many thousands of years ago a specific migration took place.

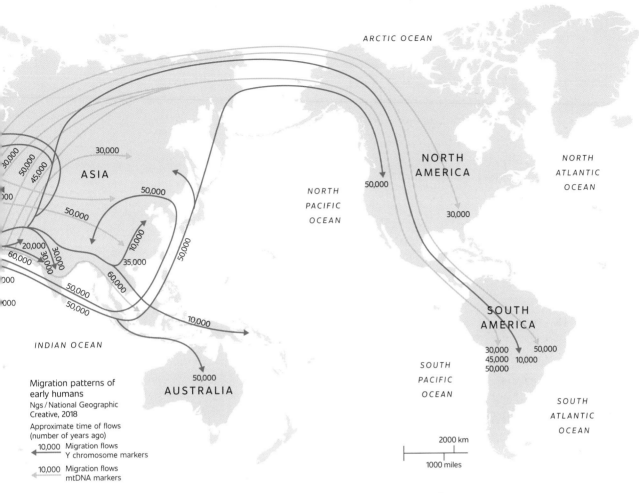

ARCTIC OCEAN

ASIA

30,000
30,000
50,000
45,000
000
50,000
50,000
10,000
20,000
30,000
30,000
60,000
000
60,000
50,000
000
50,000
50,000
35,000
50,000
10,000

NORTH
PACIFIC
OCEAN

NORTH
AMERICA

50,000

30,000

NORTH
ATLANTIC
OCEAN

INDIAN OCEAN

SOUTH
AMERICA

SOUTH
PACIFIC
OCEAN

30,000   50,000
45,000 10,000
50,000

SOUTH
ATLANTIC
OCEAN

50,000
AUSTRALIA

Migration patterns of
early humans
Ngs / National Geographic
Creative, 2018

Approximate time of flows
(number of years ago)

10,000   Migration flows
Y chromosome markers

10,000   Migration flows
mtDNA markers

2000 km

1000 miles

How migration shaped humanity

Humans have been on the move for well over 100,000 years. The map displays migration patterns of our ancestors. It is based on analysis of genetic markers, which are DNA sequences that can be used to track individual historical lineages and variations over time. These lingering traces in our gene pool are the living proof of our shared origins in Africa.[6] The desire and ability to migrate allowed our ancestors to escape famine, drought, pandemics, wars and a host of other disasters that befell them. It is precisely because they explored new opportunities and populated our planet that we have collectively thrived. As the map shows, we started our journey in East Africa before moving across the rest of the continent.

Over 50,000 years ago humans began the long journey across the fertile crescent of the Middle East into Europe and Asia. Crossing Asia, our ancestors reached Australia at the closing stages of the Pleistocene age, when ocean levels were perhaps 300 feet lower than today, allowing for a land bridge between Australia and New Guinea. In time, traversing Europe and crossing Siberia and the Bering ice bridge during the most recent ice age, which ended about 10,000 years ago, they eventually reached North and then South America through multiple migrations. Evidence of people living in the Bluefish Caves of the Yukon territory of Western Canada dates back 24,000 years. By 18,000 years ago they had journeyed the length of the Americas, as evidenced from early human settlements in the Monte Verde area of southern Chile.[7]

Compared to the past, the world is not experiencing exceptionally high levels of migration, at least as a share of our populations. The 'age of mass migration', which began around 1840 and lasted seventy years, involved much higher levels and rates of both emigration and immigration.[8] Political and economic crises, including joblessness caused by the industrial revolution, propelled millions of Europeans to move to escape food shortages, pogroms, wars and poverty. Many of them took advantage of relatively cheap, fast and safe new steamships to emigrate to the Americas, Southern Africa and Australia. By the 1850s, roughly 300,000 Europeans were emigrating a year, rising to over 3 million migrants annually until, in 1914, the First World War ended the mass migration that had seen more than 40 million Europeans sail across the Atlantic.[9]

The sheer scale of migration in previous centuries is difficult to fathom in the context of COVID-19 and today's rising anti-immigrant

Foreign born share of US population: 1820 to 2015
The figure tracks foreign-born migrant arrivals for each year as a share of the US population, from 1820 to 2015. It shows that immigration peaked at around 1.6 per cent in the middle of the 19th century and that it has subsequently fallen significantly, stabilising at around 0.3 per cent over the past decade.

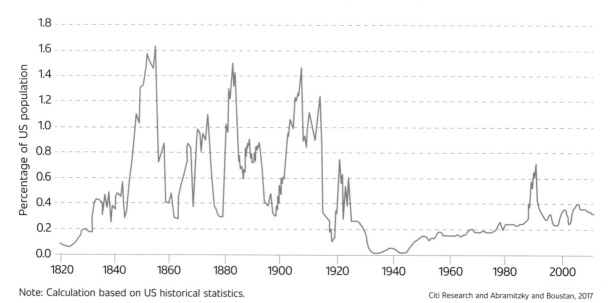

Note: Calculation based on US historical statistics.

Citi Research and Abramitzky and Boustan, 2017

sentiment and closed borders. During the second half of the nineteenth and early twentieth centuries, for example, about one in five Europeans migrated. In parts of Ireland, Italy and Scandinavia over a third of the population moved away from where they were born.[10] The first waves in the period 1800 to 1850 were mainly British and German workers who were economically dislocated by the industrial revolution.[11] They were quickly outnumbered by Irish, Italian, Spanish, Scandinavian and Eastern Europeans. In fact, European migrants accounted for at least a third of the population of North America and Australia, and half of Argentina's. In the UK, the share of migrants, including those escaping from famines in Ireland and pogroms in Eastern Europe, was also significantly higher than it is today.[12]

On balance, a significant share of migrants returned to their countries of origin once conditions there improved. In the nineteenth century, as today, the flow of migrants was not one way: about half of them eventually returned home, although the proportion that returned varied greatly by country of origin.[13] Although roughly half of the Italian and Spanish migrants went home at some point, less than 5 per cent of the Russian migrants ever returned.[14] It was not until the First World War and rising nationalism that identity documents and associated border controls were widely introduced and enforced. Previously, controls were imposed selectively, mainly to discriminate against specific ethnic groups based on their appearance or origin. For example, in 1882 the US authorities introduced regulations – the Chinese Exclusion Act – to keep out Chinese migrants and make them ineligible for naturalisation.[15]

The strict immigration and customs controls that most of us treat as normal today are a recent invention. More than 130 new nations have come into being since passports were first introduced in the early twentieth century, leading, inevitably, to many more borders and checkpoints.[16] These borders are increasingly visible and inviolate. The latest phase of globalisation did not usher in a new era of free movement of people as many expected. On the contrary, travel restrictions have grown in many parts of the world. Self-determination of the former Soviet Union republics led to the creation of fifteen new countries, each with its own immigration regime. With each new state comes new border controls, so that people who previously moved freely, now must contend with visas, immigration checks, and much more.[17]

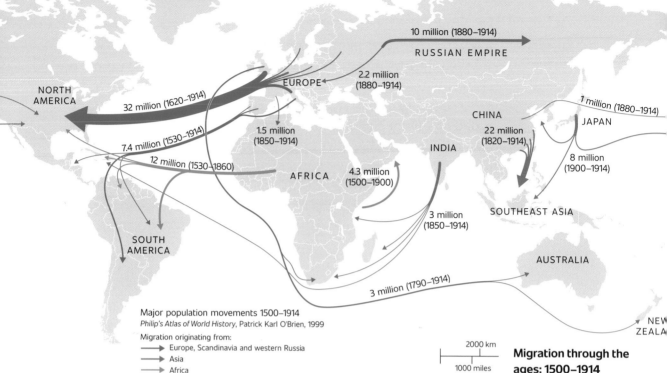

10 million (1880–1914)

RUSSIAN EMPIRE

2.2 million
(1880–1914)

NORTH
AMERICA

EUROPE

32 million (1620–1914)

CHINA

1 million (1880–1914)

JAPAN

7.4 million (1530–1914)

1.5 million
(1850–1914)

22 million
(1820–1914)

12 million (1530–1860)

INDIA

8 million
(1900–1914)

AFRICA

4.3 million
(1500–1900)

SOUTHEAST ASIA

SOUTH
AMERICA

3 million
(1850–1914)

AUSTRALIA

3 million (1790–1914)

Major population movements 1500–1914
Philip's Atlas of World History, Patrick Karl O'Brien, 1999

Migration originating from:
→ Europe, Scandinavia and western Russia
→ Asia
→ Africa

2000 km

1000 miles

NEW
ZEALA

Migration through the ages: 1500–1914
Tens of millions of Europeans left for the Americas between 1815 and 1914. The pace of migration accelerated with the arrival of steam-powered travel. European settlers also settled in Southern Africa, Australia and New Zealand. Meanwhile, millions of Chinese and Japanese migrated in search of work, many of them to Southeast Asia but also the west coast of North America. The slave trade also resulted in the massive involuntary relocation of Africans to the Americas, as well as to the Middle East.

Notwithstanding many political and administrative restraints on movement, the vectors of migration are more abundant than ever. Before the outbreak of the coronavirus in late 2019, the cost of travel had plummeted, and the availability of transportation options had dramatically expanded, not least with the proliferation of high-speed railways, low-cost airlines and competing airport hubs. Since the global population doubled over the past forty-five years, the number of migrants also rose significantly in absolute terms, to about 272 million in 2019.[18] Yet despite more borders, rising population, improved transport options and better information about destinations, as a share of the world's population, migration has remained remarkably stable.

Following the advent of steamships from around 1850 until 1914 when the First World War disrupted shipping, roughly 3 per cent of the world's population were migrants.[19] Immediately following the war, the devastating Spanish flu pandemic of 1918 added to the desire to control migration, with strict border controls becoming the norm in many countries. As the era of relatively open borders ended, a period of increased control was ushered in, establishing a new normal in which the share of people migrating hovered around 2 per cent.[20] The Great Depression and resurgent nationalism, which preceded the

Second World War, reinforced this, as did the subsequent Cold War in which national borders became ever more fortified. The end of the Cold War in 1990 and establishment of a more open global economy, together with the integration of Europe and opening up of China, has seen a return to close to the levels of migration that characterised the second half of the nineteenth century.

Migration has not exploded in the way many expected. Despite a ten-fold increase in average global incomes since 1970, combined with much cheaper air, sea and terrestrial travel and greater opportunities to move (especially from Eastern Europe and China, which until recently allowed international travel only under exceptional circumstances), the world has not experienced dramatically higher levels of migration than in the past.[21] The figure shows how, despite the absolute expansion in the number of people on the move – shown on the left hand scale – as a share of the global population that is migrating, it is not different to the nineteenth-century – hovering around 3 per cent. COVID-19 has dramatically reduced migration, but this disruption will not change the long-term trends.

Globally, one in thirty people are migrants today. Examining the hard facts rather than hyper-ventilating headlines is critical. It is certainly the case that tighter controls are keeping migrants, asylum seekers and refugees from moving. But consider the European countries that signed the Schengen Agreement: there are no border

Migrant numbers have increased, but the proportion of migrants in the world has stabilised at a new normal of around 3 per cent

The number of migrants has grown significantly over the period since 1970, as shown in the height of the bars, in millions depicted on the left hand scale. However, as a share of the global population, migration is hovering at around 3 per cent, as shown in the line and percentage indicated on the right hand scale.

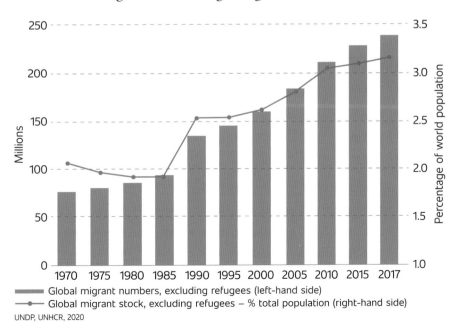

Global migrant numbers, excluding refugees (left-hand side)
Global migrant stock, excluding refugees – % total population (right-hand side)
UNDP, UNHCR, 2020

controls, yet most people have not moved despite very different levels of employment and incomes. Even when these differences widened further following the 2008 financial crisis, virtually no significant increase in migration within Europe was detected. The simple point is that most people prefer to remain at home whenever possible.[22] Migrants are 'exceptional people' in many ways.

Who are the migrants?

The sheer scale and impact of migration around the world is invisible to most people. In the US there are about 51 million migrants, constituting about 15 per cent of the country's entire population. In Saudi Arabia there are proportionately double that number, some 13 million migrants who make up over a third of the population.[23] In other Gulf countries, such as the United Arab Emirates, migrants made up as much as 88 per cent of the population in 2019. Smaller successful countries, such as Singapore, are almost 40 per cent immigrant, as is Luxembourg, which is 47 per cent foreign-born, and Switzerland, where 30 per cent of the residents are immigrants. The world's most populous countries, notably China and India, have tiny immigrant populations – about 0.07 and 0.4 per cent respectively.

International students account for a significant share of international migration. More than 5 million of them are enrolled in degrees or diplomas in foreign universities, and before the arrival of COVID-19 their numbers were growing by about 10 per cent a year. While an increasing amount of students from all countries are seeking opportunities abroad, most of them herald from the two Asian giants, China and India, as is discussed in our chapter on education. In the Anglophone world, the US notably has over a million international university students, and the UK half a million; these countries have historically recruited the greatest number of foreign students. However, Australia and Canada are no slouches, both of which now have over three-quarters of a million foreign students and have overtaken the UK.

Globally, just over 60 per cent of all migrants live in Asia (about 83 million) and Europe (around 82 million). The US hosts more migrants than any other single country which helps explain why the debate over immigration is so prolific there.[24] The United Nations projection of 27 million migrants in Africa significantly underestimates so-

Migrants Coming and Going – 2016–2019

The blue circles show countries which are receiving more migrants than are leaving, with the top net immigration countries including the US, Turkey, Germany and Saudi Arabia. The red circles show countries where more people are leaving than arriving. The main net emigration countries were Syria, Venezuela and India, The size of the circles represents the scale of net inflows or outflows of migrants over the period 2016–2019.

called 'undocumented' migration between the continent's fifty-four countries. The combination of porous borders and weak registration systems mean that the statistics on African migration are deeply flawed.[25] Even so, there is evidence that migration flows between developing countries are growing more rapidly than between poorer and wealthier countries. The red circles in this map show that between 2016–2019 Syria suffered a net loss of 2 million migrants (on top of the 6 million more who left in the previous five year period), and Venezuela had 1.9 million net emigrants, virtually all of whom were refugees. Over the same period, 1.7 million more migrants left India than entered it, mostly heading to work in the Gulf. Germany was the largest net recipient of migrants, with the blue circle showing that 2.6 million more came than left (about half of whom were refugees, with the other half entering for work), followed by Saudi Arabia (receiving 2.2 million), and the US (2.1 million net immigrants). Turkey received 1.25 million refugees, on top of 2.5 million refugees it received in the previous five year period.

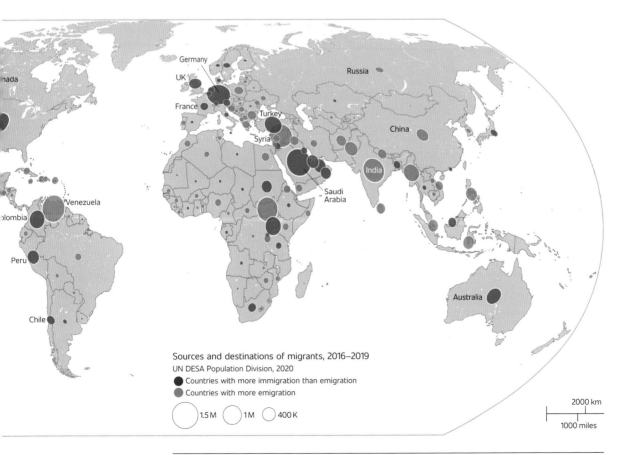

Sources and destinations of migrants, 2016–2019
UN DESA Population Division, 2020
● Countries with more immigration than emigration
● Countries with more emigration

○ 1.5 M ○ 1 M ○ 400 K

2000 km
1000 miles

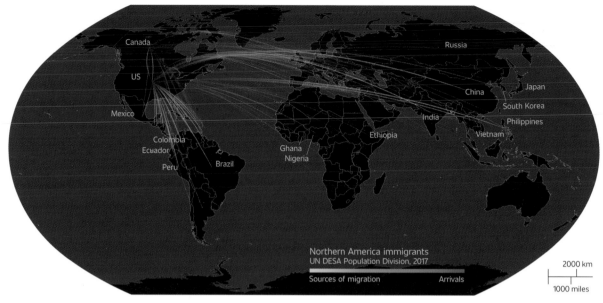

North America is a migration magnet – 2017

This series of maps highlight the dynamic flows of people between regions in 2017. The first map reveals the global attractiveness of Canada and the US to migrants from across the Americas, Europe, Asia and Africa, with the purple lines depicting arrivals and yellow showing the sources of migration.

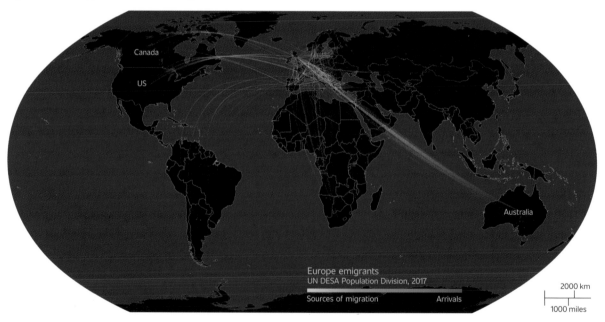

Europeans prefer Europe – 2017

The map reflects migration to Europe and within; it captures the density of the intertwined network of flows in 2017 between EU countries with Germany at its heart.[26] Europe is the second most popular migration destination after the US, with about 4 million people entering and about 2 million people leaving the region in recent years.[27]

Asian triangle – 2017

Patterns of intra-continental migration are also apparent in Asia in 2017. The extent to which the richer countries, such as Japan, Singapore and South Korea, draw in low-wage workers from surrounding countries is evident as is migration between neighbouring Philippines and Indonesia. There is also significant migration from Asian countries such as China, Japan, Philippines and South Korea to North America, and to Australia.

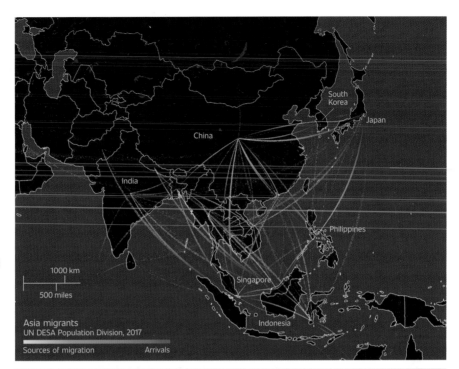

Asia migrants
UN DESA Population Division, 2017

Sources of migration   Arrivals

Long-distance African migration – 2017

The dense migration between countries like Burkina Faso, Guinea, Mali and Niger in West Africa is evident in this representation of 2017 migration patterns within the continent, with this in part being accounted for by nomadic pastoralists. Long-distance migration to work in South Africa follows patterns established over 100 years ago when repressive migratory labour systems were developed for the South African gold and diamond mines.[28]

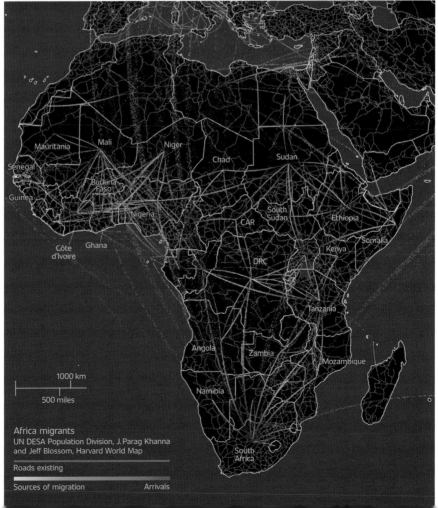

Africa migrants
UN DESA Population Division, J. Parag Khanna and Jeff Blossom, Harvard World Map

Roads existing

Sources of migration   Arrivals

Migrants fuel economic growth

Immigration – the arrival of migrants in a society – is a powerful driver of economic prosperity in most countries and cities. This type of migration generates greater prosperity on an aggregate, per capita and per worker basis. Among wealthy countries, migrants now make up between 10 and 30 per cent of the working population, in comparison to 5 per cent in 1960. Globally migrants now constitute about 3.3 per cent of the world's population, so their share in workforces is much higher than their overall share of the population.[29] Since 2000, the total number of migrants in rich countries increased by 20 per cent, with this mainly accounted for by the 70 per cent spike in high-skilled immigrants.[30] Migrants typically travel to countries to work and most of them eventually leave when their work terminates. For this reason, they constitute a much smaller share of the school age or elderly population than they do of the working population.[31]

The worsening of the geopolitical, environmental and pandemic outlook described elsewhere in this book has negative implications for migration. It also makes understanding the drivers, growth and impact of migration more pressing than ever. In many countries, migrants are blamed for 'crowding out' public services and 'draining' the public purse. But the overwhelming evidence shows the reverse is true: migration typically increases the incomes of host populations. In fact, if immigration had been frozen in 1990, the German and UK economies would have been around €155 billion (£132 or $160 billion) and £175 billion ($212 billion) smaller respectively in 2014.[32] In the US, without migration the country would have experienced a much weaker recovery following the financial crisis.[33]

One reason for the productivity of migrants is that they are typically of working age. As the figure shows, they are usually younger and work more than native populations, with far more twenty-five to forty-five year olds represented among migrants than in the local population distribution. Since fewer migrants are non-working dependants, as a group they tend to have a strong positive impact on average standards of living in most places where they reside. This also underpins their significantly greater contribution to taxes, as they contribute more and rely less on government spending, such as schooling, pensions, health and elderly care.[34]

In most countries the lifetime net financial contribution of

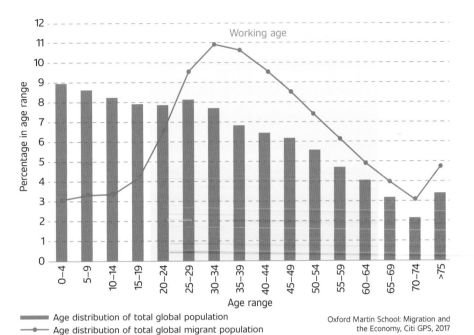

Most migrants are working age – 2017.

Age distribution of total global population
Age distribution of total global migrant population

Oxford Martin School: Migration and the Economy, Citi GPS, 2017

admitting an additional migrant under the age of forty is positive.[35] Migration tends to be more beneficial when governments, businesses and societies as a whole adopt a more comprehensive approach to absorbing new arrivals, especially in terms of integration and rights. In the short-term in the US and Europe, for example, the direct cost versus benefit of migrants in terms of taxes and expenditures is estimated at close to neutral year-on-year.[36] In the longer term, however, the positive impacts are much stronger.[37] In France, if net migration fell by half, government spending would need to rise by at least 2 per cent to compensate for their lost contributions.[38] More dramatic negative consequences are foreseen in the UK, where cuts in immigration would deteriorate public finances.[39] And in the US, the rising burden on taxpayers associated with the ageing of baby boomers could be offset by a net annual inflow of 1.6 million immigrants.[40] While more migrants would be economically beneficial to most countries, large rapid increases are nevertheless seen as politically unpalatable.

Migrants generally work longer shifts than natives.[41] They also indirectly increase the rate of employment of the native population.[42] For one, migrants exert a downward pressure on the costs of child and other care services.[43] In Europe, for example, almost 23 per cent of the population are unable to work because they are looking after other people.[44] This disproportionately discriminates against women, who are more likely to have caring jobs. In Ireland, whereas more than 55 per cent of women say they are unable to work because of their caring responsibilities, less than 10 per cent of men say this is a constraining factor.[45] In the UK about 40 per cent of women who would like to work say their caring responsibilities prevent them doing so, compared to under 5 per cent of men.[46] Migrants who can assume caring jobs at affordable prices encourage the entry of women into the labour force.[47] As childcare and other domestic services become more affordable, female participation increases.[48] So too do fertility rates, as more women are able to have children and return to work.[49]

Cutting back on migrants who engage in domestic and care-related work profoundly reduces female work opportunities.[50] The impact of migration on raising female labour force participation is most notable among highly skilled workers, and this increases the economic impact further still.[51] Not surprisingly, the UK's National Health Service (NHS) has warned that to the extent that Brexit is associated with lower migration, a shortage of care workers could force British workers (and disproportionately women) to quit their jobs in order to care for children and other dependants.[52] Lower migration has already raised the costs of the UK's health service and led to longer NHS waiting times, owing to restrictions on the recruitment of foreign nurses and doctors, as well as cleaning, kitchen and other staff who are vital to providing health care.[53] The shortages of doctors and nurses and reliance of the NHS on foreign staff became brutally apparent during COVID-19.

Inventive migrants

The transformative long-term impact of migration on local and national economic productivity is widely acknowledged by university scholars and think-tank analysts. But there is a real danger that by focusing on the immediate costs, politicians cut migration to meet short-term political objectives which have severely negative social and economic implications over the long run. Innovation and entrepreneurship are the driving forces of a dynamic economy. Two reliable ways to inject dynamism and innovation into an economy are to increase the number of highly educated workers and to introduce diversity into the workplace. Immigration provides both.

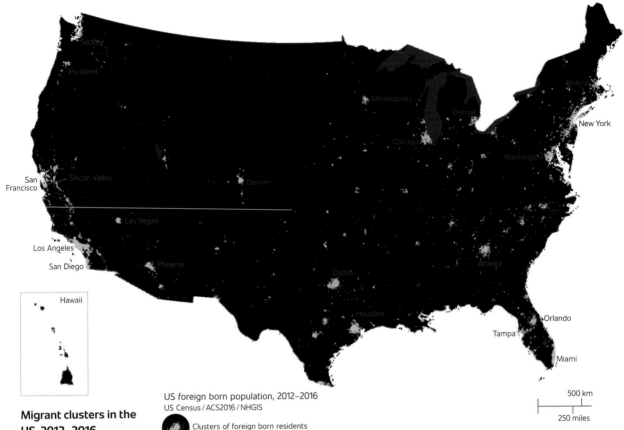

US foreign born population, 2012–2016
US Census / ACS2016 / NHGIS

500 km
250 miles

Clusters of foreign born residents

Migrant clusters in the US, 2012–2016

The blue luminescence on this map identifies areas in the US where there is a concentration of migrants, measured as the number of foreign-born people over the period 2012 to 2016. Not surprisingly, these are in dynamic metropolitan areas as migrants are drawn to job-creating centres and in turn contribute to greater wealth and create more jobs. In Silicon Valley, two-thirds of the engineers are foreign born. [54]

Migrants are more inclined to be entrepreneurs and start businesses precisely because they are willing to take more risks and have no career paths guaranteed for them in their host country. In the US, immigrants are three times more likely than natives to file patents and file around 40 per cent of global patent applications.[55] Migrants account for most of the patents filed by leading science firms: 72 per cent of the total at Qualcomm, 65 per cent at Merck, 64 per cent at General Electric and 60 per cent at Cisco.[56] In Silicon Valley, half of all venture capital-backed firms and 30 per cent of the firms taken public have at least one immigrant founder.[57] Over half of US start-ups valued at $1 billion or more that have yet to go public – the so-called unicorns with potential for high growth – are led by immigrants.[58]

Not surprisingly, the most dynamic, productive and profitable industries and geographical regions in the global economy feature high concentrations of migrants. Immigrants constitute more than three times as many Nobel laureates, National Academy of Science members, and Oscar-winning film directors than would be expected from the migrant share of the population.[59] They have won a third of all Fields Medals in mathematics. A McKinsey study of publicly listed companies found that immigrants are three times as likely as natives to start highly successful businesses.[60] Reflecting this, 40 per cent of all Fortune 500 companies were founded by first- or second-generation immigrants and they are founders of some of the planet's most recognisable companies, including Google, Intel, PayPal, eBay, Yahoo and Tesla.[61]

Harvard researchers William Kerr and William Lincoln make a direct connection between US immigration policy that is open to skilled workers and information technology innovation of the past thirty years.[62] They find that higher rates of temporary high-skilled admissions (through higher levels of H-1B visas) 'substantially increased' rates of inventiveness (measured as the number of registered patents).[63] Importantly, increased numbers of skilled migrants not only increased the contribution to innovation through their own work, but also by collaborating with natives and enhancing the dynamism of the ecosystem, they raised the contributions of non-migrants.[64] The positive contribution of migrants to the arts and sciences is not confined to the US. In the UK, one-third of all Booker Prize-winning authors have been migrants.[65] As Robert Winder noted in his book *Bloody Foreigners: The Story of Immigration to Britain*, immigrants have contributed to successive waves of innovation in politics, finance, industry and medicine.[66] Robert Guest, in his book, *Borderless Economics*, highlights the fact that migrants play a dynamic role in many countries, including China.[67]

Migrants are the source of innovation for several reasons. For one, they are more drawn to innovative sectors than is the case for the population as a whole.[68] Migrants tend to cluster in the most innovative cities, and are disproportionately represented in areas where there is a rapidly growing demand for skills, rather than in the stagnating areas of the economy.[69] What is more, migrants help fill acute skill shortages and drive productivity at a faster rate. In 2015, immigrants accounted for 45 per cent of the US workforce with a science or engineering doctoral degree.[70] The higher the level of skill required, the higher the contribution of migrants in science and engineering occupations.[71] In computer and mathematical sciences, 60 per cent of US workers are foreign-born and in engineering the immigrant share is around 55 per cent nationwide.[72]

The disconnected politics and economics of migration

If migration is so tremendously beneficial, why do so many people oppose it? In recent elections in the US, Europe, and South Asia, nationalist candidates have gained votes on the back of strongly anti-immigration platforms. For the first time in decades, the political viability of migration is under pressure. The simple answer is that anti-immigrant sentiment is heightened – especially during elections – when populist politicians stir up nativist sentiments and when voters feel anxious and afraid. The success of politicians from Austria and Italy to India and Russia in using anti-migrant rhetoric to propel themselves into power has also created a powerful mobilising narrative. This has led to a race to the bottom among politicians who try to out-do one another in showing how tough they are on migrants and refugees.

Since the 2008 financial crisis and especially the COVID-19 pandemic in 2020, growing spatial inequality has become more visible, especially as dynamic cities outperform stagnant towns. The widening income and employment gap between urban and rural areas is compounded by the concentration of new technology firms and high pay in (mostly) liberal and cosmopolitan cities. The fact that these areas are also home to the 'elites' who are blamed for the multiple crises has pushed some voters towards political parties offering to disrupt 'politics as usual'. This is not to say people are not feeling tremendous pain: austerity and reduced infrastructure spending is diminishing the quantity and quality of transport, schooling, health and other services. These trends worsened dramatically in the wake of the COVID-19 pandemic. This has provided disgruntled politicians and locals with an opportunity to blame migrants for overcrowding and queues.

Resistance to migration intensifies when scarcity, feelings of anxiety, and reactionary nationalism collide. This is reflected in the rallying cries of traditional anti-migrant parties, such as the French National Front's 1978 slogan, 'Two Million Unemployed is Two Million Immigrants Too Many!'[73] If resources and services appear to be less available, as is currently the case, hostility to new arrivals rises. In this environment even anecdotal stories about immigrants overburdening services are likely to engender opposition. As the chapters geopolitics and culture explain, support for political parties that promote conservative values has grown substantially since 2010.[74] Social media has enabled and empowered coalitions of loose constituencies to be built around ultra-right principles, ideas and misinformation. As a result, resistance to migration has become a basis for building alliances that are rattling many democracies around the world.

Certain countries, notably Greece and Spain, which suffered acutely after the 2008 financial crisis, and Italy which also was hit particularly hard by COVID-19, have been comparatively tolerant of migrants. Others, such as Hungary and Poland, where the financial crisis actually had a less dramatic impact and unemployment has come down, have seen rising anti-immigrant sentiment. Meanwhile, the much poorer countries neighbouring Syria have seen unprecedented arrivals of migrants and refugees, with some of these countries seeking to integrate the new arrivals. Clearly, politicians can exercise a powerful leadership role in reducing or inflaming anti-migrant sentiments. Radical parties in Italy have used migration to rally voters, as they have done in France, Germany, the Netherlands and UK, whereas in Greece and Spain anti-migration sentiment has not been similarly weaponised for political advantage.[75]

Campaigns against immigration are arising from changes in party politics, rather than broader shifts in social attitudes. The scapegoating of migrants reflects changes in how political parties are competing as much as anything else. Emphasising the net positive economic factors associated with migration could help offset spurious political narratives, as well as put migration policy itself on a more sustainable footing. Evidence of the aggregate gains from migration are not, however, enough to improve attitudes to migrants. A more concerted focus on sharing the benefits of growth is essential. Countries as a whole benefit from migration, but specific communities and groups – especially rural ones – tend to suffer short-term costs. Governments should ensure that communities that bear the costs of migration get more support, including by relieving the pressure on public services.

The public like migrants more than politicians do

Around the world, the public generally believes their country has more migrants than there actually are. Right-leaning politicians and news outlets frequently complain of being 'flooded' or 'inundated' with migrants in the hope of provoking anti-immigrant backlash.

Politics of migration: numbers over-estimated – 2017

The figure shows the disconnect between how many migrants national populations believe reside in their societies and how many actually do, based on 2017 surveys. In the US, respondents think there are three times as many migrants as there actually are. Even in liberal Sweden, which prides itself on receiving migrants and welcoming refugees, perceptions vary significantly from reality.[77]

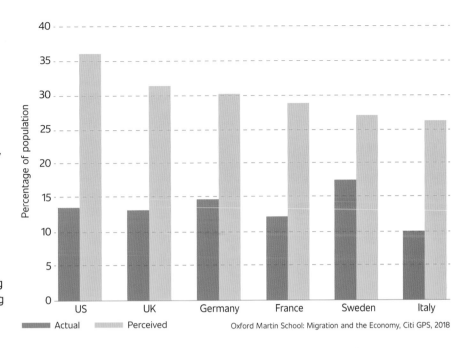

Oxford Martin School: Migration and the Economy, Citi GPS, 2018

This strategy seems to work. The perceived benefits received by migrants (in comparison to natives) are overestimated, reflecting an underestimation of their skills, incomes and employment rates.[76]

The level of acceptance of migrants varies from place to place. For example, the proportion of people who support expanding immigration is low in many European countries. Two of the countries with the lowest fertility rates in the world – Poland (roughly 1.4 births per woman) and Hungary (1.5) – are among the most opposed to migration. By contrast, other low-fertility countries such as Spain (1.2) and Germany (1.5) are more welcoming.[78] In Germany this may be due to the government taking practical steps to correct public perceptions by publicising the benefits that migrants bring and seeking to reduce fears. In many countries, however, the gulf between the evidence that migrants are economically beneficial and the polemic that they are a drain on society is widening. Surveys show that respondents drastically overestimate the proportion of migrants that are the least educated or lowest earners by a factor of three to four.[79] There is also a common misperception that migrants are more likely to be unemployed than locals[80] – with estimates as much as four times higher than the actual rate.[81] As the debate over migration heats up, economic reasoning has given way to political expediency and public negativity. Breaking this vicious circle is more important than ever.

The plight of the forcibly displaced

Although migrants clearly face considerable hardships, few population groups are more vulnerable than refugees and displaced people. They are also growing in number: one person is forcibly displaced from their country every two seconds. Refugees are not the same as migrants. The maps highlight the sheer scale of refugee flows in 2012, when more than 5 million Syrians moved into Jordan, Lebanon and Turkey, and 2015, when at least 1 million Syrians sought refuge principally in Germany, Sweden and other countries in Europe and North America. Their individual stories of struggle and survival are impossible to describe in a two-dimensional map. Indeed, each dot represents seventeen people and, as can be seen, there are tremendous flows between countries in Africa, the Middle East, Central Asia and parts of Europe. There are noticeably fewer refugees moving towards the US.

The world is experiencing an unprecedented surge of refugees. In 2005, there were just 8.4 million refugees worldwide, the lowest number since 1980. According to the United Nations High Commissioner for Refugees, in 2020 there were 26 million refugees, about half of whom were women or under the age of eighteen.[82] There are also another 3.5 million asylum-seekers who have not received refugee status, and more than 41 million more people who are internally displaced within their own borders and do not receive the same international legal protections as refugees.[83] To put these numbers in perspective, more than 37,000 people flee their homes every day because of conflict, violence and persecution. There are more refugees and displaced people at the time we wrote this book than at any other period since the end of the Second World War.

Countries unevenly share the burden of hosting refugees and displaced people. Well over 80 per cent of all refugees and asylum-seekers live in countries next door to their country of origin, and not in North America, Europe or Australia.[84] Consequently, the burden of caring for fleeing refugees is especially onerous for middle- and lower-income countries such as Turkey (with 3.7 million refugees), Pakistan (1.4 million) and Uganda (1.2 million) that are home to families fleeing conflicts in Syria, Afghanistan, South Sudan and the Democratic Republic of the Congo. In many cases, refugees may live in 'protracted situations' – for five or more years in a camp – with few solutions in

2012

US

MEXICO

GUATEMALA
EL SALVADOR
HONDURAS
COLOMBIA

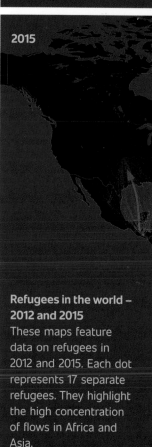

2015

Refugees in the world – 2012 and 2015
These maps feature data on refugees in 2012 and 2015. Each dot represents 17 separate refugees. They highlight the high concentration of flows in Africa and Asia.

Global refugee flows
UNHCR, 2016
Each dot represents 17 refugees

Departures   Arrivals

2000 km
1000 miles

SWEDEN
GERMANY
TURKEY
LEBANON
SYRIA
JORDAN
AFGHANISTAN
HAITI
MALI
PAKISTAN
SOUTH SUDAN
DRC
UGANDA

Global refugee flows
UNHCR, 2016
Each dot represents 17 refugees

Departures   Arrivals

Large refugee flows:
Syria to Europe – Germany, Sweden
Central America to US
Eastern Europe to Russia

2000 km
1000 miles

Syria, 2012

TURKEY

SYRIA

LEBANON

IRAQ

Syrian refugee flows
UNHCR, 2016
Each dot represents 17 refugees

Departures   Arrivals

200 km

100 miles

ISRAEL

JORDAN

**Refugees in the world –
2012 and 2015**

It is possible to see the
impact of Syrian refugee
flows, particularly in 2015
when large numbers
fled across borders to
neighbouring countries
and to Western Europe.

Syria, 2015

SWEDEN

DENMARK

NETHERLANDS

UK

GERMANY

BELGIUM

FRANCE

ITALY

Corsica

Sardinia

GREECE

TURKEY

SYRIA

LEBANON

IRAQ

JORDAN

EGYPT

500 km

250 miles

Syrian refugee flows
UNHCR, 2016
Each dot represents 17 refugees

Departures   Arrivals

1 km

0.5 miles

© KARI / ESA, 2013

Zaatari refugee camp from space – 2013[85]

The Zaatari refugee camp was established in 2012 and is still the largest camp for Syrian refugees in the world. It is home to more than 80,000 people, with tents having largely been replaced by semi-permanent structures.[86]

sight. Well over 13 million refugees are warehoused in such settlements, over half of whom have languished in poverty for decades. At least 5.4 million of them are Syrians living in informal settlements and camps or with friends and family in Egypt, Iraq, Lebanon, Turkey and Jordan.

Take the case of the Syrians fleeing civil war, including those who ended up in the Zaatari refugee camp, Jordan's largest, established in 2012. Located just a few miles from the Syrian border, the land is dusty, arid and unforgiving. The population swelled to more than 150,000 at its peak, before settling at around 80,000 today, making it one of Jordan's largest cities. While housing some of the world's most destitute, the camp also features tremendous innovation. For example, local authorities installed a large solar energy plant that, in addition to providing refugees with between twelve and fourteen hours of electricity a day (a luxury when compared to other refugee camps around the world), also reduces CO_2 emissions by 13,000 tonnes a year, the equivalent of 30,000 barrels of oil. Even cash-for-food programmes administered by international relief organisations now run on blockchain technologies. What began as a temporary refuge is fast becoming a more permanent settlement.

Discrediting anti-refugee myths

Even the best-resourced government authorities can be overwhelmed when large numbers of refugees arrive suddenly, as the recent mass movement of refugees to Middle Eastern and Western European cities shows. The sense of disorder and otherness – together with real and perceived competition over services – can fuel fear and resentment. The outbreak of sexual violence purportedly involving men of 'Arab and North African descent' during the 2015–16 New Year's celebrations in the German cities of Cologne, Dortmund and Hamburg are a case in point. Such tensions cannot be glossed over – indeed, they may be exploited by nationalist and reactionary politicians and parties, further undermining national refugee and migration policies.[87]

There are many reasons to create a welcoming environment for new arrivals. Refugees and asylum-seekers, as in the case of migrants, are generally net positive contributors to the societies where they end up living. Refugee communities – and mixed-migrant communities more generally – are typically less prone to crime than the average host community.[88] Refugee networks tend to exert greater social controls and self-restraint precisely to avoid contravening local laws and customs.[89] In fact, crime declined in nine of the ten US cities accepting the largest number of refugees between 2006 and 2015 – in some instances, dramatically so.[90] Despite the periodic (and widely reported) incidents, Syrian and Iraqi refugees reportedly commit far fewer crimes than other residents precisely because they wish to avoid jeopardising their legal status. That said, some rejected asylum claimants from Algeria, Morocco and Tunisia were slightly more likely to be involved in crime than locals.[91] From the Netherlands to Sweden, there is no hard evidence that refugee centres contribute to rising crime, despite insistent allegations to the contrary.[92]

Another mistaken assumption is that refugees are more likely to be linked to an upsurge in extremist and terrorist violence. Once again, however, this claim does not corroborate with available evidence. For example, in the US, the number of refugees arrested on terrorism-related charges since January 2015 is in low single digits. Moreover, between 1975 and 2015, not a single person in the US was murdered by a refugee from the seven countries (Iran, Libya, North Korea, Somalia, Syria, Venezuela and Yemen) against which the US administration sought to impose travel bans.[93] Put simply, refugees are much more

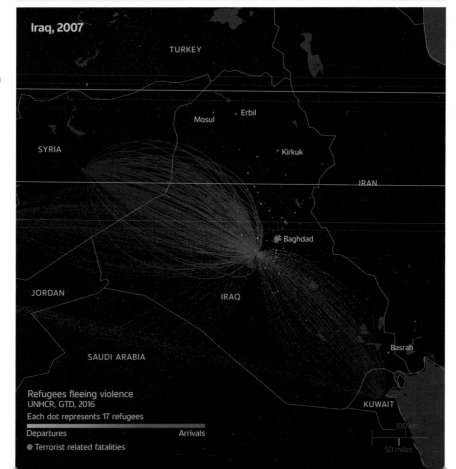

Refugees fleeing violence in Afghanistan and Iraq – 2007

Afghanistan and Iraq experienced explosive violence in the wake of the US-led interventions in 2001 and 2003. These maps show the spread of terrorist-related fatalities and the surge in refugees to neighbouring countries a few years later.

Afghanistan, 2007

Kabul
Jalalabal
AFGHANISTAN
Peshawar
Khost
Kandahar
Quetta

Refugees fleeing violence
UNHCR, GTD, 2016
Each dot represents 17 refugees

Departures   Arrivals
● Terrorist related fatalities

100 km
50 miles

PAKISTAN

Iraq, 2007

TURKEY
Erbil
Mosul
Kirkuk
SYRIA
IRAN
Baghdad
JORDAN
IRAQ
Basrah
SAUDI ARABIA
KUWAIT

Refugees fleeing violence
UNHCR, GTD, 2016
Each dot represents 17 refugees

Departures   Arrivals
● Terrorist related fatalities

100 km
50 miles

likely to be fleeing terrorism than causing it.[94] The maps on the previous page show how Afghan and Iraqi refugee numbers are surging from countries most severely affected by terrorism. The orange dots reveal how many people are fleeing and the red smudges are reported fatalities associated with terrorist violence.

Far from being a threat or burden, if properly planned for, refugees and asylum-seekers contribute to local economies. The short-term negative effects of sudden inflows are often attenuated and frequently reversed. Studies of refugee arrivals in the US revealed no adverse long-term impact on labour markets through paying taxes and entrepreneurial activity.[95] The impacts of Syrian refugees on labour markets in neighbouring countries such as Jordan, Lebanon and Turkey showed few disruptive effects on unemployment rates or labour force participation.[96]

Displacement to cities

The majority of individuals and families on the run are moving to urban areas. This is in stark contrast to the past, when most refugees were 'interned' in camps in rural areas. According to the United Nations, roughly 60 per cent of all refugees and 80 per cent of all internally displaced people are living in cities. By way of comparison, just 30 per cent of all refugees live in planned camps like Zaatari, which are typically managed by governments and international agencies. Cities have provided sanctuary to people fleeing violence for thousands of years. The practice is universal, evident in early Christian, Islamic, Judaic, Buddhist, Sikh and Hindu societies. Throughout history, city leaders have resisted demands from higher authorities – from kings to presidents – to coercively restrict the rights of people seeking sanctuary.[97]

All cities suffer from social and economic divides. These fissures can translate into racism, exclusion and marginalisation. In cities experiencing political turbulence, economic disruption and rising insecurity, refugees and other displaced populations are often targets of recrimination. The scapegoating of forced migrants is hardly restricted to cities, and they may face even more discrimination in rural areas. Complicating matters, many city authorities are struggling to address these challenges (assuming they are acknowledged at all)

Most African refugees are fleeing to neighbouring countries – 2012 and 2014

The vast majority of the world's urban refugees are not relocating to developed cities in North America or Western Europe. Instead, as the map shows, they are moving to neighbouring countries, often to poor and underdeveloped cities and slums in Africa, Asia and the Middle East. Today, Africa hosts more than 18 million refugees, roughly 26 per cent of the global total.

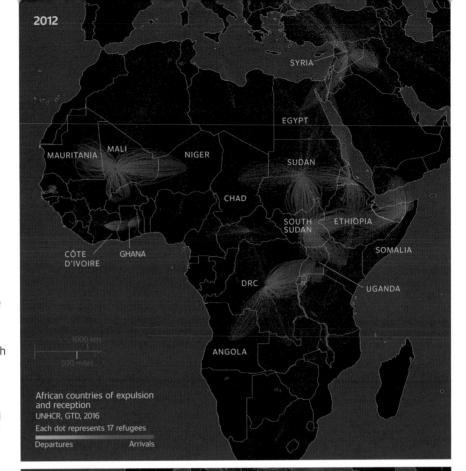

2012

African countries of expulsion and reception
UNHCR, GTD, 2016
Each dot represents 17 refugees

Departures   Arrivals

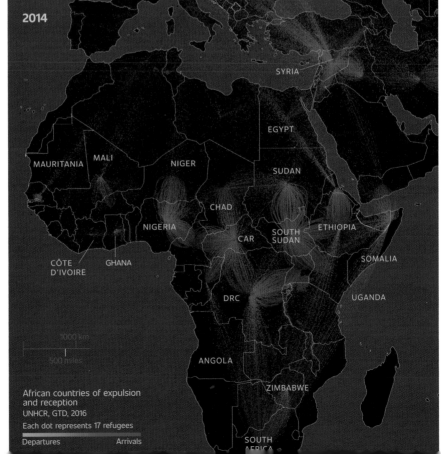

2014

African countries of expulsion and reception
UNHCR, GTD, 2016
Each dot represents 17 refugees

Departures   Arrivals

with inadequate resources. While they often work valiantly to improve the lives of refugees, they frequently lack the necessary autonomy, discretion and capacity to deliver effective services, particularly when confronted with upsurges in newcomers, which further exacerbates local grievances.

As with migrants, the backlash against refugees is global. Anti-immigration and anti-refugee politicians in Australia, Germany, Hungary, Italy, Poland, Serbia, the UK and the US are obsessed with erecting barriers to immigrants and refugees. Refugee admissions to the US are at their lowest levels since the Refugee Act was introduced in 1980. Just 33,000 refugees were accepted in 2017 and President Trump reduced the ceiling to below 18,000 by 2020.[98] And since the arrival of more than 1 million refugees in Europe in 2015, governments across the region have fortified their borders with fences and barbed wire, cut funding for assistance, and ratcheted up deportations. Similar sentiments towards refugees and migrants are on display in African, Latin American, Middle Eastern, and South and East Asian cities that host most of the world's forcibly displaced people. The extrajudicial killings of foreigners in Durban and Johannesburg over the past decade were incited by South African populists, who label immigrants and refugees as criminals. Somali refugees in Nairobi are routinely targeted by police, often at the behest of Kenyan politicians, who accuse them of being a threat to national security. Meanwhile, Bangladeshi border towns have struggled to absorb over 700,000 Rohingya refugees from Myanmar, amid complaints that they have contributed to food shortages, pushing-up prices and undercutting wages.

The United Nations, some member states and growing numbers of cities are looking for new ways to improve responses to refugees rather than simply closing borders. For example, in 2018, 176 countries approved a new Global Compact on Refugees (the US was the only country to oppose it). The deal set out a more robust and fairer response to global refugee movements, especially to ease the pressure on countries hosting large numbers of refugees and finding ways to help refugees become more self-reliant. Likewise, in 2017, the International Organization for Migration, together with the city network, United Cities and Local Governments, assembled 150 cities to sign the Mechelen Declaration, demanding a seat at the decision-making table. In 2015, a network of major European cities that represents the local governments of more than 140 of Europe's largest cities and forty-five urban centres, established Solidarity Cities in response to the influx of Middle Eastern and North African migrants.

Cities are also developing legislative and policy frameworks to welcome refugees and promote protection, care and assistance. In the US, there are more than 100 'welcoming cities' that have committed to developing institutional strategies for inclusion, building leadership among new arrivals and providing support to refugees.[99] Another 500 US jurisdictions describe themselves as 'sanctuary cities' to resist federal efforts to enforce immigration law and are on the front-line of supporting undocumented migrants and refugees – despite the threats of cuts to municipal funding.[100] Meanwhile, at least eighty 'cities of sanctuary' across the UK are committed to welcoming refugees, asylum-seekers and others seeking safety. Cities across Europe are also adopting similar strategies in co-operation with Eurocities, an inter-city network founded in 1986.[101] There are currently more than 300 inter-city networks dedicated to urban priorities ranging from governance and climate change to public safety and migration.[102] Several, including a Mayors Migration Council, have established dedicated guidelines to help cities protect and care for refugees and other new arrivals.[103]

The effective social and economic integration of refugees comes down to smart planning. It's true that newcomers can generate surging demand for housing, health, education and welfare services. But this can be partly mitigated with appropriate preparation and dispersal policies. The reality is that refugees often receive substandard accommodation, faltering social welfare support, limited access to labour rights and uneven social and disability care compared to local residents. These shortfalls were not due to the excessive needs of refugees, but rather the lack of qualified staff and adequate resources on the ground. Ultimately, refugees can have a net positive effect on revenue collection and pension systems in ways that can boost government finances in countries with ageing populations.

Rethinking policies on migration and refugees

Immigration and refugee flows have featured prominently in the media cycles and agendas of political parties in recent years. The conflation of economic migrants and refugees is at the heart of often-confused and counterproductive policy responses. Migration –

including forced displacement – is as old as humanity, with our ability to migrate defining the evolution and success of *Homo sapiens* and our peopling of the planet. Yet there is a profound disconnect between the overwhelming positive economic contribution of migrants and refugees and the negative perceptions and politics that conspire to keep them out. Migrants and refugees are mistakenly accused of overwhelming services, depressing wages and compounding shortages in public services. And, worse still, they are unjustly decried as criminals and terrorists.

Aside from much-needed political leadership, there are a wide range of actions that should be taken to improve migration.[104] The costs of migration tend to be borne at the community level, through increased pressures on schools, housing, health or transport systems. Meanwhile the benefits accrue elsewhere, to companies and to society, through higher profits and taxes, and lower costs and improved services for citizens. The uneven geographical impact points to the need for national and regional governments to pay particular attention to supporting local communities in integrating migrants and providing for the communities' needs. For migrants themselves, proximity to each other brings benefits, but risks creating ghettos, reinforcing segregation and slowing integration. Designing our cities to be more sensitive to the needs of migrants and displaced people, especially in the COVID-19 era, is critical.

Although the international framework governing refugees is in need of renewal, the one established for economic and other migrants is virtually non-existent. The evolution in 2016 of the International Organization for Migration into a fully-fledged United Nations agency has at last created an international legal framework for migration, but this remains narrow in its coverage and has minimal enforcement powers. Countries are unwilling to allow international organisations to limit their choices when it comes to migration policy. As a result, migrants have little in the way of international law that can protect them from abusive practices. Greater clarity is required with regard to their rights to migrate, and to safe passage, and their treatment in their destination countries. The legal limbo extends to many practical areas and is highly detrimental to migrants. Among the areas that require clarity is pension portability to ensure that the benefits they have accrued while working, through pension, national insurance and other contributions, can be claimed when they leave. The extension of political rights and professional qualifications is also subject to different national rules and as a result millions of migrants have no political voice or representation.

More migration would be highly beneficial for the host countries as well as the migrants themselves. To make this possible, both the rights and responsibilities of migrants should be clarified and secured. The rights include many of those enjoyed by citizens, including the full protection and freedoms that come with the rule of law, including those covering employment and human rights. The responsibilities include being documented, paying tax and abiding by the law of the land. This grand bargain, of accepting more migrants with better rights in return for an acceptance by migrants of more responsibilities and stricter controls on undocumented entry and unsafe passage, could help provide for a transition to a more positive virtuous circle in which migrants are seen as an opportunity to be embraced by countries. Migrants are key to the dynamism of countries, cities, companies and societies as a whole. Whereas economic factors should be at the fore in policies on migration, ethical and legal considerations should govern our policies towards refugees. Our maps highlight the evidence on migration and refugees. Our hope is that this provides the basis for greater clarity in perceptions and policies, for the benefit of migrants and refugees, and for us all.

Palm oil mainly comes from two countries, Indonesia and Malaysia, where palm plantations have replaced tropical forests

UN Comtrade, OEC, 2017

Food

Predictions of running out of food are wrong

Overeating causes more deaths than starvation

Unsustainable agriculture causes climate change

Changes to diets, with less meat eating, are required

Introduction

Palm oil is the most widely consumed vegetable oil on the planet. It is used in everything from bread, chocolate and peanut butter to shampoo, cosmetics and cleaning products. The production and trade has soared, particularly to the growing consumer markets of Asia, Africa and Latin America. Its rapid expansion threatens some of the world's most important and sensitive habitats.[1] This map exposes the scope and scale of its trade, particularly from producers like Indonesia and Malaysia. It also shines a light on the re-export trade from Europe and North America by big processing and trading companies based there. In Latin America, Colombia is the largest exporter followed by Ecuador, both of which have plans to expand their plantations at the expense of their precious indigenous forest. The destructive impact of our daily use of palm oil is illustrative of the tensions that need to be resolved in agro-industrial food systems.

The notion that 'you are what you eat' probably came from Jean Anthelme Brillat-Savarin, a French lawyer, politician and famed gastronome, whose book, *Physiologie du Goût* (the physiology of taste), was published in 1825. He famously said 'Dis-moi ce que tu manges, je te dirai ce que tu es,' which translates into 'Tell me what you eat, and I will tell you what you are.' With the passage of time his words seem ever more prescient. Many of us are becoming aware that our diet shapes not only our own individual health but that of our planet. Yet not enough of us are changing our behaviours quickly enough.

The bottom line is that the global food systems that shape our eating habits are unsustainable. The COVID-19 pandemic exposed the risks of both our habits, and the supply chains that distribute our food, in a matter of months. For most of history, humans had too little food. We spent most of what few calories we consumed scavenging for more calories simply to survive. While this is still tragically the case in some parts of the world, in others the problem is that we have too much food. In this chapter we show how feeding a rapidly growing population healthily, while simultaneously reducing the negative impact of food production, processing and consumption on other species and on our climate, is going to require a fundamental transformation in how we grow food and in our diets.

Life expectancy is closely linked to diet. One of the diets commonly associated with good health comes from the Mediterranean. This is

Palm oil – destroying forests for food
UN Comtrade, OEC, 2017
Each dot represents $10,000 traded in palm oil

Origin   Destination

Palm oil – destroying forests for food, 2017
Almost 90 per cent of palm oil comes from two countries, Indonesia and Malaysia, where palm plantations have replaced tropical forests. Each dot on the map shows $10,000 in palm oil traded in 2017, with the white parts of the lines representing the source country and the redder parts showing destinations.

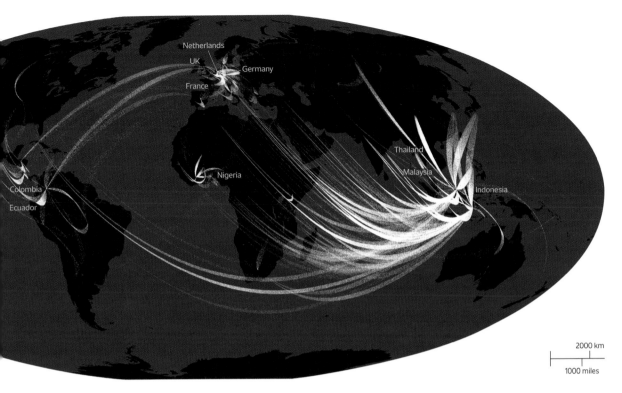

Netherlands
UK
Germany
France
Thailand
Malaysia
Nigeria
Indonesia
Colombia
Ecuador

2000 km
1000 miles

high in fruits, vegetables and (unsaturated) olive oil, and with fish rather than red meat as the primary source of protein.[2] Likewise, the average Japanese person consumes far more fish than red meat and lives longer on average than their US counterparts who eat processed foods that are high in saturated fats. These fatty sugary diets are a major contributor to obesity, diabetes and other chronic diseases. Poor diets are also associated with declining life expectancy in both the US and UK in recent years.[3] While hard to swallow – what you put in your mouth is a decent predictor of how long you'll live.

Throughout history, a key to human success has been our ability to efficiently cultivate land and domesticate animals. This has enabled our species to specialise in tasks other than scrambling for fruits and berries, and, through more predictable food availability and improved nutrition, develop our cognitive and physical capabilities. Yet, agriculture, the very thing that helped us thrive, is now a major threat to our planet. Food production accounts for over a third of all the greenhouse gases that are contributing to climate change.[4] The mass production of food such as palm oil, soy beans, and beef is leading to the extinction of millions of wild mammals, fish, birds and insects.[5] What is more, food production accounts for well over three-quarters of all freshwater use (of which more than half is wasted)[6] and to the degradation of rivers and oceans through waste products and fertiliser run-off.[7]

Food glorious food

The nature of food production has changed dramatically in recent decades. Its expansion and diversification are being propelled by advances in technology and the rising demand from rapidly growing populations and incomes in developing countries. Whereas most countries used to produce the majority of their own food, a growing share of what is on our plates comes from elsewhere. This has allowed many of us to consume a greater variety of foodstuffs throughout the year and at much lower costs. It has also given jobs and incomes to people in far-away places. But it is wreaking havoc with our planet too.

The extent to which globalisation has contributed to changing patterns of food production and consumption is evident in the extraordinary diversity of what we eat and buy. Although major crises like COVID-19 can trigger food shortages, it is no longer a surprise to see asparagus, avocados, bananas, flowers, lamb or strawberries on sale in supermarkets and even small grocers throughout the year. These maps highlight the dramatic increase in food trade around the world between 2000 and 2017. Whereas in 2000 food was mainly traded between the Americas, Europe, Japan and Australia, by 2017 China's intense trade is evident. Deepening supply chains and trade are also apparent in Europe and North America as free trade agreements and reductions in tariffs and other barriers encouraged greater movement of goods and services.

Global food trade
The maps show global food trade in 2000 and 2017 with each dot representing $10 million of trade in food, including live animals. The rapid growth in trade, not least with Asia and within Europe, is striking.

Too little food for some and too much for others

In early 2020, even before COVID-19 greatly increased food insecurity in many countries, over 821 million people faced chronic hunger owing to lack of food, the highest level in a decade. And one in five people around the world died prematurely because of poor diet.[8] Poor nutrition – especially inadequate protein, fats and oils intake – caused more than 151 million children to suffer impaired physical

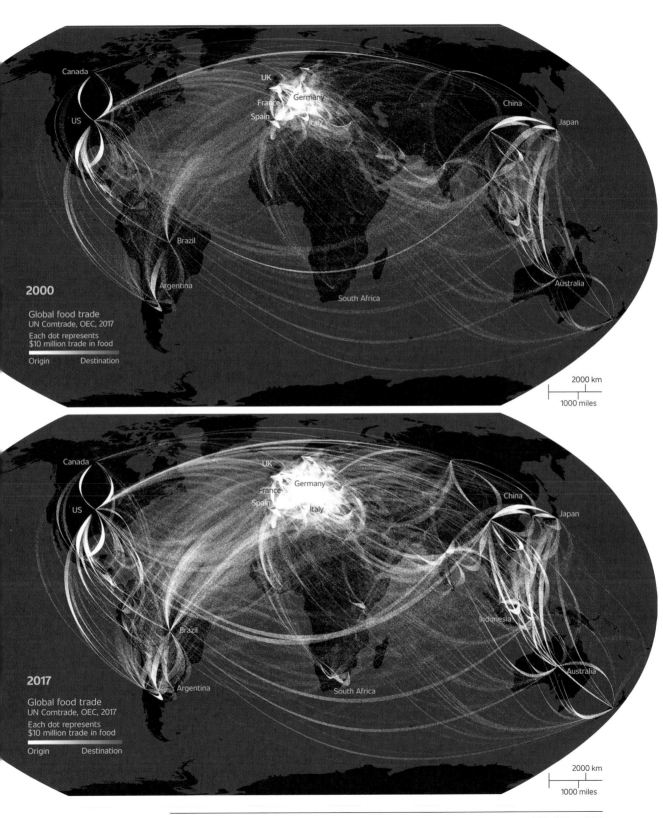

2000

Global food trade
UN Comtrade, OEC, 2017

Each dot represents
$10 million trade in food

Origin   Destination

2000 km
1000 miles

2017

Global food trade
UN Comtrade, OEC, 2017

Each dot represents
$10 million trade in food

Origin   Destination

2000 km
1000 miles

FOOD 331

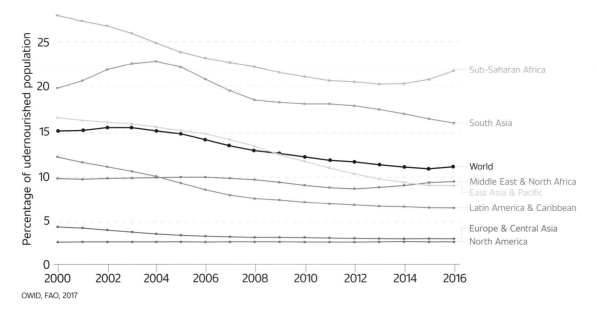

Sub-Saharan Africa

South Asia

World
Middle East & North Africa
East Asia & Pacific

Latin America & Caribbean

Europe & Central Asia
North America

OWID, FAO, 2017

and mental development and an additional 51 million children to be underweight.[9] Remarkably, however, for the first time in human history, more people are dying from eating too much rather than too little food.[10] Overeating is now a bigger problem than starvation.[11] Fixing the twin perils of rising obesity and rising undernourishment requires wide-ranging action from all of us individually, as well as our governments and the food industry.

Even before the devastating impact of COVID-19, at least 2 billion people around the world – more than one in four people on Earth – were micronutrient deficient, with inadequate intakes of key vitamins and minerals, such as iron. This made them more prone to disease, an issue we explore in the chapter on health.[12] More nutritious food is needed by the roughly one-quarter of the world's population that suffers from undernutrition, overnutrition, or malnutrition. And as we discuss in the chapter on demography, there will be at least another 2 billion more mouths to feed before our planet reaches peak population late in the 21st century. The good news is that there are many good examples of how to avoid the unnecessary pain and suffering that arises from eating too little, too much, or even the wrong food.

Starvation, as the Nobel Prize winner Amartya Sen brilliantly exposed in his essay *Poverty and Famines,* is virtually always man-made. It is almost never the result of prolonged droughts, flash floods and failed crops but rather the outcome of structural inequality, exclusion, conflict and the skewed distribution of political power.[13] Notorious

Share of the population that is undernourished: 2000–2016
The figure shows the estimated share of undernourished people in different regions of the world. While the global average hovers just over 10 per cent, in Sub-Saharan Africa it is over double this. The dramatic improvement in South Asia is evident as are the improvements in Latin America and the Caribbean.

famines in which millions of people died, including those in India (1769–92), Ireland (1845–1849), China (1959–61), Bangladesh (1974) and Ethiopia (1984–85), were catastrophic not because there was inadequate food available, but because food was hoarded and not distributed at affordable costs to starving and malnourished people.

Famines are still distressingly common, however, especially in Africa and parts of the Middle East. Between the 1870s and 1970s, famines purportedly killed close to 1 million people annually. Since the 1980s, annual deaths have dropped to well under 100,000.[14] However, COVID-19 threatens to lead to a sharp spike in malnutrition. Africa is particularly vulnerable, with tens of millions of people at risk in the Sahel, including in Niger, Chad, Sudan and Nigeria. A particularly disturbing feature of famines is that they are preventable, in that food can be provided to those in need if the governments and international agencies are determined and have the capacity to do so. Even worse, famines are at times deliberately created or exacerbated by governments in order to undermine communities who are not aligned with those in power, be it for ethnic, political, religious or other reasons. It is also noteworthy that there has never been a famine in a country – even an impoverished one – with a democratic government and free press, although COVID-19 could pose the greatest challenge to this record.[15]

Feeding people who don't have enough food, let alone catering for the nutritional needs of future generations, is a monumental challenge. If we are to reach this goal while reducing our negative impact on the planet it will require that those who eat too much adopt healthier diets and that everyone consumes food which has a much lower carbon and environmental footprint. A green revolution in Africa to improve agricultural yields and the nutritional content of crops will be a necessary part of all this. In Asia, such a revolution in the 1960s and 1970s transformed agricultural production. Within twenty years over 90 per cent of existing wheat fields were planted with high-yielding varieties, as were two-thirds of rice paddy fields. This, together with improved farming methods and the increased use of fertilisers, pesticides and irrigation, led to a doubling of yields per hectare. In China and India today, wheat and cereal yields are approaching levels found in the US and Europe. Meanwhile, African yields have remained stubbornly low and lag far behind other regions', and Africa's challenges were compounded in 2020 by not only COVID-19 but also unprecedented plagues of locusts that have devastated crops.

Obesity a growing problem

Obesity has tripled worldwide since 1975. Today, an astonishing 2.1 billion adults are believed to be overweight or obese for one simple reason: they are eating too much unhealthy food.[16] Among the many consequences of the rising consumption of fatty and processed food is rising mortality and morbidity from diabetes, which has almost doubled in the past thirty years. Other negative outcomes of obesity are a higher incidence of heart disease, stroke, cancer, gout and breathing problems such as sleep apnea.[17] Amazingly, bad diets now pose a greater risk of ill health and death than do unsafe sex, alcohol, drugs and tobacco use combined, and are placing a growing strain on national health systems.[18]

The number of calories any one individual requires differs according to their age, the extent to which they are sedentary or active, and whether or not they are pregnant. Countries with the most obesity in 2016 are indicated in purple on the accompanying map. Over one-third of all residents in the US, Kuwait, Saudi Arabia and a number of Caribbean and Pacific Island states are classified as obese, implying that they eat at least a third too many calories.[19] In these countries the average daily calorific intake is over 3,400, whereas a diet of around

The growing problem of obesity – 2016
The most obese countries in 2016 are indicated in purple below, with over a third of the US, Kuwait, Saudi Arabia and some Caribbean and Pacific Islands obese.[22] The UK is the 'fattest' country in Europe, and in Australia, Argentina and Mexico over a quarter of the population is similarly obese, whereas in Egypt and Algeria over 30 per cent of the population is obese.[23]

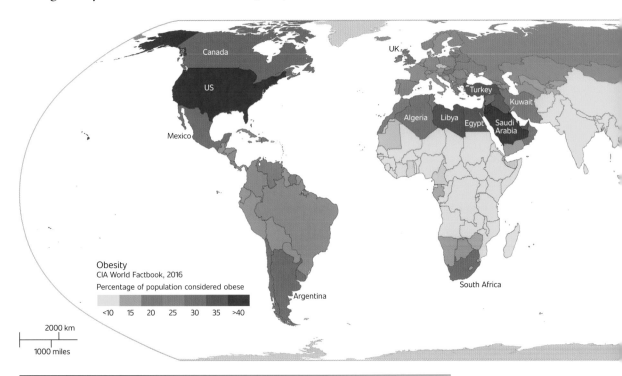

Obesity
CIA World Factbook, 2016
Percentage of population considered obese

<10   15   20   25   30   35   >40

2000 km
1000 miles

2,500 calories for an active 70 kg individual is more than adequate.[20] The UK is the 'fattest' country in Europe, depicted in dark blue-green, with obesity levels rising even more rapidly than in the US, and in Australia, Argentina and Mexico over a quarter of the population is similarly obese, whereas in Egypt and Algeria over 30 per cent of the population is obese.[21]

Agriculture globally: how much more can be grown?

Increases in global food production have more than matched rapidly rising demand.[24] Despite the gloom and doom predicted by the Club of Rome which in 1972 published a report predicting high food prices and famines, a claim which has been repeated regularly since, more recently by the British NGO Oxfam and Stephen Emmott, the world has not run out of food.[25] On the contrary, before the COVID-19 pandemic food prices had fallen. The expansion of food production is due, in large part, to technological innovations such as new crop varieties and improved inputs that have raised yields. It also is due to globalisation which has brought more farmers around the world into global markets.

The map on the next page collates data on crop-lands around the world and illustrates how most land is already being farmed. It shows that Argentina, China, India, Mexico and countries in Europe have devoted more than half their land to agriculture. Thirty years ago, just over a quarter of all land on Earth was used for agricultural purposes; today the proportion is well over one third and rising.[26] What's left is typically tropical forests, mountains, deserts, tundra or otherwise unusable, as is the case in the Arctic circle or Sahara.

The past may, however, be a poor guide to the future of agricultural production, not least as most arable land globally is already in use. To avoid deforestation, more food will need to be grown without more land, implying existing arable land needs to be more productively farmed, yields need to be improved and the use of the land will need to change, in terms of the choice of crops or livestock. Sustainable intensification requires growing more food from the same or less land without harming the environment. Avoiding greater destruction of forests, savannah and vulnerable ecosystems means that new types

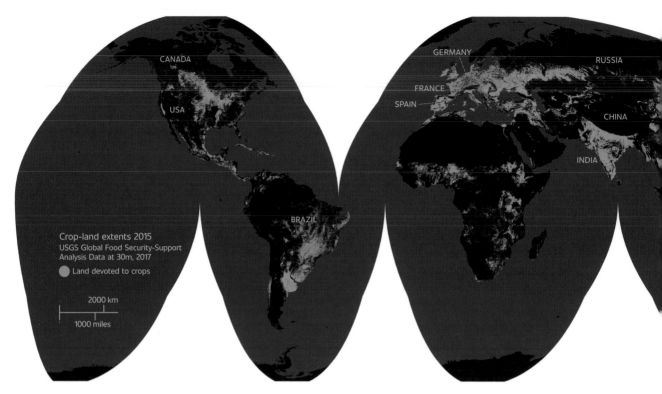

Crop-land globally – 2015

The composite 2015 satellite image shows in green the land devoted to crops globally. It highlights the extent of farming across much of India, Brazil, the great plains of the USA and Canada, the southern coast of Australia and across temperate parts of Europe.

of production and distribution systems will be required for food. In addition to improvements in yields and the nutritional content, greater reliance on hydroponic, drip feeding and other intensive cultivation methods, as well as the removal of the harmful subsidies which encourage over application of chemicals and production in unsuitable soils, is required. Changing food systems requires a wide range of interventions, including technological innovation and the reform of economic, regulatory and trading frameworks, as well as the provision of technical assistance and crop insurance to farmers, not least in poor countries.

The satellite image of crop-land, above, shows where food is grown globally, depicted in green. The total geographic area for food production has remained relatively stable since the mid-twentieth century.[27] However, this trend masks substantial reductions in agricultural land in the temperate regions of Europe, Russia, and North America, which has been offset by the substantial expansion of agricultural land in

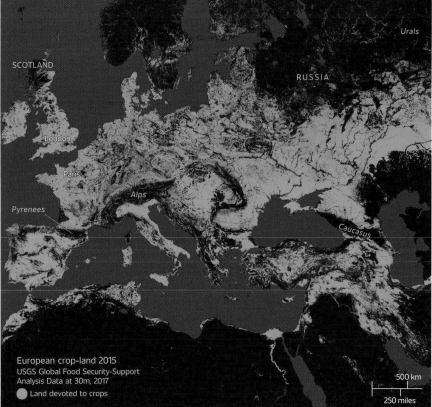

European crop-land 2015
USGS Global Food Security-Support
Analysis Data at 30m, 2017
● Land devoted to crops

500 km
250 miles

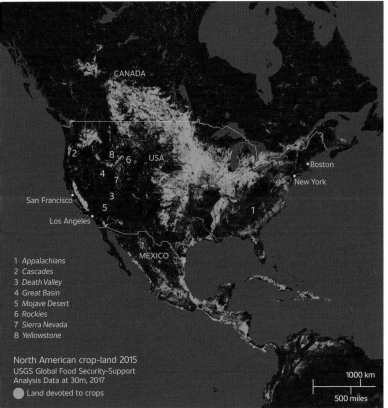

1 Appalachians
2 Cascades
3 Death Valley
4 Great Basin
5 Mojave Desert
6 Rockies
7 Sierra Nevada
8 Yellowstone

North American crop-land 2015
USGS Global Food Security-Support
Analysis Data at 30m, 2017
● Land devoted to crops

1000 km
500 miles

European crop-land – 2015 *(above)*
Zooming in on Europe, it is clear that
by 2015 there was no more potential
for 'new' arable land. The few remaining
dark areas are either already heavily
urbanised, such as London and Paris,
are mountainous, like the Pyrenees,
Alps and Scottish Highlands, or due to
the cold conditions are unsuitable for
agriculture, other than the grazing of
sheep and goats.

North American crop-land – 2015
In the US, we can see how by 2015
only deserts, such as the Mojave and
Great Basin, natural parks, including
Death Valley and Yellowstone, mountain
ranges including the Rockies, Sierra
Nevada, Cascades and Appalachians,
and large metropolitan areas, such as
those centred on New York, Boston, San
Francisco and Los Angeles, remained
free from arable agricultures.

biodiversity-rich tropics.[28] Food production is the largest driver of land use and land-use change, mainly through the clearing of forests and burning of biomass. Some parts of the world have experienced monumental shifts in land production, with devastating effects on primary forests. Between 2000 and 2014, for example, Brazil lost on average 2.7 million hectares of forest a year, the Democratic Republic of the Congo lost 0.57 million hectares a year, with this accelerating by a factor of 2.5 since 2011, and Indonesia lost 1.3 million hectares a year, with 40 per cent occurring in primary forest.[29]

Climate change and the food, water and energy nexus

Food, water and energy are inextricably linked, with food accounting for over 70 per cent of water use and over a third of diesel fuel use globally.[30] This dependency cannot last, as water resources are becoming depleted and urban requirements for water are increasing, while diesel-based energy which previously was subsidised in most countries is being increasingly phased out as it is particularly polluting and carbon intensive. Meanwhile, current agro-industrial farm practices are leading to the degradation of farmland, water reserves and biodiversity, which in turn is reducing agricultural potential, and is expected to lead to sharp declines in yields.[31] Already vulnerable land, on which many of the poorest people depend, is being irreparably degraded by overgrazing and the erosion and leaching of soils, with the devastating consequences for subsistence farmers being compounded by increasingly extreme weather.[32] In temperate regions like North America and Europe, which serve as the bread-baskets of the world, changes in climate are expected to lower production of major crops such as wheat, rice and corn.[33] A combination of temperature and rainfall changes may even lead to a 70 per cent decline in yields by 2050.[34]

Greenhouse gas emissions are also dangerous for food production, although higher concentrations of carbon dioxide (CO_2) provide more fuel for photosynthesis and can enhance growth of plants. This, however, is likely to be more than offset by the impact of rising CO_2 contributing to excessive peak temperatures, more extreme weather, such as destructive storms and winds, and to the spreading of plant diseases.[35] Devastating crop diseases can emerge and spread with astonishing speed. In the 1840s, for example, a previously unknown fungus from Mexico destroyed the Irish potato crop, which combined with the hoarding of stockpiles by the ruling British government and wealthy traders, induced a famine killing more than a million people.[36] Now the growing risk arising from global trade, increased plant stress and the migration of insects and parasites, owing to climate change, could lead to rising food insecurity plunging large parts of the world into crisis.[37]

Industrial-scale food production involves a series of dangerous feedback loops. First of all, the growing of crops and livestock releases greenhouse gases into the atmosphere directly. It also drives land use change that generates additional carbon dioxide when forests are cleared, wetlands are drained, and soils tilled. And when vegetation decomposes or burns, it also produces methane, which is around thirty-four times more damaging than carbon dioxide when it comes to raising atmospheric temperatures.[38] The degradation of land accelerates these processes, which means that inefficient farming practices and advancing desertification are leading to soil eroding at a rate that is ten to 100 times faster than it can recover.[39] Nitrous oxide, which has almost 300 times the impact of carbon dioxide, is also released from soil microbes in crop-lands and pastures and in the application of fertilisers.[40]

As we discussed in the Climate chapter, there are already changes in rainfall patterns, temperatures and, by extension, growing seasons. Climate extremes such as droughts and floods are among the key drivers of rising hunger, food insecurity, economic instability and even conflict, as the chapter on violence shows.[41] The consequences of global temperatures rising by two degrees, thought to be the lower bound of future warming, would be dramatic, leading to increasingly severe droughts in North Africa, the Sahel and the Middle East. These are already among the most conflict-prone regions on Earth and would greatly increase the risk of poverty and malnutrition across Africa and in South and Southeast Asia.[42] Countries in these regions depend on large-scale and subsistence agriculture for a significant share of their national income and employment. Small-scale farmers there already have limited access to fertilisers and irrigation, which undermines their ability to adapt to changing rainfall and withstand the increased variability in growing conditions.[43]

While some places may benefit from higher temperatures and more rainfall, this may not be beneficial as farmers do not care about decade-long averages. A single minute of temperature that is too high or too low, or hailstones that are too big, or wind that is too severe, or torrential downpours, or rains that come a few days too late or too early, can destroy their crops and livelihoods. It is the extremes that matter to farmers, and which create the greatest risk. The scientific consensus articulated in the Intergovernmental Panel on Climate

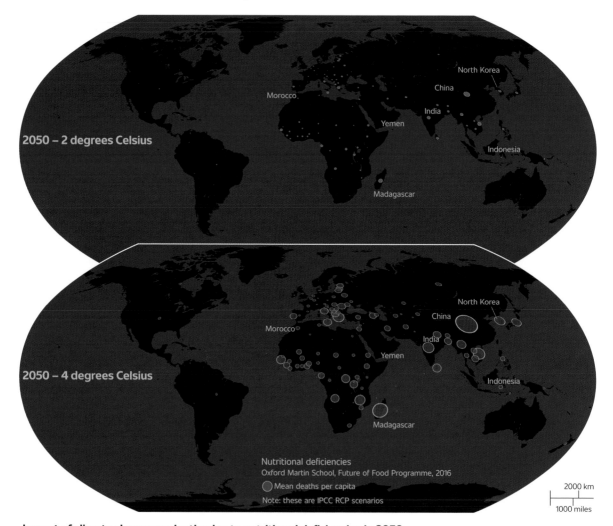

Impact of climate change on deaths due to nutritional deficiencies in 2050
The maps show the projected additional deaths annually from malnutrition resulting in 2050, under scenarios in which the average global temperature rises by two and four degrees Celsius. An estimated 30,000 additional deaths are forecast in China, and over 10,000 in Madagascar, and hundreds of thousands globally a year, notably in Asia and Africa, arising from 4 degrees of global warming.

Change (IPCC) is that much greater variability is to be expected and this is likely to impact negatively on farmers globally.[44]

The extent to which climate change will impact on malnutrition and induce famines depends in no small part on the responses of governments and the companies involved in food trade and distribution. The maps reveal the potential impact of global temperature rises of two and four degrees Celsius on nutrition and excess deaths. The size of the circles on the maps reflects the scale of possible deaths above the baseline due to climate-related food crises. It shows that climate change is forecast to result in hundreds of thousands of additional deaths annually, notably in Asia and Africa.

Meat matters

Huge swathes of land are cleared every year to feed rising demand for meat, sugar, and soy and palm oil. The rapid growth in the consumption of beef is especially damaging to the environment and is now estimated to contribute to as much as three-quarters of all agriculture-related greenhouse gas emissions.[45] After rapidly growing in the decades following the Second World War, North Americans and Europeans have already reached peak demand for meat. While consumption per head is declining in these regions, this has been more than offset by the growing appetite for meat among Asians and Latin Americans, who can finally afford to buy it. Globally, about 80 per cent of all agricultural land is devoted to feeding more than 4 billion cattle, goats and ruminant livestock and more than 2.5 billion chickens.[46]

The rapid change and growth in food production and trade reflects the pace at which global diets are changing. As countries get richer, what people eat changes. As a result of their ability to afford improved nutrition, the average urban Chinese boy has grown by nine centimetres, or three and a half inches, and the average Chinese girl by seven centimetres (two and three-quarter inches) since 1985.[47] The average Chinese citizen consumes around 136lb (62kg) of meat a year, seventeen times more than fifty years ago.[48] As a result, China's 1.4 billion citizens eat more than one-quarter of all meat, double the share of their US counterparts.[49] Meanwhile, China produces about half of the world's pork, a quarter of the poultry, and 10 per cent of all beef.[50] But the mismatch between having 20 per cent of the global

World meat production, 2017
FAO

● Meat production (tonnes)

2000 km
1000 miles

Global meat production

The map shows meat production in 2017, with the size of each circle indicative of the millions of tonnes produced. That China has overtaken the US as the world's top producer is evident in the larger purple circle, with Brazil the third largest. The wide global distribution of meat production is also clear to see.

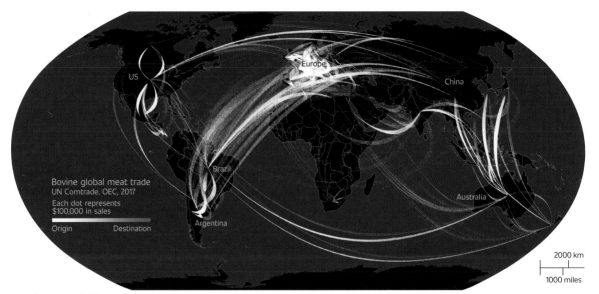

Bovine global meat trade
UN Comtrade, OEC, 2017
Each dot represents
$100,000 in sales

Origin   Destination

2000 km
1000 miles

Beef trade – 2017

Global beef trade in 2017 is shown with each dot representing $100,000 in sales. China is now far and away the biggest meat importer, especially of beef and pork, mainly from Australia. Argentinian and Brazilian sales into Europe and Asia, and the intense trade within Europe, is also strikingly visible.

population and just 7 per cent of the arable land means that surging demand requires meeting the deficit through imports.[51] Simply to feed its domestic livestock, China has to import more than 100 million tonnes of soybeans, accounting for 60 per cent of the global trade. This in turn is fuelling rampant deforestation and the proliferation of monoculture soybean farms, including in Brazil, Argentina and Paraguay.[52]

The Chinese shift from pork to beef, aggravates climate change. Pigs do not require pasture and are more efficient than cattle in converting feed to meat. Cattle produce five times more greenhouse gases per kilogram of meat and require 2.5 times more water than pigs. The transition may come sooner than we think. In 2019, the trade war between the US and China increased the latter's demand for soya from Brazil. In the same year the devastation of pig farming owing to the spread of African swine flu accelerated the Chinese transition to beef consumption. As the Climate chapter makes clear, the Amazon rainforests and the world's atmosphere and ecosystems are casualties of these changes in food consumption and trade.

Indonesia: from forest to palm oil plantations – 1984 and 2019

The satellite images from 1984 and 2019 show how the forests of Indonesia were cleared to make way for palm oil.

While in South America the ravaging of tropical rainforests and savannah is mainly for rearing cattle and growing soya, in Indonesia and Malaysia it is primarily the result of the rampageous expansion of palm oil plantations to meet the needs of the food processing industry. These two Southeast Asian countries account for 90 per cent of all palm oil production globally.[53] In Indonesia alone the area cleared for palm oil has grown twenty times over the past three decades, to more than 10 million hectares.[54] The satellite images from 1984 and 2019 show how the forests of Indonesia were cleared to make way for palm oil over this 35 year period. The clearing of the peat land that is particularly suitable for palm oil leads to more than 1,600 tonnes of CO_2 being released for every hectare cleared.[55] Worldwide, rainforests the size of Israel or Rwanda are being destroyed every year.[56]

Fishy business

Fish have been part of the human diet for tens of thousands of years. But industrial-scale fishing emerged only in the last century. This was made possible by the deployment of technologies originally developed for warfare such as satellite positioning systems, radar and echo sounders, as well as bigger maritime fleets. As the industry grew and subsidies increased so did the scale of the catch. As with

Fishy trade – 2017
The global fish trade in 2017, with each dot representing $1 million worth of fish products, shows white dotted exports to red-dotted import destinations. China is the largest exporter of fish. followed by Norway and Vietnam.[57] The intensity of trade in fish in Europe and across the Mediterranean as well as the scale of exports from South America is clear.

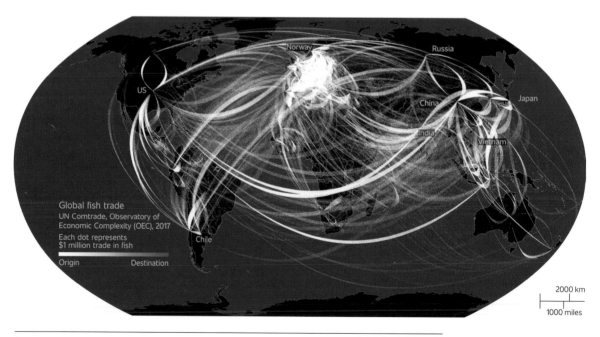

Global fish trade
UN Comtrade, Observatory of Economic Complexity (OEC), 2017
Each dot represents $1 million trade in fish

Origin   Destination

2000 km
1000 miles

beef, rising incomes have increased demand for fish, not least in Asia. The rapid rise in industrial fishing is exhausting fish stocks around the world. As the catch waned off the shores of rich countries, their fleets, powered by cheap diesel fuel, have gone to greater lengths to increase their catches, including Arctic and Antarctic waters. No one knows exactly how much overfishing is occurring. Countries with big fleets – China, Indonesia, India, US, Russia and Japan – are repeatedly accused of overfishing and breaching regulations in distant waters.[58]

Fish are not just destined for human consumption, as about a third of all marine fish catch is used for animal feed. This demand is further accelerating the exhaustion of stocks, not least as small plankton and infant fish, which are too small for human consumption, are vacuumed out of the ocean or dredged in fine nets.[59] The stocks of many fish species are already reaching critical levels owing to overfishing and the rising acidification and degradation of the oceans.[60] Taken together, the annual growth in fish consumption has been twice as high as population growth since the 1960s.[61] The map shows the global fish trade in 2017, with each dot representing $1 million worth of fish products and the lines going from the white dots showing exports to red-dotted import destinations. As was the case with the trade in meat, the rapid integration of Asia into global fish markets is striking, with the largest fishing fleet in the world, with more than 3.5 million vessels (or three-quarters of the global fleet). China since

The growth of aquaculture
The capture of fish in oceans and rivers peaked in the late 1980s and has been diminishing ever since. Since then, all the growth has been accounted for by fish farming, which now exceeds the capture of wild fish.

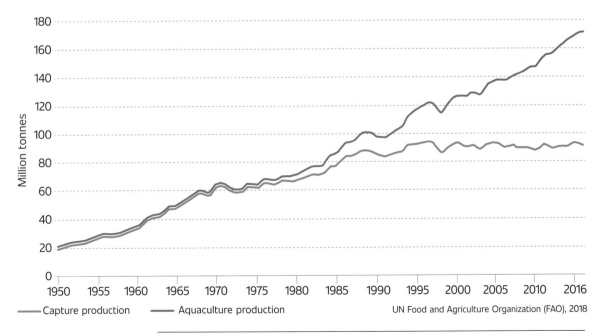

Capture production   Aquaculture production   UN Food and Agriculture Organization (FAO), 2018

**Fish farming in Asia –
1990 and 2018**

The dramatic increase
in fish farming in Asia
is apparent when
comparing 1990 and 2018
production. The size of
each circle represents
the scale of aquaculture
production in millions of
tonnes. In 2018 China's
output is 48 million
tonnes, followed by
India (7 million)
Indonesia (5 million),
and Vietnam (4 million).

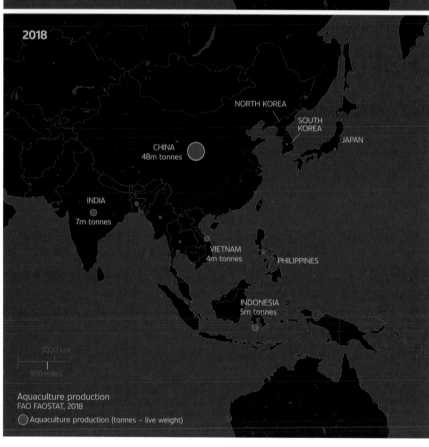

2002 has been the world's top fish producer as well as the largest exporter of fish and related products, followed by Norway and Vietnam.[62] The intensity of trade in fish in Europe and across the Mediterranean as well as the scale of exports from South America, notably anchoveta and salmon from Chile, is also evident in the map.

While most fish we consume are sourced from wild fisheries, fish farming, or aquaculture, already accounts for almost half of total fish used for food and feed products.[63] As shown in the figure on page 345, the amount of wild fish captured in oceans and rivers peaked in the late 1990s and has stayed the same ever since.[64] All the growth in production since then has been accounted for by fish farming, with the amount of fish farmed exceeding that of the capture of wild fish.[65]

Biodiversity benefits

Maintaining the diversity and richness of all living organisms on land and in water is necessary for the stability of ecosystems and the productivity and resilience of food production systems. All human and economic activity depends on services provided by nature, estimated to be worth as much as $125 trillion per year.[66] Yet the critical importance of these natural systems is not close to being reflected in the amount of attention and care given to biodiversity. As the system is being plundered, it is close to collapse with some scientists speaking of 'biological annihilation'.[67] The highly respected EAT-Lancet Commission claims we are entering the sixth mass species extinction, losing flora and fauna at a rate 100 to 1,000 times faster than is typical[68] of the current geological period that began about 12,000 years ago.[69] Human activity has led to a reduction of roughly 60 per cent in the size of animal populations since 1970.[70] Species extinction could permanently disable our ability to sustainably feed the global population.

A wide range of human activities contributes to biodiversity loss. Terrestrial and aquatic habitat loss and the breaking up of migratory patterns destroys natural systems, particularly through farming, but also as a result of urbanisation and the building of infrastructure like roads, railways, dams and ports. Biodiversity is further threatened by climate change which compounds other threats to biodiversity such as the impact of industrial and other pollution, the spread of invasive species, unsustainable harvesting of wild species and the spread of agriculture and infrastructure. According to the International Union for Conservation of Nature, 80 per cent of all extinction threats to mammal and bird species are due to agriculture.[71] Before extinction, species' population sizes are reduced and the species no longer exists in areas where previously it was to be found. Insect biomass has reduced by 75 per cent over the past thirty years and farmland birds by 30 per cent in fifteen years.[72]

The loss of bees offers an example of how reductions in biodiversity will impact all our lives. Bees gathering pollen from a variety of plants have healthier immune systems than those taking pollen from less diverse sources.[73] Healthy immune systems are vital for bees to be able to create food for the colony. Yet bee and insect populations are falling alongside the diversity of plants they pollinate and feed on. The collapse of bee colonies is widespread, but in the UK it is more severe than elsewhere in Europe, with a possible explanation being the decline in biodiversity. The epidemic was described as a 'bee-pocalypse' and is referred to by entomologists as 'colony collapse disorder'. In the UK the estimated value of bee pollination exceeds £200 million per year and in the US it is more than $14 billion. The consequences of a complete collapse in global bee populations would far exceed these estimates.[74] So far, this has yet to occur, though this could easily happen.[75]

Healthy diets for people and planet

The combined pressures of declining productivity and increasing demand for food were building for some time.[76] To reverse the course, several changes are needed immediately. First, we need to close the gap between global obesity and hunger through radical changes in our diets. We have to do this in a way that is sustainable and slows climate change. Because food production is one of the largest contributors to climate change, diet, health and sustainability are fundamentally inter-linked. The failure to change our diets will lead to devastating increases in greenhouse gas emissions, growing pollution from fertilisers, bio-diversity loss and accelerating degradation of water and land. This not only threatens food production, but also our ability to contain climate change, much less meet the Paris Agreement targets.[77]

A widely cited *Lancet* study assessed the environmental implications of various dietary options.[78] Its authors concluded that to achieve improvements in our diets and health, we need to reduce the consumption of unhealthy foods – especially red meat and sugar – by more than half.[79] Meanwhile, the consumption of healthy foods, such as nuts, fruits, vegetables and legumes, needs to be more than doubled. The changes required differ greatly by region, with the greatest reductions in meat and sugar consumption needed in the richest countries. Basic dietary changes could stabilise and reduce greenhouse gas emissions and greatly benefit human health, averting an estimated 11 million deaths per year arising from poor nutrition, a reduction of more than 20 per cent.[80]

The IPCC has also stressed how more productive land use could significantly reduce global warming.[81] As the climate chapter makes clear, land masses are often natural carbon sinks, absorbing greenhouse gases through photosynthesis.[82] Sustainable food production requires that gaps in yields between high and low production regions be closed.[83] In order to ensure that this does not result in more fertiliser use and related pollution, the former needs to be reallocated globally from countries where this is excessively applied to those countries that suffer deficits. Agriculture needs to be transformed to guarantee that the twin health challenges of undernourishment and obesity can be met while simultaneously ensuring that food systems swing from destroying the environment and emitting carbon, to acting as carbon sinks and restoring biodiversity.

Change starts with all of us. So-called 'flexitarian' diets are compatible with a wide variety of agricultural systems, cultural traditions and individual dietary preferences. They can be combined with various types of omnivorous, vegetarian and vegan diets. The simple truth is that the regular (much less growing) consumption of meat is incompatible with good health and environmental sustainability. For your own sake, and that of our environment, cut back on meat and replace it with sustainable fish, or better, plants. Try going vegan for at least a meal a day and buy food with less packaging and a lower environmental footprint. You will be doing yourself, and the planet, a favour.

The global spread of COVID-19 as of June 2020

Health

Pandemics and superbugs are super threats

Life expectancy has risen virtually everywhere

Infant, child and maternal mortality rates are falling

Mental illness and suicide are major risks

Technology is transforming health care

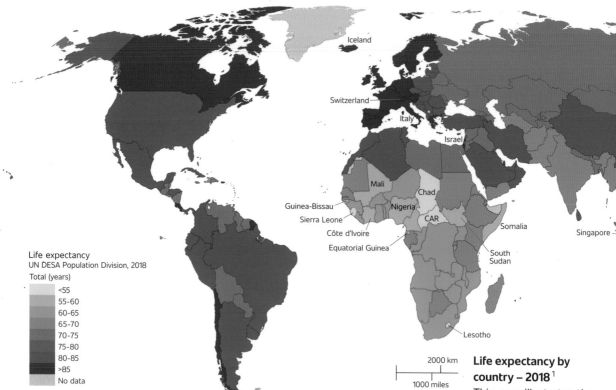

2000 km

1000 miles

Life expectancy by country – 2018[1]
This map illustrates the average life expectancy for virtually every country in 2018. The darker the colour, the longer citizens are expected to live. People are living longer in North America, Western Europe and Australia. Africa, however, is the outlier, where life expectancies can be between twenty and thirty years lower.

Introduction

More people are living longer and healthier lives than at any time in history. The oldest person who ever lived was a Frenchwoman by the name of Jeanne Calment. She was born in 1875 and died in 1997 at the age of 122 (and 164 days). Although her record still stood when this book went to press, it won't last. Jeanne is part of an exclusive club: only a few thousand people have earned the distinction of being 'super-centenarian', living more than 110 years. By way of comparison, there are several hundred thousand centenarians – people over 100 – many of them Japanese.[2] While still comparatively rare, soon century-long lifespans won't seem so special. Depending on where they live, people born in 2000 can expect to live for at least one hundred years. Some scientists believe that human life could be extended by an order of magnitude longer. If you think this all sounds like science fiction, you're not alone. For years, the accepted view was that we had reached the outer limits of human longevity.[3] But this may soon

Japan

South Korea

Hong Kong

be a minority opinion. A wave of new research led by biologists and gerontologists suggests that the maximum age may be much higher than earlier believed.[4] Advances in technology, especially gene editing and regenerative medicine, mean that average life expectancies could be elongated to 150 years by the end of this century.[5]

It may sound obvious, but one of the central reasons we are living so long is because human health has dramatically improved. Owing to advances in everything from medicine and lifestyles to the quality of nutrition and care, health has ameliorated virtually everywhere over the past century. The clearest indicator of this is global life expectancy at birth. For virtually the entire time *Homo sapiens* have inhabited this planet – give or take 200,000 years – life was short, often brutally so. Then something extraordinary happened. Global life expectancy doubled in less than a century, the blink of an eye in evolutionary terms. Not every part of the world moved ahead at the same speed. In Monaco, people live on average to eighty-nine, while in Chad life expectancy is closer to fifty-three. Notwithstanding these disparities, the prolongation of life is one of our greatest achievements. And this is just the beginning. Advances in Artificial Intelligence (AI) and biotechnology and the ubiquity of patient data are transforming how we think about health. Although the upper age limit for humans continues to inspire debate and controversy, everyone agrees that we are on the precipice of a monumental revolution in medicine.[6]

In this chapter, we explore how we humans defeated some of our biggest killers. A big reason most of us are healthier comes down to the spread of hygiene, nutrition, antibiotics and professionalised medicine that helped us beat back our mortal enemies, especially bacteria and disease. The spread of education and changes in mindsets about personal and public health changed the game. Straightforward and common-sense ideas – that smoking kills, universal vaccination is smart, national health systems can save lives – generated staggering improvements in population health. And while contagious and drug-resistant viruses such as COVID-19 are a reminder of our vulnerability, we must not forget that we've come a long way in an incredibly short period of time. From declining mortality rates and the rise of antibiotics to the growing threat of pandemics and mental health-related illness, we explore some of the big transformations in health care that changed the world and will define our future.

Living longer – how we doubled our lifespan in a century

Life expectancy more than doubled over the past century. The first map reveals how average lifespans in 1960 were around fifty years. The global average was pulled up by North America, Western Europe, Japan and Australia and was dragged down by large swathes of, Asia, Africa, Latin America and the Middle East. Today, the global average lifespan is closer to 71.5 years with dramatic improvements all over the world. Because males are generally less healthy and adopt riskier habits than females,[7] they tend not to live as long.[8] Today, Japanese men live on average to 81 while women live on to 87. In Sierra Leone, one of the world's poorest countries, life spans are closer to 52 for males and 53 for females.[9] Lifespans are still climbing, especially in wealthy countries. By 2030, US men could live to around seventy-nine while females may live to eighty-three.[10] A baby girl born in South Korea this decade can expect to live until almost ninety-one as compared to eighty-four in the case of boys.[11] Until recently, scientists believed that an average life expectancy of ninety was impossible. Yet astonishing advances in biomedicine, nutrition and access to health care are turning such prognostications upside down. We will explore the future in a moment – but let's first attend to the past.

Statistically speaking, for 99.99 per cent of humanity, we lived on average twenty to twenty-five years. Skeletal remains recovered from the Neolithic period suggest that average life expectancy hovered closer to twenty-one. Unluckily for our ancestors, it stayed this way for another 10,000 years. Burial inscriptions dating back to the Roman Empire indicate that most people didn't live past their mid-twenties.[12] The English philosopher Thomas Hobbes was on point when he described life in nature as 'nasty, brutish and short'. When he wrote that memorable passage in *Leviathan* in 1651, average life expectancy was just thirty-five in England and twenty-five in its newest acquisition, the US colony. In the nineteenth century, life expectancies climbed up to forty in Belgium but still hovered closer to twenty-three in India and Korea. Then everything started to change. Life expectancies began shooting up from the late nineteenth and early twentieth centuries onwards. In less than 100 years we tripled human lifespans in many parts of the world. So, what happened?

1960

2020

Life expectancy at birth
UN DESA Population Division, 2019

Total (years)

<55 55-60 60-65 65-70 70-75 75-80 80-85 >85 No data

2000 km

1000 miles

Life expectancy at birth – 1960 and 2019 [13]

These maps reveal the dramatic increases in life expectancy around the world
between 1960 and 2019. The transformation is remarkable in Latin America and
Asia, in particular, where life expectancies increased by on average two decades or
more. Improvements have been much slower in parts of Africa. Moreover, there is
still considerable variation in life expectancies within countries, especially between
cities and the rural hinterland.

Surviving birth – strongly improving odds

The first hurdle involved surviving childbirth and the first five years of life. For most of our existence, new-borns had even odds that they would last until five years of age. Human remains gathered from hunter-gatherer societies show child mortality rates as high as 70 per cent.[15] As recently as the middle of the nineteenth century, one in two children died before their fifth birthday. A hundred years later, the proportion had shrunk to one in five. Today, child mortality is a comparatively rare event, involving roughly one in twenty-two births globally, and one in 100 in wealthier countries. While this sounds high (and it is), the chart highlights the steady decline over the past 200 years. Social scientists such as Max Roser, Hannah Ritchie and Bernadeta Dadonaite have shown how child mortality started falling earlier in parts of Europe and North America, but then started dropping sharply in middle- and lower-income countries during the second half of the twentieth century. Brazil and China registered a tenfold decline in the past four decades alone. Even in Sub-Saharan Africa child mortality rates – while still more than twice the global average – dropped dramatically to one in ten.

Declining child mortality rates – 1800–2015
(selected countries)[14]

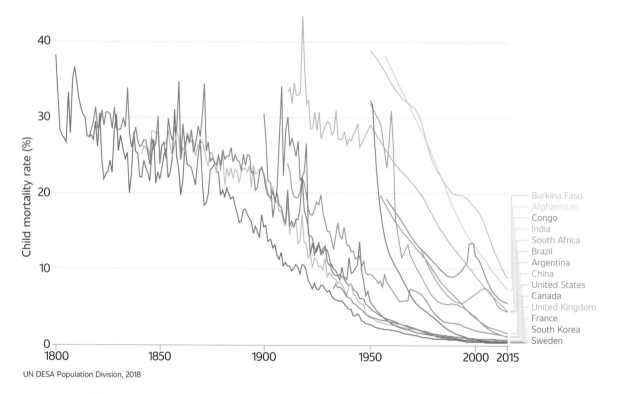

UN DESA Population Division, 2018

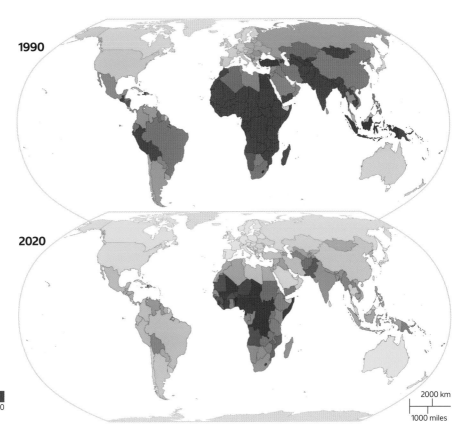

1990

2020

Mortality rate, under-5
(1990 and 2020)
UNICEF, WHO, World Bank,
UN DESA Population Division

Per 1,000 live births

1   100

2000 km

1000 miles

Falling child mortality rates: 1990 and 2020[16]
These maps visualise changes in the share of children who die before the age of five around the world. Darker colours represent higher child mortality rates. While challenges persist in Africa, by 2020 more children are living past five than at any time in history. Great strides in hygiene, sanitation and education have improved rates in the world's less-developed countries, but inequality persists.

What accounts for the striking decline in child mortality? A principal factor is basic improvements in hygiene, sanitation and education.[17] The associated decline of infectious diseases like pneumonia, malaria, measles and diarrhoea, and birth-related (or intrapartum) complications, also helped immensely.[18] During the nineteenth and twentieth centuries, countries like Canada, France, Sweden and the US took between eighty and 100 years to reduce child mortality from 30 to less than 5 per cent. By contrast, Brazil, China, Kenya and South Korea achieved similar results in twenty-five to fifty years. In the 1960s, as many as 20 million under-fives died every year; today, the number is closer to 6 million.[19] As the map shows, there are still persistent disparities in child mortality rates around the world. Although the so-called 'fast catch-ups' in Asia and South America extended global life expectancy, there are still stark differences between the northern and southern hemispheres – and especially in Africa, where countries exhibit poorer social and economic determinants of health.[20] The spread is enormous: infant mortality rates range from just 2.3 per 1,000 live births in Singapore to above 100 in Somalia.[21]

Falling maternal mortality

Another factor explaining the prolongation of life is the stunning drop in maternal mortality – defined as the death of a woman while pregnant or within six weeks of termination of a pregnancy. Surviving birth has virtually always been a harrowing affair for infants *and* their mothers. While precise numbers are hard to come by, as many as one-third of all mothers died in childbirth for most of human existence owing to pregnancy-related complications (like haemorrhaging, infections and pre-eclampsia, a condition that involves high blood pressure and excess protein in the urine). Less than a century ago between 500 and

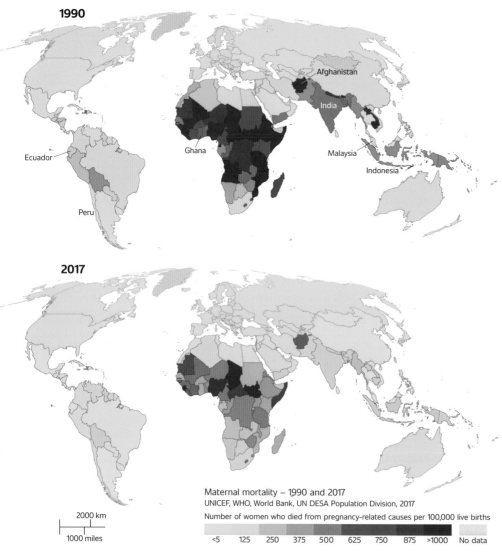

1990

Afghanistan
India
Ghana
Malaysia
Indonesia
Ecuador
Peru

2017

Maternal mortality – 1990 and 2017
UNICEF, WHO, World Bank, UN DESA Population Division, 2017

2000 km
1000 miles

Number of women who died from pregnancy-related causes per 100,000 live births

| <5 | 125 | 250 | 375 | 500 | 625 | 750 | 875 | >1000 | No data |

Falling maternal mortality – 1990 and 2015[24]

These maps reveal the estimated number of women who died while pregnant per 100,000 live births in 1990 and 2015. The most dramatic improvements in recent decades have been in parts of Latin America and East and Southeast Asia. However, there are still deep challenges in Africa and South Asia, where women continue to face a significant risk of maternal mortality.

1,000 of every 100,000 births ended with the death of the mother.[22] As the map shows, by 2015, the overall rate fell to roughly 240 per 100,000 in poorer countries and twelve per 100,000 in wealthier ones.[23] A big reason for this is that women started having fewer babies and began spacing out their children. The decline in births, and improved access to maternal health care and better information for young mums (and dads), changed the game.

There are reasons to be optimistic about continued improvements in maternal health. Over the past twenty years, maternal mortality fell roughly 40 per cent around the world. Yet sharp gaps persist in outcomes between rich and poor countries. For example, women in wealthy high-income nations are twenty times less likely to die during childbirth as women in poorer low-income countries.[25] Today, an astonishing 99 per cent of all maternal deaths occur in poorer parts of the world, especially Sub-Saharan African states. Notwithstanding these disconcerting gaps, the African novelist Chinua Achebe reminds us that 'no condition is permanent'.[26] The spread of maternal and infant care together with specialised procedures to manage complications are closing these divides.[27] And while deaths occurring before, during and shortly after childbirth remain unacceptably high, especially in countries lacking universal health care, they are still less prevalent than at any time in history.[28]

Battling bacteria – the rise of the superbug

For most of human existence our most dangerous enemy was invisible – bacteria. Throughout history, most people did not die of heart disease or cancer for the simple reason that they did not live long enough to experience these ailments. Instead, they died of rudimentary infections arising from minor injuries. For thousands of years, bacterial infections spread by cholera, pneumonia, smallpox, tuberculosis and typhoid killed hundreds of millions of people. Making matters worse, no one had the faintest notion where these quiet killers came from. Most people didn't even know that bacteria existed until the late nineteenth century. This is not to say that early civilisations didn't try to stop disease or resort to home-grown antibiotics. All manner of healers and homoeopaths applied natural moulds and plant extracts to treat infections.[29] They very seldom worked.

We started winning the fight against bacteria about a century ago. In the late 1800s, the French chemist and bacteriologist Louis Pasteur and his wife and assistant, Marie, proved that diseases were not spontaneously occurring, but rather a product of living germs that multiplied. In 1910, the German physician Paul Ehrlich invented arsphenamine (later marketed as Salvarsan), which was used to treat syphilis that had ravaged Europe for centuries. Arsphenamine became one of the most widely prescribed drugs in the world for the next thirty years. Ehrlich's approach to designing and testing drugs was as important as his discoveries. In addition to coining the term 'chemotherapy', Ehrlich and his colleagues pioneered a systematic approach to developing, screening and producing drugs that gave rise to a new era of mass-produced antibiotics.[31]

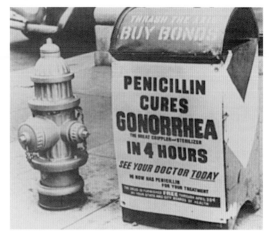

The spread of penicillin, the world's first wonder-drug.[30] Poster c.1944. *Source*: NIH, Wikimedia

The rise of antimicrobial resistance – projected deaths in 2050 [34]
The map depicts predicted deaths due to antimicrobial resistance in 2050. While a serious threat, there are comparatively few deaths predicted in North America, Western Europe and Australia. Risks are much higher in Africa and Asia, and to a lesser extent South America, where hundreds of thousands, even millions, are projected to die prematurely.

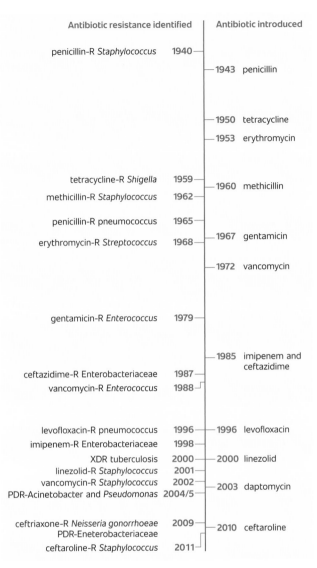

Antibiotic resistance identified			Antibiotic introduced
penicillin-R *Staphylococcus*	1940		
		1943	penicillin
		1950	tetracycline
		1953	erythromycin
tetracycline-R *Shigella*	1959	1960	methicillin
methicillin-R *Staphylococcus*	1962		
penicillin-R pneumococcus	1965		
erythromycin-R *Streptococcus*	1968	1967	gentamicin
		1972	vancomycin
gentamicin-R *Enterococcus*	1979		
		1985	imipenem and ceftazidime
ceftazidime-R Enterobacteriaceae	1987		
vancomycin-R *Enterococcus*	1988		
levofloxacin-R pneumococcus	1996	1996	levofloxacin
imipenem-R Enterobacteriaceae	1998		
XDR tuberculosis	2000	2000	linezolid
linezolid-R *Staphylococcus*	2001		
vancomycin-R *Staphylococcus*	2002	2003	daptomycin
PDR-Acinetobacter and *Pseudomonas*	2004/5		
ceftriaxone-R *Neisseria gonorrhoeae*	2009	2010	ceftaroline
PDR-Eneterobacteriaceae			
ceftaroline-R *Staphylococcus*	2011		

Spread of antibiotics and resistance: 1940–present. Recreated based on *Antibiotic Resistance Threats in the US*, CDC, 2013.

The discovery of sulfanilamide drugs and then penicillin in 1928 was game-changing.[32] After the microbiologist Sir Alexander Fleming, accidentally discovered penicillin, it was mass produced at the University of Oxford in the early 1940s. This new wonder-drug went on to tame many infectious diseases that were the leading causes of mortality and morbidity at the time.[33] But its widespread use had a dark side. As the popularity of antibiotics grew, so did over-prescription. Almost as quickly as penicillin and other antibiotics spread around the world, they met bacterial resistance. In fact, microbial resistance to penicillin was discovered even before it began being commercially distributed in 1943.[35] Another drug, tetracycline, launched in 1950, faced resistance in 1959. Methicillin was introduced to the world in 1960 and resistance was detected just one year later. Virtually every antibiotic released on to the market became redundant within a few years.[36] Today, there are more than 100 antibiotics available and just a handful still seem to work. The World Health Organization (WHO) and the US Centers for Disease Control and Prevention (CDC) have found infectious bacteria that are resistant to all but two remaining antibiotic drugs.[37] Both organisations are sternly warning against non-essential use for humans and livestock.[38] The truth is that none of this would have surprised Fleming, who issued several warnings about the potential resistance to his wonder-drug if overused.[39]

Communicable disease specialists believe we are on the cusp of a post-antibiotic era.[40] Warnings about antibiotic and antimicrobial resistance are reaching fever pitch.[41] A world without antibiotics is terrifying, especially for anyone with weakened immune systems or exposed to intrusive surgery. Simple exposure to hospitals could prove fatal. The UK's former chief medical officer, Sally Davis, until September 2019, Sally Davies, believes that such resistance poses a 'catastrophic threat'[42] that could 'kill humanity before climate change'.[43] According to the CDC, every year at least 2 million Americans already experience infections that resist the antimicrobial drugs used to treat them – resulting in an excess death toll of at least 23,000.[44] Another study determined that 700,000 people already die annually around the world from drug-resistant infections.[45] If trends continue as they are, this could rise to 10 million deaths a year by 2050, more than the projected deaths from cancer (8.9 million), diabetes (1.5 million), diarrhoeal diseases (1.4 million) or road traffic accidents (1.2 million).[46] The potential economic burden of a mass superbug outbreak is frightening – as much as $10 trillion annually.[47]

Preventing pandemics – humankind's biggest killer

Although the COVID-19 pandemic has taught an entire generation about the risk of killer viruses, the deep anxieties about plague and pestilence are deeply-seated. And for good reason: pandemics are among the single biggest killers throughout history. Technically speaking, pandemics are epidemics that cross international boundaries and affect large numbers of people. Most, but not all, are communicable. Some of them can be incredibly destructive, with lasting effects on societies that can span generations. A typhoid outbreak in 430 BC killed a quarter of the Athenian army, permanently crippling it. The bubonic plague from AD 541 to 750 eliminated between 25 and 50 per cent of the planet's human population. The Black Death, caused by the bacterium Yersinia pestis, lasted from 1331 to 1353, extinguishing around 75 million lives. Meanwhile, smallpox, measles and influenza, introduced by European explorers to Latin America and the Caribbean, extinguished roughly 95 per cent of indigenous populations there. And in the twentieth century these same diseases killed up to 500 million people, especially infants and children.

One of the first known uses of maps to analyse the relationships between diseases, their underlying causes and outcomes, occurred in London in the 1850s. A physician, John Snow, refused to believe that

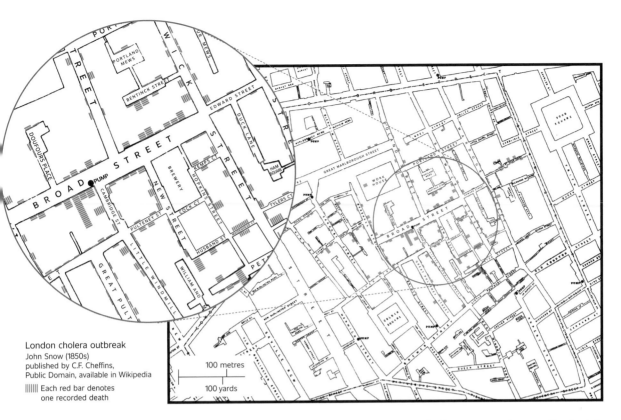

London cholera outbreak
John Snow (1850s)
published by C.F. Cheffins,
Public Domain, available in Wikipedia

||||| Each red bar denotes
one recorded death

Cholera clusters during the London epidemic of 1854 [48]

diseases such as the bubonic plague or cholera were due to 'bad air', as was widely assumed at the time. Working with Reverend Henry Whitehead, Snow created the map above and traced the source of a massive cholera outbreak to a single water well. His careful scrutiny of the disease's trajectory – the red bars denote reported deaths from the disease – eventually convinced the local council to disable the specific pump linked to the infected water source. The incidences of cholera immediately plummeted. By using data and maps to show the connection between water quality and cholera, Snow revolutionised public health in London and gave birth to the field of epidemiology in the process.

Maps help illustrate the way geographic, demographic, social and economic factors can influence the incidence and spread of disease outbreaks.[49] Today, they are used to track a wide range of pathogens such as COVID-19 and influenza, chikungunya, Ebola, tuberculosis, yellow fever and Zika. Take the case of the H1N1 influenza virus, also known as 'swine flu', which can be transmitted by human to human contact.[50] Unlike other influenza strands that tend to infect older adults, H1N1 is particularly dangerous because it infects children and younger adults – groups who usually have the strongest defences against seasonal flu. Symptoms typically graduate from fever and headaches to breathing difficulties and in some cases pneumonia, acute respiratory distress and death. The H1N1 is particularly chilling since it is a virus associated with the Spanish flu outbreak that infected 500 million people between 1918 and 1920, killing between 3 and 5 per cent of humanity at the time.[51]

While overshadowed by the 2020 COVID-19 outbreak that started in Wuhan, China, the 2009 H1N1 outbreak was already a stark reminder of just how risky these contagious outbreaks can be. The first documented H1N1 case was detected in the town of La Gloria in Veracruz, Mexico in April 2009.[53] Newly infected patients quickly emerged in other parts of the country and eventually the capital, Mexico City.[54] Because it was a new strain there were no vaccines and the virus spread quickly around the world. It was upgraded to pandemic status in June, before slowing by November after newly developed vaccines were distributed to sixteen countries. The WHO declared the outbreak officially over in August 2010,[55] but not before an estimated 18,000 people were presumed dead. This may have been a significant undercount: researchers later determined that as many as 248,000 people were likely to have been killed by the disease

Deaths per 100,000 associated with H1N1, 2009–2010[52]
This map tracks the number of H1N1-related deaths between 2009 and 2010 globally. The first reported outbreak occurred in Mexico, but then spread rapidly across North America, Western Europe, South and East Asia, and Australia.

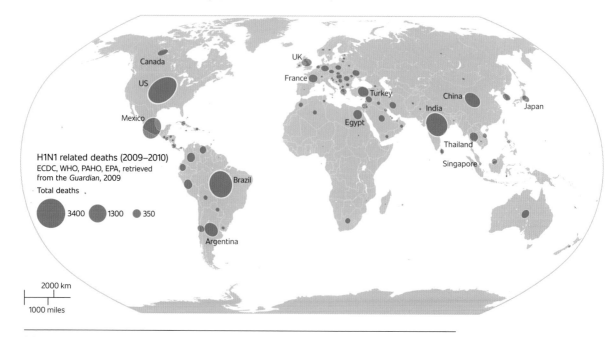

worldwide.[56] Put another way, three times more people died of H1N1 between 2009 and 2010 – a flu spread by coughing and sneezing – than all wars combined that same year. Since then H1N1 outbreaks have been reported in India, the Maldives, Malta, Morocco, Myanmar and Pakistan, though these are less worrying now that H1N1 is considered part of the regular flu strain.

Even so, the sheer speed and scale of the COVID-19 pandemic – and its spread to over 190 countries in three months – caught many decision-makers by surprise.[57] But COVID-19 is hardly the only virus keeping health experts awake at night. Infectious disease specialists worry that other strains of influenza, including the H7N9 virus (also known as 'avian flu'), could spread from chickens to humans.[58] The CDC, for example, has warned that H7N9 has a 'high likelihood' of evolving into a widespread pandemic in the short to medium term.[59] What makes these new strains particularly disconcerting is their high kill rate. During the last H7N9 outbreak in 2013 in China, roughly 88 per cent of those diagnosed with the virus developed pneumonia and 41 per cent of everyone who contracted the disease died prematurely.[60] The map below simulates the trajectory of a possible H7N9 pandemic. There have been at least five known outbreaks of H7N9

Simulated spread of an H7N9 pandemic emerging in China [64]
The map highlights the possible spread of a human-transmissible virus from China to major population centres across Asia, Europe and North America. The estimated numbers of people living within two hours' travel time from an airport was calculated using gridded population-density maps and a dataset of global travel times.[65]

A. J. Tatem, Z. Huang and S. I. Hay, 2013

Residents within 2 hours of the airport (million)
- <1
- 1–5
- 5–10
- 10–20
- >20

Passengers per month
- <1,000
- 1,000–2,500
- 2,500–5,000
- 5,000–15,000
- >15,000

2000 km
1000 miles

since the virus was first discovered in humans, all of them in China. Cities where it has been reported – including Beijing, Fujian and Shanghai – are commercial and transport hubs, with transportation links to population centres across Asia, Europe and North America.[61]

Although the map depicts a fictitious outbreak, its potential scope and scale echoes the COVID-19 outbreak. Researchers estimate that more than 131 million people, 241 million domestic chickens, 47 million domestic ducks and 22 million pigs lived within a 30-mile radius of each of the sixty H7N9 human cases that occurred in 2013. More-over, roughly a quarter of the global population outside of China lives within two hours of an airport with a direct flight from the outbreak regions, and almost three-quarters if a single connecting flight is included.[62] And it's not just H7N9 or COVID-19 that pose a threat. Between 1980 and 2013 there were more than 12,000 recorded disease outbreaks involving 44 million cases around the world.[63] Today, the WHO tracks roughly 7,000 new signals of potential outbreaks every month. In a single month in 2018, the WHO counted – for the first time – outbreaks of six of the eight 'priority diseases'. The next year, COVID-19 made its appearance. It killed hundreds of thousands of people, infected potentially millions and triggered massive global disruption affecting politics, economics and social life across the globe.

What explains the steady increase in pandemic outbreaks? Anti-microbial and viral resistance discussed earlier are certainly a big part of the problem. So is the refusal of some segments of the population to take precautions, including vaccinations.[66] But there are other factors at play such as global warming and accelerating trade. It is worth recalling that all vector-borne diseases – illnesses transmitted by a living organism between humans or from animals to humans – are not necessarily spread in similar ways. Meanwhile, the potential of vector-borne diseases such as malaria (which is transmitted by the mosquito) to cause a pandemic depends on the survival and reproduction capacity of the vector itself. Vectors, pathogens and hosts all survive and reproduce within a specific range of conditions related to everything from temperature and precipitation to elevation and wind speed.[67] So as temperatures rise, forest penetration expands, meat consumption grows or commerce picks-up, so too does the potential for pandemics. The bottom line is that we are no longer safe from diseases because of our location – in today's globalised world, the

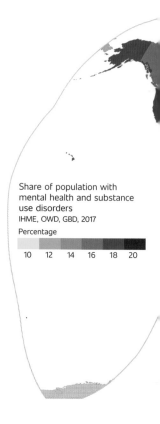

Share of population with mental health and substance use disorders
IHME, OWD, GBD, 2017
Percentage

10 12 14 16 18 20

Share of population experiencing mental health and substance abuse disorders – 2017[68]

Mental health and substance abuse disorders are hiding in plain sight. This map shows how parts of North America, South America, Western Europe and Australia experienced comparatively high rates of such disorders in 2017, despite ranking better on other health indicators. Even so, these variations may also be a result of under-reporting owing to weak data collection and stigma.

source of the next pandemic could quite literally be sitting (and sneezing) right next to you.

Mapping mental health disorders – more than a billion people with symptoms

Many diseases are hiding in plain sight. Take the case of mental health disorders – including anxiety, dementia, depression, eating and bipolar disorders, and schizophrenia – that are routinely under-diagnosed. As a result, they tend to receive less attention and treatment than other forms of illness – especially in low-income settings. Yet mental health disorders are among the largest contributors to global mortality and morbidity.[69] A whopping 1.1. billion individuals – one in six people on the planet – suffers from one or more mental health and substance abuse disorders.[70] About one in four people will experience a mental disorder in the course of their life.[71] As the chart overleaf shows, the incidence of depression is widely distributed across Africa, the Americas, Asia and Europe. But overall, mental disorders are still very

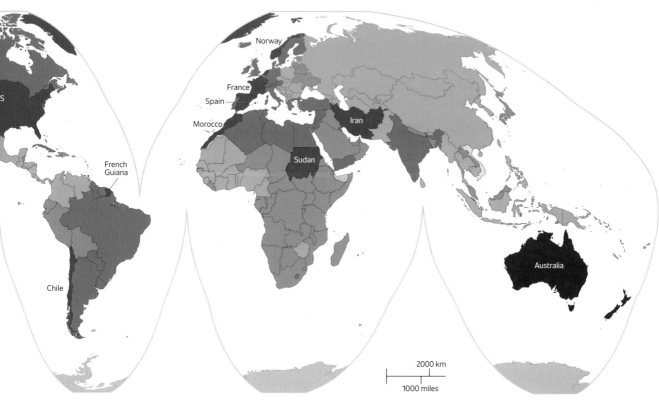

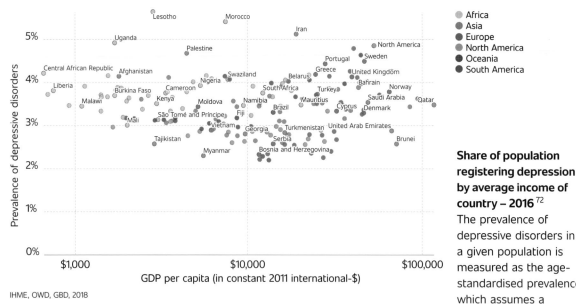

Prevalence of depressive disorders

- Africa
- Asia
- Europe
- North America
- Oceania
- South America

GDP per capita (in constant 2011 international-$)

IHME, OWD, GBD, 2018

Share of population registering depression by average income of country – 2016 [72]
The prevalence of depressive disorders in a given population is measured as the age-standardised prevalence, which assumes a constant age structure to compare between countries and through time.

likely to be under-reported, suggesting the problem may well be much bigger than we realise.

Mental health disorders are present across all geographies and income categories. Overall, around half of the general population of middle- and high-income countries will suffer from at least one mental disorder at some point in their lives.[74] In the United Kingdom, for example, an estimated one in four people will experience some kind of mental illness this year.[75] But some population groups are more vulnerable than others. The poor are twice as likely to experience mental disorders as the rich. The relationships are dangerously self-reinforcing – poverty increases the risk of mental disorders and having a disorder increases the likelihood of experiencing poverty.[76] Yet in low- and middle-income countries, as few as 15 per cent of the population receive treatment as compared to up to 65 per cent in higher-income ones.[77] While significant progress is being made to identify, treat and destigmatise mental illness, it continues to be taboo in some parts of the world.[78] The economic burden of mental health disorders is astonishing. Notwithstanding potential classification and reporting errors, a Lancet study determined that they could cost the global economy as much as $16 trillion by 2030, including at least 12 billion working days lost every year.[79]

The majority of deaths arising from mental health disorders occur indirectly, a result of suicide.[80] People diagnosed with clinical

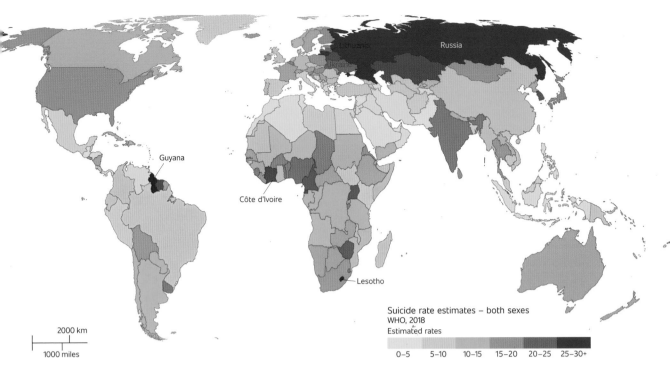

Suicide rate estimates – both sexes
WHO, 2018
Estimated rates

| 0–5 | 5–10 | 10–15 | 15–20 | 20–25 | 25–30+ |

Age-standardised suicide rates (per 100,000) for both sexes – 2016 [73]

This map provides an overview of suicide rates per 100,000. Countries in Eastern Europe – especially Russia and Lithuania – suffer exceedingly high levels of suicide. Meanwhile, wealthy countries such as Australia, Canada and France have rates that are on a par with many low- and middle-income countries.

depression[81] are on average about twenty times more likely to die by suicide than people who are not.[82] Intriguingly, depression is evenly distributed across most countries regardless of their level of economic development, accounting for about 2 to 6 per cent of a given population. But the extent to which depression is connected to suicide varies from place to place.[83] Very generally, in wealthier countries, roughly 90 per cent of all suicides can be traced to mental health and substance abuse problems. Meanwhile, in poorer ones, the proportion is lower, closer to 60 per cent, and is also connected to underlying social and economic conditions and possibly also cultural stigma and under-reporting.[84]

Suicide affects people from all countries, cultures and classes. As the map illustrates, the countries registering the highest suicide rates are incredibly diverse. In 2019, for example, Eastern European states such as Belarus, Kazakhstan, Lithuania and Russia registered among the highest rates of suicide. Yet these countries were rivalled by Guyana, South Korea and Suriname. Even Bhutan, famous for its gross happiness index, suffers from a high suicide rate. War-torn countries like Afghanistan, Iraq and Syria exhibit comparatively low rates of suicide, while the countries with the lowest rates are clustered in the Caribbean, including the Bahamas, Grenada and Jamaica. Meanwhile, Belgium, the Netherlands and Sweden report relatively high rates, but this may have something to do with the high prevalence of physician-assisted suicide, which is legal there.[85]

While it is typically more common at older ages, suicide can occur across a person's lifespan. In fact, suicide is today the second leading cause of death of fifteen to twenty-nine year olds worldwide. A person takes his or her own life every 40 seconds, more than 800,000 deaths a year, making it a bigger killer than murder and manslaughter.[86] Although over three-quarters of all suicides typically occur in lower- and middle-income settings (because of their larger combined population), it is hardly the preserve of the poor and destitute. Suicide typically comes down to an interplay of personal attributes – including genetic traits – alongside a mixture of social, cultural and economic conditions and environmental factors.[87]

There are signs that suicide rates are worsening in some parts of the world. In the US, for example, suicides have risen by almost a third since 1999. At least 47,000 Americans took their own lives in 2017 and there were another 1.4 million suicide attempts. In North America, mental health, souring relationships, loneliness and financial troubles seem to play a disproportionate role in the decision of people to attempt suicide. As the map shows, many western, Midwestern and Rust Belt states were particularly severely affected in recent years. Rural states and counties are especially susceptible to economic downturns and

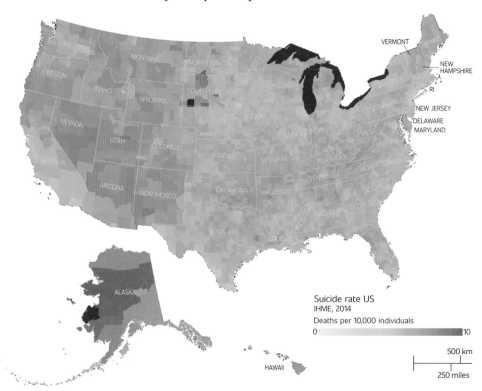

Suicide rate US
IHME, 2014

Deaths per 10,000 individuals

0   10

500 km

250 miles

Suicide death rate in the US (per 10,000) – 2014
There is considerable variation in the distribution of suicide rates between and within countries. This map displays suicide rates per 10,000 individuals at the county level across the US in 2014. It shows how comparatively high levels of suicide cluster in Alaska, the Southwest and Midwest, as well as parts of the eastern Rust Belt.

many people living there are more isolated and lack access to proper care.[88] Populations there have experienced a surge in what Nobel Laureate Angus Deaton and Anne Case have called 'diseases of despair', including deaths from liver disease and addiction to and overdoses from alcohol and opiates.[89] This map highlights the clustering of overdose deaths in Arizona, California, Colorado and New Mexico, but also in Alabama, Kentucky, Tennessee, and West Virginia. In 2017, there were an estimated 158,000 deaths of despair. More than 700,000 U.S. citizens have died from diseases of despair since 2000, more than the total number of U.S. combat deaths in both world wars.

Different factors are at play in shaping the decision of people to kill themselves in Japan and South Korea, two nations with exceedingly high suicide rates. In both countries, children have long been expected to care for ageing parents. But as young people moved in larger numbers to cities and traditional support systems came unstuck, the elderly have taken to ending their own lives to avoid imposing a financial burden on their families. Japanese and South Korean students also register a higher-than-average suicide rate due in part to the cultural and social pressures of having to succeed academically, as the education chapter makes clear. In Japan, where

Overdose death rate in the US (per 10,000) – 2014

The US has faced a wave of drug overdose-related deaths in the past decade. This map highlights the prevalence of overdose deaths per 10,000 in 2014, the latest year in which disaggregated data are available.

There is a clustering of overdose deaths owing to overuse of opioids in the Appalachia region (Alabama, Georgia, North Carolina, Pennsylvania, South Carolina and West Virginia) as well as the Southwest (Nevada, Arizona, New Mexico).

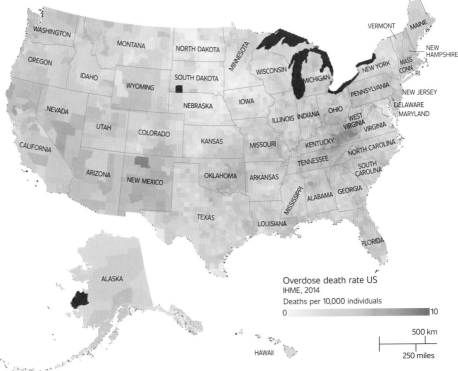

Overdose death rate US
IHME, 2014
Deaths per 10,000 individuals
0   10

500 km
250 miles

suicide is the leading cause of death for men aged twenty to forty-four and women aged fifteen to thirty-four, it is often motivated by the shame of divorce, lost employment and other forms of perceived dishonour. Child suicide is also extremely high there – the most common cause of death for people aged ten to nineteen – and peaks in September each year, at the start of school terms.[90]

Rethinking ageing – is sixty the new thirty?

The world's population is ageing more rapidly than ever. As the maps imply, the number of people aged sixty and over will more than double in the next three decades, from roughly 962 million today to 2.1 billion by 2050.[92] For the first time in history, the over-sixty set are among the fastest-growing age group in the world.[93] One of the reasons for this is that people are simply living longer, healthier lives. An equally important factor is that fertility rates are declining, which is reducing the stock of young people, as the demography chapter makes amply clear. Around the world, the average woman has just 2.5 children, compared to five a century back.[94] It is not just longer life and fewer children that explains ageing in some parts of the world; large-scale migration is also changing age structures in certain countries, particularly since those who leave tend to be younger.

All of these structural changes in population health are forcing national and municipal governments, health-care providers and insurance companies to fundamentally rethink the politics and economics of ageing. Traditionally, youth and fertility were associated with economic vitality while ageing and shrinking populations were a sign of stagnant and moribund economies. That view needs updating: an older and ageing population may be a valuable asset for societies. Not only can an elderly population lend perspective, experience and wisdom, but they will play a growing role in the labour force as income generators.[95] Some societies are already adjusting to the social and economic implications of a growing number of older people living alone, by changing retirement ages, developing life-long learning options and rethinking urban design.[96] The development of housing, employment and services that are more accommodating of the elderly, alongside the fostering of social networks, are all essential to avoiding loneliness and low-productivity that is associated with old age.

Predicted percentage of the population aged sixty and over – 2014 and 2050 [91]

These maps underline the speed at which populations are ageing around the world. They show the proportion of a country's citizens who are sixty years of age or over in 2014 and 2050. In Japan, addressing ageing is one of the most urgent policy priorities facing the country today. But notice how Canada, China, Russia, parts of Europe and also some countries in South America and Southeast Asia are projected to have a much older population by 2050. Countries such as Australia, Brazil and the US are not far behind. By contrast, Africa is projected to be by far the youngest continent in 2050.

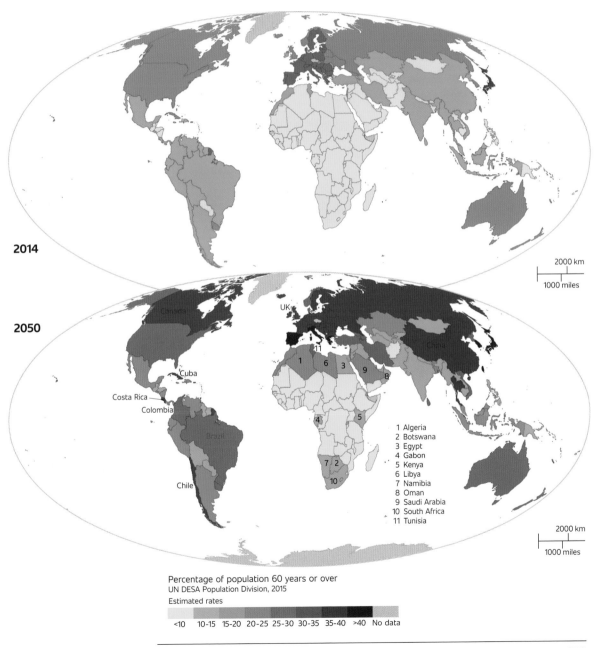

2014

2050

Percentage of population 60 years or over
UN DESA Population Division, 2015

Estimated rates

<10   10-15   15-20   20-25   25-30   30-35   35-40   >40   No data

Projecting fertility rates – 1990, 2020 and 2050 [97]

Fertility rates have dropped considerably in most parts of the world. These maps display past, current and future fertility rates by country in 1990, 2020 and 2050. Around the world, women are projected to have dramatically fewer children in 2050 than in previous years. The starkest differences are in Sub-Saharan Africa, South America and Central Asia, where the darker colours (representing more children) lighten over time.[98]

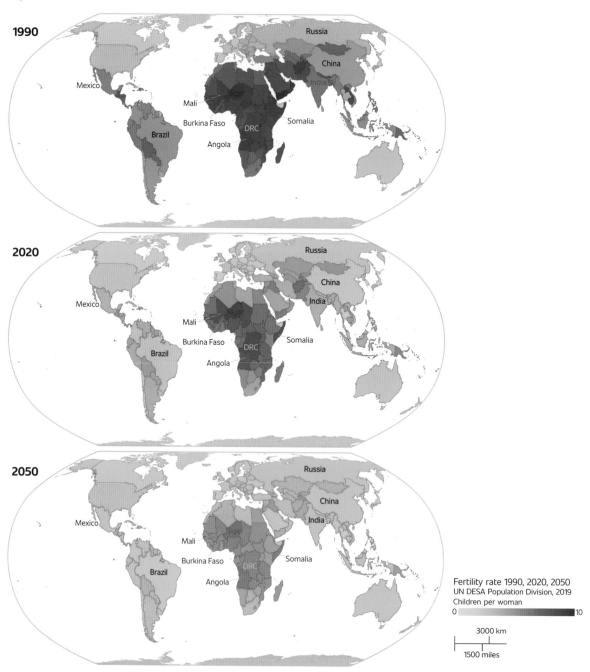

Fertility rate 1990, 2020, 2050
UN DESA Population Division, 2019
Children per woman

0 ▭▭▭▭▭ 10

3000 km
1500 miles

What can be done – the future looks healthier

The mesmerising achievements in global population health must not mask sharp inequalities in health between and within countries. Many countries where health improvements have slowed or stalled are bedevilled by their own tragic circumstances. Consider Russia, which has among the lowest life expectancies in Europe owing in part to binge-drinking habits among men (referred to locally as 'zapoi').[99] In the US, overall life expectancies also recently declined owing to unprecedented drug overdoses and suicide crises.[100] An even more dramatic example is Southern Africa, where HIV/Aids shaved an average of fifteen years off life expectancy in a single generation,[101] even as the rest of the world had fifteen years added to average lifespans.[102] Fortunately, antiretrovirals have helped reverse this trend, contributing to significant improvements in life expectancy.[103] Or consider the dramatic collapse of health systems – and life expectancies – in countries suffering from armed conflicts and extreme violence. Syria and Yemen each experienced unprecedented declines in life expectancy, especially among young males.[104] In Brazil, life expectancy is just under seventy-four years and improving, but not for everyone. Among young, under-educated black males living in the urban periphery of cities like Rio de Janeiro and São Paulo, life expectancy can drop to below sixty.[105]

Although pandemics remain a clear and present danger, over the coming decades human longevity is likely to be extended further. Many of the successes of the past century are due to the invention and application of tried and tested public health initiatives and infection-control measures, the spread of antibiotics and vaccination programmes, better education, and astounding medical, surgical and technological breakthroughs.[106] The extension of life has also coincided with a massive downward shift in fertility rates in most parts of the world. International efforts to improve health have helped. Indeed, the impressive decline in maternal mortality in lower-income countries is at least partly connected to global efforts catalysed by the United Nations Millennium Development Goals of 2000 and a host of other efforts supported by international philanthropic organisations.

In the future, humans will not die for the same reasons as in the past. In 2050, death rates from maternal and perinatal causes and non-communicable diseases should still be falling. With improvements in mobility solutions, smarter eating habits and improved screening and detection techniques, deaths due to road traffic injuries, heart disease, diabetes and breast cancer should also still be declining.[107] Meanwhile, mortality and morbidity associated with pandemic infections and mental health will probably increase. Of course, nothing is written in stone. All projections, including these ones, suffer from uncertainties, such as those caused by major new discoveries in medicine, economic shifts, the rise of superbugs, catastrophic climate change, and war. But taken together, and as the maps show, the prognosis for human health and longevity in most parts of the world looks good, and potentially might get radically better. There is also a possibility that within our lifetime we may see the rise of 'super-humans' with prolonged lives, an upgrade available only to those with the financial means to pay for it.

There are some low-hanging solutions for improving global population health. For one, universal health care coverage is the most obvious way to reduce the gap in health outcomes between and within countries and population groups. The provision of basic hygiene and health care for infants, children and mothers – especially in parts of the world that are chronically disadvantaged – is essential. Eradicating malaria, which infects almost 220 million people and kills 400,000 every year, would be immensely beneficial, and help towards eradicating poverty and reducing ill health in the poorest countries where it is endemic.[108] Another goal must be to eliminate especially devastating diseases and viruses – and minimise the threat of superbugs. We have done this before. Take the case of smallpox. The last reported cases were in Bolivia, Colombia and Somalia in 1977 – the first time in history that a disease was completely eradicated.[109] There are today just two known places that still hold samples of the virus for research purposes – in Atlanta (US) and Novosibirsk (Russia). Smallpox was not wiped out without a fight.[110] The first wave of eradication occurred in Scandinavian countries, followed by the rest of Europe, Australia and North America. The next wave occurred in the wake of massive vaccination programmes across Africa, Asia and Latin America, including a $300 million intervention overseen between 1966 and 1977 by the WHO. Yet the costs of these prevention efforts are tiny in comparison to the savings in lost lives.

Reducing the threat of superbugs requires the same kind of dedicated global response as past efforts to eradicate smallpox.[113] We must start by identifying, researching and developing responses for the top priority pandemic risks.[114] Equally important is preparing

Reported cases of smallpox by country – 1943 and 1977[111]
Smallpox was a vicious killer throughout history and into the twentieth century. These maps depict known cases in 1943 and 1977, the year smallpox was officially declared eradicated. Notice the dramatic decline in known cases, a tribute to the combined efforts of governments and international and civic organisations around the world.

1943

1977

2000 km
1000 miles

Number of reported smallpox cases
Fenner et al., WHO, 1988
(includes endemic and imported cases, therefore some countries still
recorded cases after the official year of eradication)

0   1   10   50   100   500   10,000   50,000   >100,000

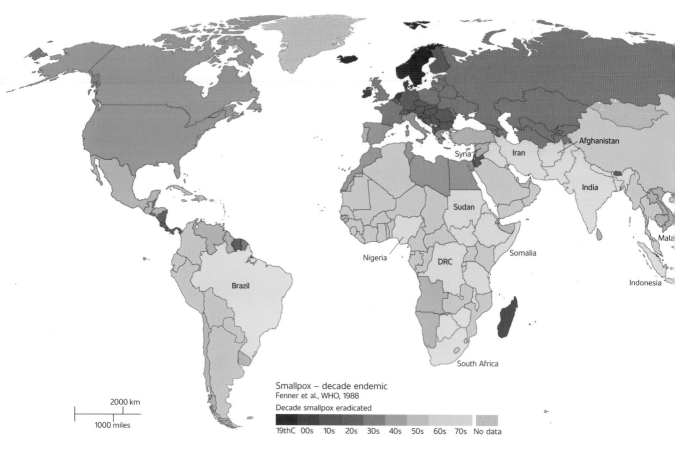

Smallpox – decade endemic
Fenner et al., WHO, 1988
Decade smallpox eradicated

19thC 00s 10s 20s 30s 40s 50s 60s 70s No data

Decade in which smallpox ceased to be endemic [112]

Smallpox was not wiped out all at once. This map highlights when specific countries eradicated the virus – with lighter colours representing more recent decades. Iceland, Norway and Sweden were among the first to successfully eradicate smallpox, followed much later by countries in Latin America, Africa and South and East Asia.

for lesser known threats – a possible 'disease x' – that would generate catastrophic outcomes should (or rather when) it make the 'jump' to humans such as COVID-19. Concerted efforts are also needed to radically accelerate antibiotic and vaccine development and deployment. Take the case of Ebola, where preliminary vaccines were developed in twelve months rather than the usual development cycle of five to ten years. There are some signs of progress: experimental vaccines used to contain the Ebola virus are up to 97.5 per cent effective. [115] These innovations will not come cheap even if the costs are coming down. The price tag for developing a vaccine for each of the eleven most infectious diseases runs into the billions of pounds. But it is possible that with the advent of new biotechnologies, testing techniques, and

even 3D printing facilities, these costs could drop considerably. In the end, the old adage – an ounce of prevention is worth a pound of cure – is more appropriate than ever.[116]

Preparedness and co-operation are the watchwords for the twenty-first century. While there are encouraging technological advances, the reality is that many countries and cities are struggling to implement even the most rudimentary measures to prevent and contain these threats.[117] As the COVID-19 pandemic makes clear, one weak link can put the entire world at risk. And while progress is clearly being made to build awareness about the risks, most governments are still failing to meet minimum international standards to detect, assess, report or respond to major public health threats, including superbugs. When infectious disease outbreaks hit, as they will with increasing frequency, responses will be absent, delayed or stretched too thin. The cycle of neglect, carelessness and panic that follow pandemics is fatal – not just on the front-line, where outbreaks occur, but to populations across the planet. It is possible to change course and bolster our collective defences, but this will require significantly expanding global partnerships between governments, investors, vaccination producers and health experts.

The good news is that a new generation of drugs (and methods to produce them) is appearing on the horizon. These developments can be accelerated with discovery grants, extended patents, and incentive competitions to, for example, lure private companies into making antibiotics and specialised drugs for especially high-risk bacteria and viruses. There are grounds for hope. As we were writing this chapter we were aware of dozens of clinical trials testing for a COVID-19 vaccination, and over 30,000 peer-review scientific articles on every conceivable facet of the newly discovered virus. Already there are thousands of new antibiotic combinations emerging that could reinforce human defences and target drug-resistant pathogens.[118] Entirely new classes of antibiotics are being tested with potentially game-changing properties.[119] To improve the speed at which they are brought to market, greater investment is needed to harvest and analyse data on antibiotic use and bacterial and viral response. And to limit the possibility of more drug resistance, governments will need to restrict and potentially end the livestock use of antibiotics (and help nudge consumers to adopt more plant-based diets). Ultimately, all of us will need to change our mindsets and decisions – including limiting antibiotic use for minor infections and avoiding any food products that use antibiotics.

The advent of Big Data and analytics, along with the Internet of Things, is profoundly changing the way we track patient and vital statistics. Wearables help detect how people sleep, their heart rates and their exercise routines. Innovations are allowing physicians to monitor blood pressure and glucose levels remotely, helping them to identify potential health concerns and provide life-saving treatment before they worsen. The availability of this data is also set to drastically improve how health-care facilities predict admission rates, allocate resources and ensure home care for ageing populations. Many new technologies are being adopted and applied in middle-income countries such as China and India, with virtual care already being deployed to reach remote populations. The insurance sector also stands to gain handsomely by promoting wearables, health trackers and data analytics to keep patients from spending excessive time in hospitals and health-care facilities. New medical technologies will not only increase efficiency, they can also help prevent human error – for example, the prescription of the wrong medicine or the incorrect diagnosis of an X-ray. This of course raises a new set of questions related to data privacy and protection, issues picked-up in the chapter on technology.

Runaway advances in neurology, stem cell research, gene editing and other fields are about to upend our conception of health and ageing – and potentially even what it means to be human. Some biologists and geneticists regard ageing not as a mysterious or inevitable condition, but as the undesirable side-effect of being alive. Over time, the body suffers from metabolic damage, and physical and psychological pathologies accumulate. In their view, ageing will increasingly be 'healed' through occasional cellular and molecular-level repair and replacement.[120] While there is still an element of fantasy to the cure for ageing, advances in personalised medicine that pair genomic mapping with targeted drug treatments and upgrades are set to revolutionise the field of health. Make no mistake: future humans will be cyborgs.

One of the most important things all of us can do is continue educating ourselves about how to stay healthy, something we discuss at length in the chapter on food. This requires continually tracking diseases, constantly improving our knowledge about how to prevent and treat them, and passing on insights so as to improve health outcomes and reduce disparities between and within countries. Data sharing is still a major roadblock, though technology can help improve this. The advent of new drugs and a greater appreciation of the determinants of population health are going to generate new breakthroughs. Whether humans live to 100 or 1,000, we would all be advised to prepare our societies for an ageing population. This means investing in appropriate health and social-care systems to assist populations with multiple health needs. It also requires rethinking the nature of insurance and pension systems, and the duration and meaning of work itself. The social, economic and ethical implications are daunting. For example, who gets to live longer? How do we manage population growth? What about immortal dictators? While these changes are hard to fathom, this is precisely why we need to start debating them today.

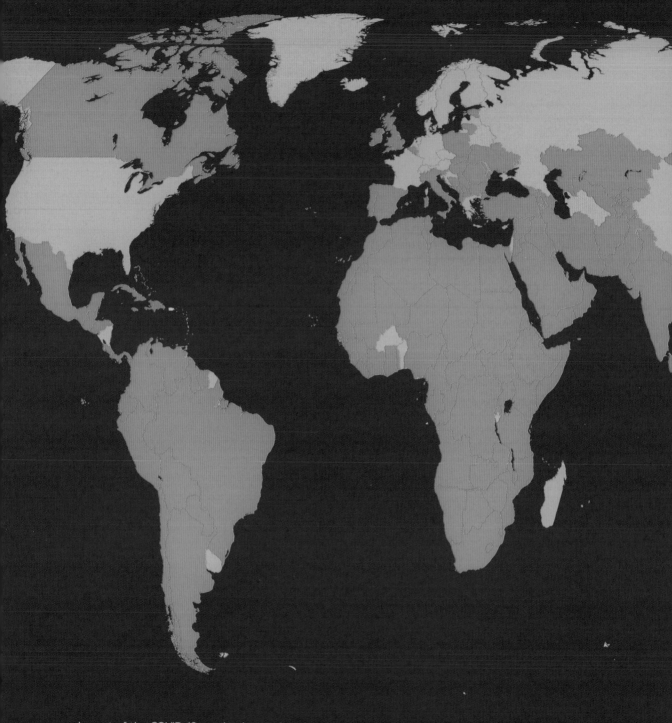

Impact of the COVID-19 pandemic on education

Wikimedia Commons, 2020

Country-wide school closures ☐ Localized school closures ☐ No school closures ☐ No data

Education

Universal education is a new idea

Access to education is improving

Improving education requires smart investment

Better education equals more economic growth

Education systems are in need of an upgrade

Introduction

The COVID-19 pandemic is a reminder that the world our children will occupy in the future will be radically different to the one we live in today. The only constant will be change. Digital and virtual tools and remote learning will be omnipresent. The availability of unlimited content will for all intents and purposes eliminate the need to remember facts. To navigate this bewildering and fast-changing environment, future generations will need to be literate in science, technology, engineering and maths, subjects increasingly taught using Artificial Intelligence, or AI.[1] In some primary schools, children are already learning to code before they can read or write. This will soon be the norm. But to genuinely thrive, they must also be creative, empathetic, critical thinkers and problem-solvers. All of them will require the mental fortitude and stamina to navigate tremendous complexity. School will not end in their early adulthood: life-long learning and constant reinvention will be the norm. Our children will acquire all these skills without having a clue about what jobs will exist by the time they graduate.[2] No wonder parents everywhere are feeling anxious and unprepared.

While there are many reasons to be nervous about the future, especially as the world struggles to recover from multiples waves of COVID-19, there are also grounds for optimism. After all, we have come an extraordinarily long way in a short time. As the map shows, most people living outside of North America and Northern Europe had fewer than four years of education in 1950. By 2017, the average years of schooling around the world had more than doubled. The number of children and adolescents out of school has also fallen spectacularly in recent decades. In 1970, around 27 per cent of primary-school-aged children were out of school: today, it is closer to 9 per cent. Most parts of the world are also closing in on gender parity at the primary school level. These extraordinary gains have propelled economic growth and have even helped democracies to flourish, not least because of the strong association between access to quality education and broader improvements in social and economic development.[3]

Mean years of schooling – 1950 and 2017

More people than ever are accessing education. These maps display the mean years of education of adults aged twenty-five and older. Notice the stunning increase in mean years of education between 1950 and 2017, illustrated with darker colours.

1950

2017

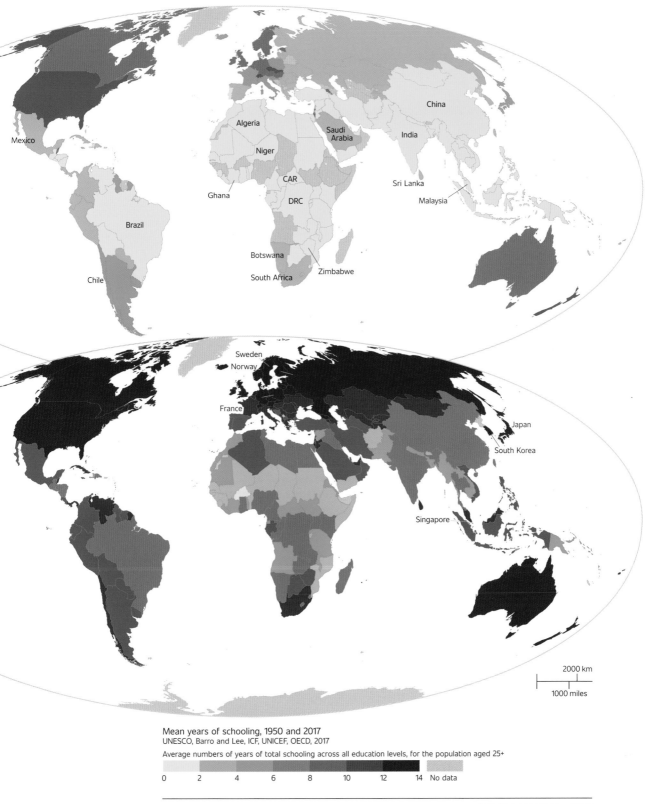

Mean years of schooling, 1950 and 2017
UNESCO, Barro and Lee, ICF, UNICEF, OECD, 2017

Average numbers of years of total schooling across all education levels, for the population aged 25+

0 2 4 6 8 10 12 14 No data

Not everyone is benefiting equally from recent transformations in education. At least 120 million children are still out of school.[4] Another 250 million young people have yet to acquire rudimentary skills in literacy and numeracy. There are at least 750 million youths and adults who cannot read or write, which excludes them from full participation in their societies.[5] Despite tremendous advances, younger girls are still held back from attending school or taking certain types of courses in some parts of the world. As the map shows, the mean years of schooling for many Africans, Asians and Latin Americans in the past decade is still considerably lower than for North Americans, Europeans, Australians and New Zealanders. Yet despite these stubborn disparities, the story has been a largely positive one. While pockets of disadvantage persist in wealthy and poorer countries and between males and females, educational inequality has steadily declined.[6]

There's just one big problem: the world's education systems are not remotely prepared to deal with the dramatic transformations ahead. Seismic changes such as the rise of the knowledge economy, structural changes in the nature of work, and the exponential spread of new technologies have tremendous implications for what, where and how we educate the next (and current) generations. In some parts of the world, educational reforms and opportunities for AI-powered instruction are accelerating.[7] In others, more traditional models of education using blackboards, notepads and pencils prevail. What is more, the COVID-19 pandemic will probably accentuate inequalities between and within countries and hasten the digitisation of education. In the COVID-19 era, it is not at all clear whether curricula, teachers, schools and universities can change fast enough to adapt to this uncertain future.

In this chapter we chart the remarkable spread of education. Our point of departure is the unprecedented expansion of primary, secondary and tertiary education over the past century. We take stock of the education dividend – including the steady improvement in the quantity and quality of teaching – and how it has helped power growth and even democratic governance in most parts of the world. We then consider new experiments in education in places like Finland, Singapore and South Korea, and the disruptive impact of new educational technologies. The power and potential of new digital tools are real. But one thing is for certain: we have virtually no idea what comes next.

The rise of literacy – from 1 per cent to 86 per cent in 500 years

In many countries we take education for granted. This may change in the wake of the COVID-19 pandemic, which has rapidly introduced billions of parents to home-schooling. It is hard to believe that the education revolution is only a few hundred years old. The idea that children, much less girls, should be entitled to a full day in school is a radically new idea. For almost all the past 5,000 years, 'education' was restricted to the privileged few – almost exclusively wealthy males. Some of the first dedicated schools and libraries were built on Egyptian soil. Most of them were designed to instruct scribes, healers and temple bureaucrats in basic reading and writing. While different civilisations developed their own language systems, the oldest known alphabet – hieroglyphics – emerged just a few thousand years ago, around 3,000 BC.[10] Various forms of education spread across the Middle East, Europe, China and India, all of them catering to a tiny elite. Until a few hundred years ago, less than one per cent of the global population could be considered literate by today's standards.

The concept of universal education took off during the European Enlightenment in the seventeenth and eighteenth centuries. The explosion of book publishing, public libraries and a print culture

Global literacy rates between 1475 and 2015 [8]

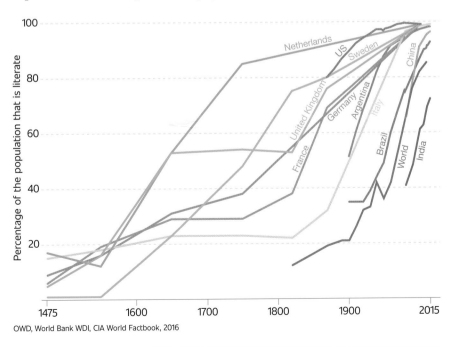

OWD, World Bank WDI, CIA World Factbook, 2016

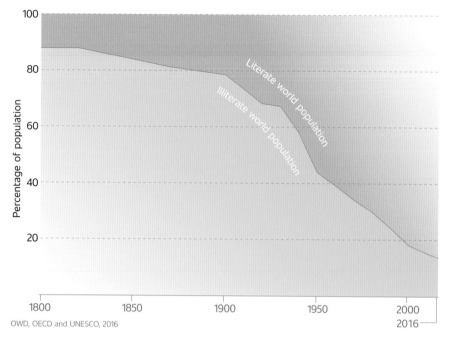

Literate and illiterate world population 1800–2016 [9]

all played a part. Popular philosophers such as John Locke and Jean-Jacques Rousseau advocated passionately for public and private investment to shape young minds as early as possible. Although literacy rates started rising in Europe from the 1500s onwards,[11] global literacy rates grew comparatively slowly before accelerating in the 1900s. The figure shows how at the beginning of the nineteenth century, global literacy rates hovered at around 12 per cent. By the twentieth century, the percentage had risen to 21 per cent. Over the past six decades, literacy rates increased every year to the point where today, an astonishing 86 per cent of the world is considered literate.[12]

An educated world – how South Korea went from 5% to 100% enrolment in half a century

While it may not feel like it sometimes, the world is more educated than it's ever been. This is chiefly a result of laws guaranteeing a citizen's right to education. In most countries, children are legally required to attend primary school. The inspiration for compulsory primary education goes back to Prussia's Frederick the Great, who set up the first primary education system in 1763. He insisted that all Prussians – boys *and* girls from five to fourteen – attend publicly funded schools.

Population having attained at least basic education – 1870–2010 [13]

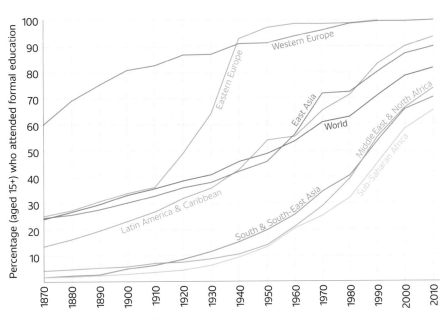

OWID, Clio-infra.eu via Van Zanden et al., 2014

He also invested in secondary schools, known as 'gymnasiums', and university pre-prep schools. By the early nineteenth century the Prussian system included everything from mandatory kindergarten, specialised teacher training and a nationwide curriculum, to national testing for students and certification requirements for teachers. Similar systems spread throughout Europe and North America in the nineteenth century and the rest of the world thereafter. Today, the right to education is enshrined in international treaties that require free education, accessible secondary education for all, and access to equitable higher education.[14] As the figure shows, some countries took longer to make this transition than others. For example, India introduced universal, free and compulsory primary education only in 2009. Several countries in Africa, the Middle East and Southeast Asia have yet to make the leap. The worst performers are in the world's poorest countries – Burkina Faso, Chad, Ethiopia, Mali and Senegal – where children still average only three years of school.

Although some children are still denied access to education, the global rate of educational improvement overall in recent decades is awe-inspiring. School enrolment and attendance have soared while the proportion of people without formal education has plummeted. The world's two largest countries by population – India and China – have achieved spectacular improvements over the past half-century.

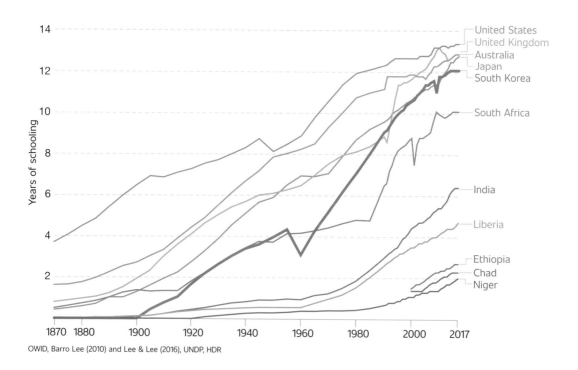

OWID, Barro Lee (2010) and Lee & Lee (2016), UNDP, HDR

In India, for example, 12 per cent of the population was literate in 1947, as compared to 74 per cent today. And China has graduated from 20 per cent literacy in 1950 to 85 per cent today. Around the world, younger generations are progressively better educated than older ones – which means that overall improvements are likely to persist into the future.[15] To put all this in perspective, in the coming three decades more people will obtain formal primary, secondary and university education than in all of human history combined.

School attendance is one thing, but educational attainment is another. When it comes to educational performance, some countries vastly outpace others. The South Korean experience stands out. In 1950, just 5 per cent of Koreans had a high-school diploma. After introducing universal education and a host of other reforms, it now registers among the highest literacy rates globally.[17] South Korea's school system is widely hailed as one of the world's finest and the country's teenagers routinely outpace their counterparts in educational performance.[18] And it's not just primary and secondary schools that have registered improvements: South Korean universities are consistently climbing up the international rankings.[19] Before the COVID-19 pandemic, South Koreans also made up the third largest group of overseas students in the US after Chinese and Indians.[20] South

Mean years of schooling in selected countries, 1870–2017[16]
The sharp differences between the years of schooling in richer countries and the poorest countries is starkly apparent. Note the rapid rise of South Korea.

Share of population with no formal education – 1970 and 2050 [21]

These maps illustrate the share of a country's population with no formal education in 1970 and projections for 2050. Some of the most impressive achievers include Brazil, China, India and Mexico. Significant improvements are also likely across many African countries as well.

Korea has benefited tremendously from this educational dividend. The so-called 'Miracle on the Han River' – the country's remarkable economic boom that kicked off in the mid-twentieth century – comes down to smart investments in education and a culture that rewards scholastic achievement. Today, virtually everyone finishes high-school and three-quarters of them go on to university.

Stark inequalities in educational achievement persist. As the maps show, despite improvements, Sub-Saharan Africa ranks poorly. Countries such as Burkina Faso and Niger are at the bottom of the world tables with numeracy and literacy rates hovering below 30 per cent. Although most African countries expanded primary school

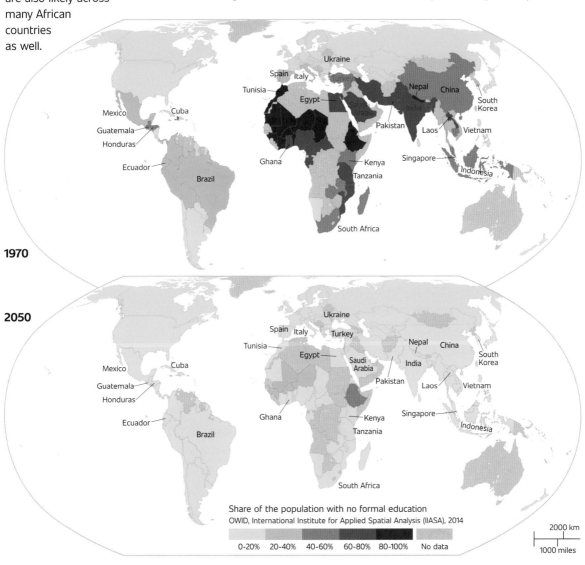

1970

2050

Share of the population with no formal education
OWID, International Institute for Applied Spatial Analysis (IIASA), 2014

0-20%   20-40%   40-60%   60-80%   80-100%   No data

2000 km
1000 miles

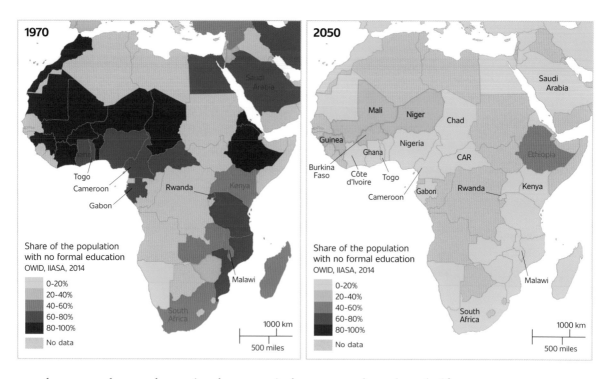

1970

Share of the population
with no formal education
OWID, IIASA, 2014

0-20%
20-40%
40-60%
60-80%
80-100%
No data

Togo
Cameroon
Gabon
Rwanda
Kenya
Malawi
South Africa

1000 km
500 miles

2050

Share of the population
with no formal education
OWID, IIASA, 2014

0-20%
20-40%
40-60%
60-80%
80-100%
No data

Saudi
Arabia
Mali
Niger
Chad
Guinea
Nigeria
Ghana
CAR
Ethiopia
Burkina
Faso
Côte
d'Ivoire
Togo
Cameroon
Gabon
Rwanda
Kenya
Malawi
South
Africa

1000 km
500 miles

enrolment and attendance in the twentieth century, less than half of all Chadian, Liberian and Nigerois school-age children show up for class. Some African nations have even started regressing in core measures of educational outcomes.[22] Making matters worse, most graduating students are woefully unprepared for tertiary education, or a competitive labour market. Less than half of all high-school graduates in Sub-Saharan Africa actually achieve the minimum threshold of educational proficiency (a global benchmark for a basket of skills). And as the maps above show, these trends may not dramatically improve in the Sahel, Great Lakes or Southern African regions in the next three decades. It is likely that the COVID-19 pandemic and its aftermath will also slow educational achievement, especially in poorer parts of the world with more limited educational opportunities.

Education does not magically improve just by throwing more money at it. In fact, material investment in education has risen just about everywhere yet improvements have been far from universal. This is partly because there are sharp differences in how societies prioritise spending. In wealthy societies, for example, households typically spend proportionately more on higher education than on primary or secondary education. This is not surprising: in cities like Toronto, London, New York, Paris or Sydney, the availability and quality of

Education deficits in Africa – 1970 and 2050
The education revolution will spread to Africa over the coming decades. These maps highlight how the proportion of the population receiving no formal education is expected to decline by 2050, but it is also a reminder that major shortcomings remain.

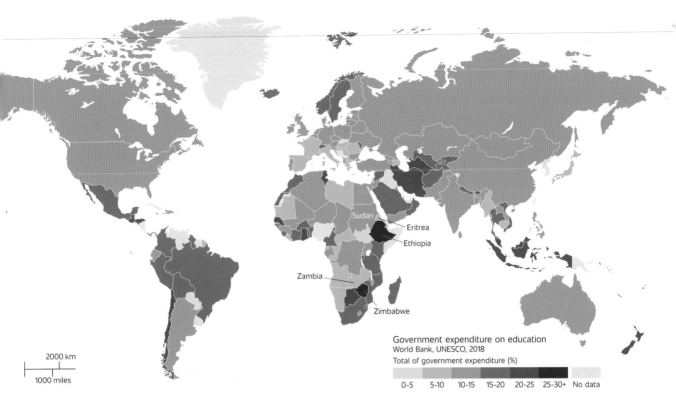

Government expenditure on education
World Bank, UNESCO, 2018
Total of government expenditure (%)

| 0-5 | 5-10 | 10-15 | 15-20 | 20-25 | 25-30+ | No data |

Government expenditures on education – 2018 [23]
More funding of education does not necessarily translate into higher educational achievement. This map displays overall public expenditures on education (current, capital and transfers) expressed as a percentage of total government expenditure on all sectors. Darker shades represent a larger portion of government expenditure on education.

public education is generally high. The outcomes are progressive to the extent that many lower-income students receive subsidies to attend higher education and students from better-off households are more likely to be able to afford to go to university. In poorer countries the situation is reversed: households spend more on primary education than on high schools or universities. Even when state schools in lower- and middle-income settings are free, many parents prefer to put their children in higher status or better performing private institutions instead.

Even so, improving both the availability and quality of education will require greater investment by governments. Achieving rapid gains in low- and middle-income countries will require increasing global annual spending from around $1.2 trillion today to $3 trillion by 2030.[24] The map shows how the proportion of public spending on education is not the most trustworthy measure of better educational outcomes. Some countries in Central America, Sub-Saharan Africa and the Middle East devote as much as a quarter of their budgets to education, with very mixed results. In recent years, some of the highest spenders as a proportion of overall government spending were Ethiopia and Zimbabwe and the lowest spenders were South Sudan, Zambia and Eritrea. Studies by the World Bank suggest that there is in fact a weak correlation between public spending and learning outcomes.[25] Of course, correlation is not causation. But it turns out that the way financing is delivered to schools, together with certain types of educational reform, are key to making a positive difference in completed years of education, future wage earnings and even reducing adult poverty. As with all social policy, it is not just the amount of money that is spent that matters, but how it is spent.[26]

Spending more money, building more schools and hiring more teachers is necessary but insufficient in low- and middle-income settings. Good performance depends on many inputs ranging from the types of skills being taught and the quality of teacher training, teaching and facilities, to the extent of school attendance, incentives for academic effort and a host of social and economic factors that may enhance or inhibit educational outcomes. There is a real opportunity here for innovation. In some (but not all) countries, for example, programmes like conditional cash transfers (which provide funds and other incentives to mothers to keep children in schools),[27] offering free breakfast and lunch for students, adjustments to school day schedules, and targeted scholarships for poorer households to keep children in school (rather than working), have had as positive an effect on overall educational achievement as the number of teachers or size of classrooms.

An education windfall

Education is first and foremost about helping everyone live to their full potential. The good news is that basic education – especially the development of cognitive and life skills – usually confers economic benefits on individuals, households and societies.[28] Universities and think-tanks have generated countless studies showing the economic rate of return of primary, secondary and tertiary education over the past half-century.[29] A common refrain is that regardless of gender or age, people with more education tend to report earning higher wages, lower mortality, and more pro-social behaviour.[30] For example, the average rate of financial return for one year of schooling is between 8 and 13 per cent. One study looking across 139 countries from 1950 to 2014 found that each additional year of education is associated with, on average, a 9 per cent increase in hourly earnings.[31] And the differences in earnings between individuals with varying degrees of education expands as they advance in their careers.

More and better education is correlated with higher incomes and broader economic growth.[32] Put simply, improving education is good for the economy. According to one study, if every child were provided with basic education it could boost GDP by an average of 28 per cent per year in low-income countries and 16 per cent in high-income countries for the next eight decades.[33] Increases in the number of universities in a country is associated with GDP per capita growth as well.[34] So why have school-based development strategies under-delivered?[35] One of the reasons is that the expansion of school enrolment and attainment does not on its own translate into improved economic performance.[36] Take the case of Latin America, which registered adult school attainment levels far ahead of those in East Asia and the Middle East in the 1970s. Yet economic growth in Latin America lagged far behind these two regions over the past forty years. This is because it is not just the quantity, but the quality of education that matters. While Latin America may register decent attainment in schools, what and how students are taught is often of comparatively lower quality.[37]

Education can also generate positive knock-on effects because of the way it forges social capital and other forms of well-being. For example, educational attainment is correlated with higher self-reported levels of trust.[38] Specifically, individuals with university and college education tend, on average, to trust one another more than those with primary or secondary education. Overall, adults reporting higher educational qualifications are also typically more invested in voluntary activities and are more politically active in their communities. And because it leads to more empowerment, healthier habits and a wider array of choices, women and girls' education in particular is strongly associated with lower child mortality and positive household earnings as the demography chapter makes clear.[39]

There is also evidence that educational improvements are associated in certain settings with the growth and sustainability of democratic governance.[40] This is because, for the most part, educated populations are typically more involved politically and exhibit a more developed sense of civic duty.[41] Notwithstanding notable exceptions like China, Cuba and Singapore, countries reporting the highest average levels of adult education in 1970 are more likely to have democratic governments today.[42] Social scientists have also determined that more educated populations are typically less likely to fracture and tilt towards populism and authoritarianism.[43] It is no coincidence that many of the people most in favour of populist and authoritarian parties register comparatively lower levels of education than their fellow citizens. In the US, for example, a survey of more than 5,000 people in 2017 found that support for democracy was statistically lower among people with less formal education, who did not follow the news and who said they did not vote.[44]

Rethinking education – what Finland and Singapore can teach the world

Despite the remarkable impacts of twentieth-century education systems, they are due for a major overhaul.[46] A major reason for this is that current educational models are ill-suited to the massive technological and labour market disruptions on the horizon.[47] In the long run, many of today's jobs are going to disappear and the future economy will be fundamentally different from the one we are presently teaching for.[48] Virtually everyone agrees that sweeping educational reforms are long overdue. Schools have not changed much since the first industrial revolution, when young people were being prepared for a future in manufacturing and public service. With exceptions, education and training systems have lacked investment for decades and are not primed to prepare young people for the robots that will render many of today's jobs obsolete.[49] While there is consensus that schools and universities need reforming, there is less agreement about what tomorrow's educational system should look like, much less how to pay for it.

Several educational experiments are in motion and offer a hint of what's to come. Their principle innovations are to dramatically improve the quality of teaching and rethink the delivery of school-based learning. One widely heralded example is from Finland, which has registered dramatic improvements in maths, science and reading among primary and secondary students.[50] This did not happen overnight – it started more than four decades ago.[51] These stunning achievements come down primarily to advances in the quality of teaching. Leading up to the COVID-19 pandemic, Finland had roughly 62,000 teachers in

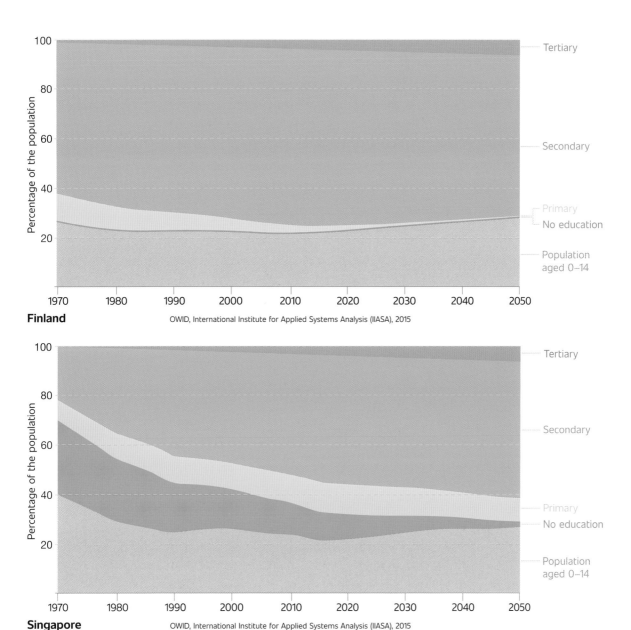

100 — Tertiary

80

60 — Secondary

40

20 — Primary
— No education

— Population aged 0–14

Percentage of the population

1970 1980 1990 2000 2010 2020 2030 2040 2050

Finland OWID, International Institute for Applied Systems Analysis (IIASA), 2015

100 — Tertiary

80

60 — Secondary

40

20 — Primary
— No education

— Population aged 0–14

Percentage of the population

1970 1980 1990 2000 2010 2020 2030 2040 2050

Singapore OWID, International Institute for Applied Systems Analysis (IIASA), 2015

Finland and Singapore: Population breakdown by highest level of educational achievement: 1970–2050 [45]

more than 3,500 schools. Acceptance into teacher-training programmes is a major accomplishment there: candidates are selected from the top 10 per cent of the nation's university graduates. Since salaries are competitive, around 90 per cent of Finnish teachers are life-long educators.[52] After-school tutorial programmes have also helped. Nearly a third of all Finnish students receive additional assistance during their nine years of schooling. And all of this pays off. Since 2000, Finnish students have ranked at the top of the Programme for International

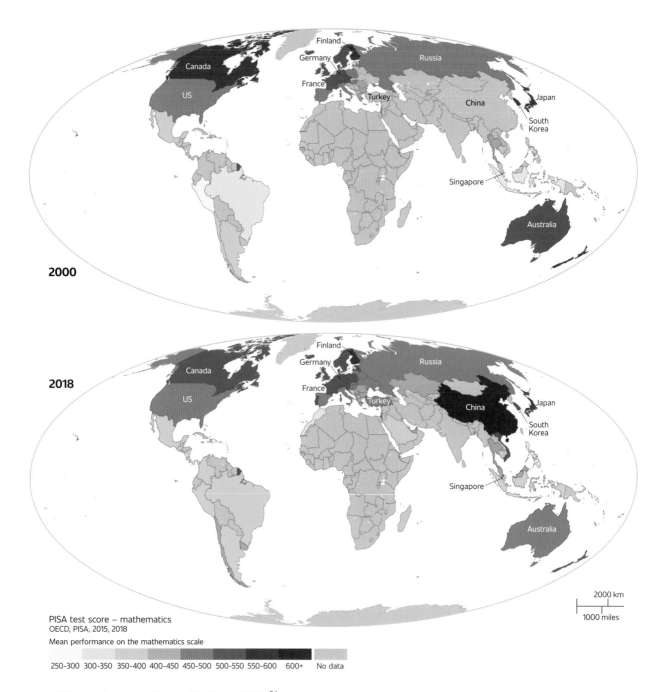

2000

2018

PISA test score – mathematics
OECD, PISA, 2015, 2018

Mean performance on the mathematics scale

250-300 300-350 350-400 400-450 450-500 500-550 550-600 600+   No data

PISA worldwide rankings – 2000 and 2018 [54]

Some students are consistently performing much better than others. This map
highlights the results of PISA test scores in seventy countries in 2000 and
2018. PISA measures the scholastic performance of fifteen-year-old students in
mathematics, science and reading. Darker colours on the map represent higher
average test scores, with China, Finland, Japan and Singapore ranked at the top of
the class.

2000 km

1000 miles

Student Assessment (PISA), a standardised test for fifteen-year-olds in more than forty countries. About 93 per cent of them graduate from academic or vocational high school – 17 per cent more than in the US. And 66 per cent of them go on to university.[53]

What makes the Finnish experience so astonishing is that all of this was accomplished without students having to do homework or take mandated standardised tests. Finns take just one exam at the end of their senior year of high school. There are no rankings or comparisons between students. What is more, the country's educational facilities are all publicly funded – they do not involve vouchers, private tutoring or charter schools. About 97 per cent of all six-year-old children attend a state preschool, where they are also provided with food, medical care, counselling and transport. As a result, almost every Finnish child has an equal shot at getting the same quality education no matter where she or he lives, something unheard of in most parts of the world. And the difference between the weakest and highest achievers is one of the smallest in the world, according to the Organisation for Economic Co-operation and Development, or OECD. Incredibly, Finland achieves all this at a cost that is 30 per cent lower per student than in the US.[55] And there are signs that social outcomes are improving across the board, from increases in overall happiness,[56] to reductions in alcohol consumption[57] and decreasing levels of suicide.[58]

Given these remarkable outcomes, the next question is whether Finland's educational experiment can achieve similar outcomes in other countries. The answer is maybe, but only if several conditions are met. The first is the extent to which a society values education. Finland's current educational system emerged from a culture that sees education as a fundamental right and not a privilege for the select few. For around 100 years, Finland's constitution has required that 'everyone has the right to basic education free of charge', which guarantees a life free of economic hardship.[59] By way of contrast, despite amendments to the Constitution, education is still not a fundamental right in the US. Another requirement is the extent to which there is equity in educational services. In Finland, funding for education is issued according to the needs of schools, not their ranking. Even the tiny number of private schools that exist do not charge fees and are prohibited from enforcing selective admission criteria. And students are offered both general and vocational training throughout their lives at no cost. In the end, the goal of the Finnish

Share of top performing students – 2017

Another measure of education attainment is the share of students who are performing at a high standard. This map reveals a country's share of top performing students based on average maths and science test scores – from primary through to the end of secondary school. The darker colours represent larger portions of top performing students. As can be seen, Finland, South Korea, Singapore and Japan all score extremely highly.

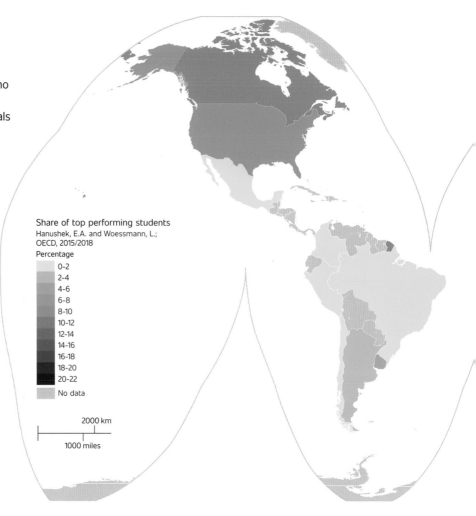

Share of top performing students
Hanushek, E.A. and Woessmann, L.;
OECD, 2015/2018

Percentage

- 0-2
- 2-4
- 4-6
- 6-8
- 8-10
- 10-12
- 12-14
- 14-16
- 16-18
- 18-20
- 20-22
- No data

2000 km
1000 miles

education system is to support the full development of students and to establish a wide range of professional and life skills.[60] This is a high bar, but a necessary one.

Tellingly, nations that have risen to the top of the educational pyramid are constantly striving to improve. Take the case of Singapore, where schools have delivered stunning academic results despite criticism for being highly scripted, overly-dependent on rote learning and lecture-based.[61] In contrast to Finland, teachers there rely heavily on textbooks, drills and testing.[62] Yet today, Singapore's educational system is considered among the world's best: pupils are nearly three years ahead of their US peers in maths. While these gains are impressive, they come at a cost. From a very young age, Singaporean students sit tests that will determine their prospects in adulthood.

Parents are determined to ensure their children receive after-school tuition and enrichment classes to give them a competitive edge. By the time they complete primary school at twelve years of age they face a decade of intensive tests.[63] In Finland, the equivalent national exam only comes at eighteen. One result is a society that is fixated on grades and focused on educational achievement as the basis of self-worth.[64]

Singaporeans, like their Japanese and Chinese counterparts, are extraordinary test takers, but not necessarily the best risk-takers. This approach may deliver stunning results in algebra, calculus and physics, but it may not be the best for preparing a workforce that is creative, innovative and adaptive. This is starting to change. Singapore is upgrading its education system to prepare for the rapid transformations ahead. For example, its 2018 Smart Nation Initiative is encouraging greater adoption of new technologies and the acquisition of digital skillsets.[65] The Ministry of Education is undertaking sweeping reforms for primary and secondary students, including measures to reduce the emphasis on grades as the measure of progress and self-worth. The idea is to reduce the focus on *kiasuism* – the fear of losing out or under-performing – and the perennial need to get ahead.[66]

The future of schooling – why our ten-minute attention span matters

As the experiences of Finland, South Korea and Singapore illustrate, there is no one-size-fits-all approach to education. They also remind us of the way education is changing. At least three trends – continual learning, distributed learning and better learning – will reshape the future educational landscape. Just as medicine was transformed by the advent of X-ray machines, biochemistry and genetic research, so too will education be reimagined by insights from neuroscience, machine learning and AI. For example, some scientists believe that human beings have an average attention span of just ten minutes (which means we might have lost you a few pages back!), not forty-five or sixty, which is the length of a typical class or lecture. Researchers are also discovering that learning is dramatically amplified by immersion, active involvement and processing ideas in groups. The act of learning fundamentally depends on curiosity since it can trigger dopamine response mechanisms with positive feedback loops. New discoveries in learning disabilities are also opening up possibilities for targeted and personalised instruction using a combination of sensors and highly tailored instruction.

A dollop of caution is called for. This is not the first time there have been calls to reboot education. During the 1990s dot-com bubble, technologists spoke breathlessly of the coming revolution in education, but the promised uprising never arrived. One of the reasons was that the conversation was too technology-centric, focusing more on designing (and selling) hardware than changing the software of teaching itself. There are good reasons to believe that this time is different. The world really does appear to be on the cusp of the transformations envisioned in the 1990s. The pace of digitisation has sped up just about everywhere, especially since the COVID-19

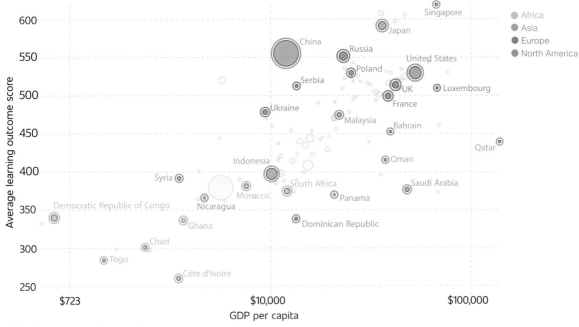

- Africa
- Asia
- Europe
- North America

Singapore
Japan
China
Russia
United States
Poland
Serbia
UK
Luxembourg
France
Ukraine
Malaysia
Bahrain
Qatar
Indonesia
Oman
Syria
South Africa
Saudi Arabia
Morocco
Panama
Democratic Republic of Congo
Nicaragua
Ghana
Dominican Republic
Chad
Togo
Côte d'Ivoire

Average learning outcome score: 600, 550, 500, 450, 400, 350, 300, 250

GDP per capita: $723, $10,000, $100,000

OWID, Altinok, Angrist and Patrinos, 2018

National average learning outcomes vs GDP – 2015 [67]

On average, countries with higher education learning outcome scores also report higher GDP per capita.

pandemic.[68] China is already leading the way in AI-powered education in and outside the classroom. AI is being used to teach rote tasks, freeing up teacher time to focus on individual students. New applications are also dramatically improving user test scores in the college entrance exam, *gaokao*. Chinese investors poured over $1 billion into AI-driven education platforms in 2019, and experiments there could potentially transform education globally.[69] But investments in ed-tech are hardly restricted to China: global spending is growing fast and is expected to double to over $340 billion by 2025.[70]

One of the most disruptive forces in education is the internet for the simple reason that it connects more people to information and opportunity. In the early 1990s, just a few million people had broadband internet access. Today, the number is closer to 5 billion. The map captures the dramatic worldwide transformation in access since 2000, even if swathes of Africa and Asia are still catching up. Put another way, just under 2 per cent of the world's population were internet users in 1995 as compared to over 59 per cent today. A similar number of people have access to smartphones and tablets, giving them instant access to the internet, and new opportunities for in-class and distance learning. And if you think the web is a chaotic place right

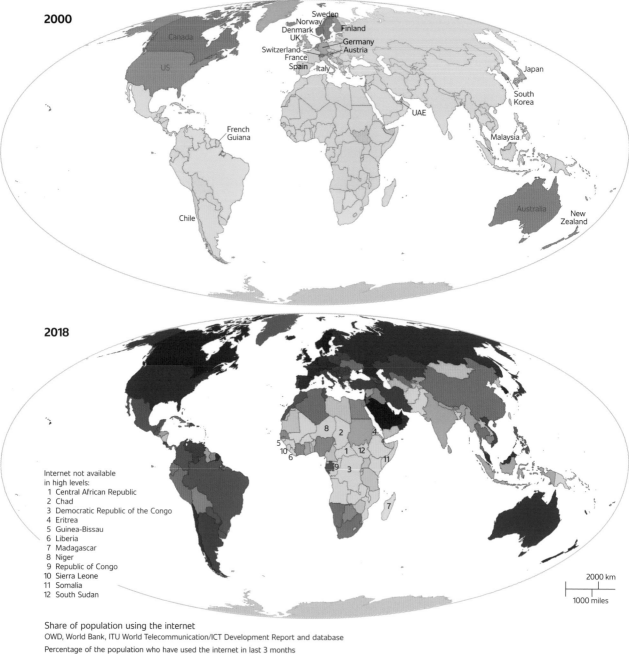

2000

2018

Internet not available
in high levels:
1 Central African Republic
2 Chad
3 Democratic Republic of the Congo
4 Eritrea
5 Guinea-Bissau
6 Liberia
7 Madagascar
8 Niger
9 Republic of Congo
10 Sierra Leone
11 Somalia
12 South Sudan

2000 km

1000 miles

Share of population using the internet
OWD, World Bank, ITU World Telecommunication/ICT Development Report and database

Percentage of the population who have used the internet in last 3 months

0-10 10-20 20-30 30-40 40-50 50-60 60-70 70-80 80-90 90-100 No data

Share of the population using the internet – 2000 and 2018 [71]

The penetration of the internet is changing the way we communicate and learn.
These maps explore the share of the population using the internet in the previous
three months. Darker colours represent a higher percentage of internet users.
Notice the global increase from 2000 to 2018.

now, just wait until the next few billion people come online. More fundamentally, the way information is digitally organised, packaged and transmitted has fundamentally changed. In a world of Amazon, Apple, Google and Wikipedia, algorithms are supplanting teachers and librarians to guide our access to knowledge in the most profound ways imaginable.[72] Younger people today are comfortable navigating the information age in ways that were inconceivable just a few decades back. Unsurprisingly, a backlash is also spreading over how our privacy is being steadily eroded.

Over the past decade there has been a shift in what kind of content is being prioritised in schools. There is already widespread consensus on the importance of STEM – science, technology, education and mathematics – skills. Most education professionals agree that immersion in STEM subjects is essential for young people to compete in the knowledge and gig, or platform, economy – though in a world of rapidly accelerating technologies and an expanding skilled workforce, these skills alone are insufficient. Not surprisingly, most major technology firms – from Google to Microsoft and IBM to Sony – have launched education platforms to promote STEM and collaborative learning.[73] Increasingly, primary and secondary schools that can afford it will provide learning in AI, digital manufacturing, genetic engineering, augmented reality and robotics. Learning analytics, mobile learning, virtual laboratories, 3D printing, gamification, virtual assistants and wearable technologies are already popping up in classrooms around the (mostly rich) world.[74] There is also widespread agreement that to genuinely compete in tomorrow's economy, students also need to be equipped with training in humanities and social sciences. Put another way, creativity and critical thinking, together with hard skills, are what will give people an edge.

All of these discoveries are profoundly reconfiguring the pedagogy and methods of teaching. Teaching will no longer be confined to the classroom or lecture hall – but will become increasingly multimodal and immersive. In the COVID-19 era, classes delivered on virtual platforms such as Zoom will be the rule, not the exception. New technologies are helping spread adaptive and personalised teaching, allowing students to determine their own pace, pathway and destination. Around the world universities and colleges are rolling out online undergraduate degrees.[75] Massive open online courses, or MOOCs, have opened up educational opportunities to millions.[76] These new approaches have their critics: some studies suggest that MOOC participants do not complete their courses and a majority of them are from affluent countries.[77] Regardless, we can expect an explosion of self-directed learning platforms and apps ('ed-tech', in the vernacular) designed to stream educational content in schools, companies and homes in ways that are tailored ever more specifically to learner needs.[78] In other words, teaching and learning are literally being designed into everyday life. These trends may be accelerated with the rise of climate change and pandemic risks.

Even before the COVID-19 pandemic, new technologies were already disrupting the business model of many post-graduate education providers. Leaving aside their positive impacts, the spread of

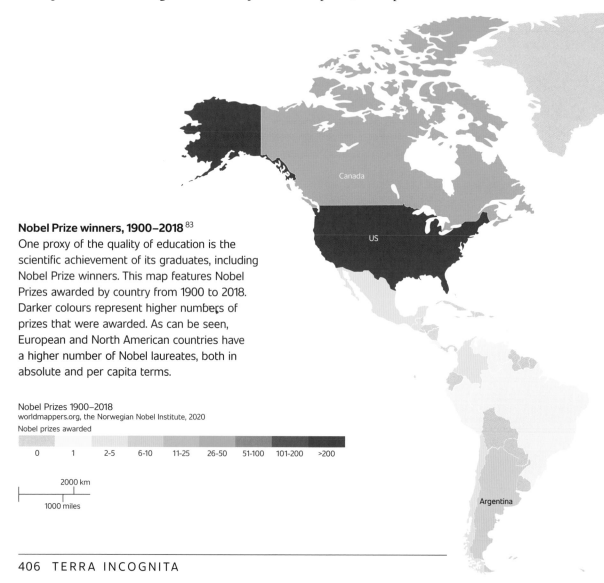

Nobel Prize winners, 1900–2018 [83]
One proxy of the quality of education is the scientific achievement of its graduates, including Nobel Prize winners. This map features Nobel Prizes awarded by country from 1900 to 2018. Darker colours represent higher numbers of prizes that were awarded. As can be seen, European and North American countries have a higher number of Nobel laureates, both in absolute and per capita terms.

Nobel Prizes 1900–2018
worldmappers.org, the Norwegian Nobel Institute, 2020
Nobel prizes awarded

| 0 | 1 | 2-5 | 6-10 | 11-25 | 26-50 | 51-100 | 101-200 | >200 |

2000 km
1000 miles

online education poses an existential threat to many universities and colleges, especially second-tier ones. In the US, about half of all tertiary institutions are at risk of closing in the next couple of decades as digital alternatives proliferate.[79] Since the COVID-19 pandemic struck the US in 2020, several smaller colleges have permanently closed their doors. While universities obviously vouch for the value of in-person learning and personal contact, in the future people may not necessarily need to physically attend classes, go to physical libraries, or live on university campuses. Many would-be students already baulk at the price of admission: in the US, student loan debts of more than 44 million borrowers amount to $37,000 per person, or a total of $1.45 trillion in current dollar values.[80] In Africa, it costs twenty-seven times

more to finance a place in university than one in primary school.[81] As students shop around for different courses from a variety of universities, we can expect more competition, lower costs and diminished university revenues. We will also see more and more online courses providing free content. An example of this is the education charity, core-econ.org, that has already been adopted by 150 universities and reaches more than 200,000 students. A small cluster of high-achieving universities will survive, developing their own international partnerships, online instruction companies and remote learning facilities.[82] The business case for the rest is harder to discern.

For the foreseeable future, North America, Western Europe and East and Southeast Asia will keep dominating the education sweep-stakes. The 2019 rankings of the world's top universities featured more than 1,200 institutions from eighty-six countries. The top 200 were virtually all Western European and North American, though several Chinese and Singaporean institutions have registered sharp improvements. High achievement at the tertiary scale translates into the concentration of innovation. As the map on pages 406–7 shows, more than 80 per cent of all Nobel Prizes (across all categories – from physics, chemistry and medicine to economics, literature and peace) have been awarded to recipients from just these three regions. This is down to their considerable investment in education together with brain drain (and brain gain), since high achievers are drawn to higher quality research institutions. Getting young people into high quality and affordable tertiary schooling is critical to driving growth. At the moment, just 9 per cent of young Africans who complete high school go on to a college or university, far lower than their counterparts in South Asia (25 per cent), East Asia (45 per cent), Latin America and the Caribbean (51 per cent) and North America and Western Europe (78 per cent).[84]

As in the past, stunning transformations in governance, business and society are forcing changes in education. The difference today is that for the first time in history, most people don't have a clue about what tomorrow's job market will look like. Even as recently as the twentieth century, people could reliably predict where they might land with a secondary school education or a degree in law, medicine or literature. Today, some futurists predict that roughly two-thirds of children entering primary school will work in jobs that don't yet exist.[85] A casual glance at online networking sites will soon reveal job titles that come straight out of a science-fiction novel – virtual habitat designer, ethical technology advocate, freelance biohacker, space tour-guide, personal content curator, human body designer, Internet of Things creative, and planetary urban planner.[86] But as more younger and older people from around the world are educated, the potential for innovation – including some that results in Nobel Prizes – will also accelerate.

Educators and parents everywhere know they need to prepare their young for an uncertain and turbulent world. In some parts of the planet, the priority is to get boys and girls into school for the first time. In others, there are ambitious plans to change the mission, culture and underlying technology of schooling itself. They also recognise to a greater extent that schools must build hard skills while also fostering critical thinking, communication, collaboration and creativity. To be sure, schooling will increasingly draw on tools and technologies that address the highly differentiated learning needs of students.[87] It is also likely that the entire notion of 'school' will be rethought. We are now in a world where learning is not confined to a set number of years, but one where learning lasts a lifetime. While daunting, the explosion of education innovation has the potential to significantly improve the human experience and unleash a new era of progress.

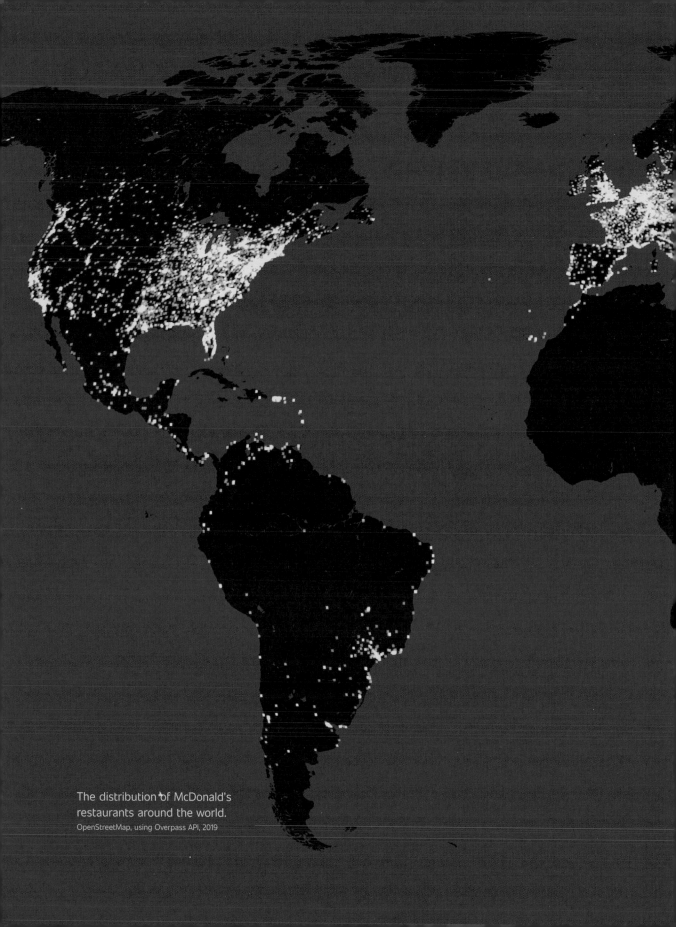

The distribution of McDonald's
restaurants around the world.
OpenStreetMap, using Overpass API, 2019

Culture

Culture is the glue holding societies and nations together

Globalisation leads to cultural homogeneity and heterogeneity

New technologies are preserving and spreading culture

Some countries are weaponising culture

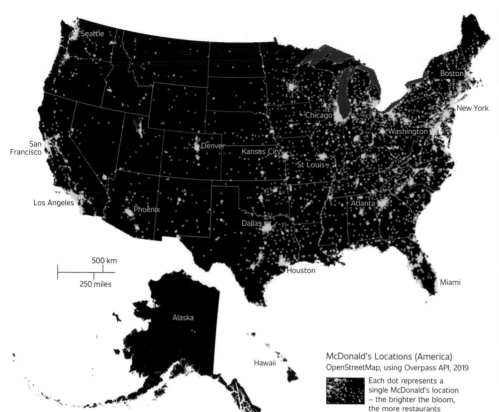

Introduction

Before the COVID-19 pandemic, an inescapable feature of modern travel was the sensation of 'sameness' in cities around the world. It started in the airport duty-free shops and continued on to the main streets. From New York and London to Mumbai and Shanghai, it was only a matter of time before you stumbled across a Starbucks, the latest hit by Billie Eilish or adverts for a new Marvel super heroes blockbuster. The forces of globalisation are bringing more and more objects, ideas and values into direct contact. These encounters are not coincidental – market forces and advertising shape how culture spreads. The highest-grossing movie of all-time was made in the US and came out in 2019 – *Avengers: Endgame* raked in $2.8 billion in earnings within three months of being launched.[1] The bulk of these receipts did not come from North American cinemas, but rather Chinese ones. The intermingling of different cultural products has many consequences, not all of them positive. There is understandable anxiety in many societies about the ways in which the spread of certain cultural objects, ideas and values can collide with, distort and even extinguish local customs, traditions and livelihoods.

Some readers are probably convinced that globalisation has made the world more homogeneous – more the 'same'. They are right. In recent centuries, some dominant cultures – including the language, art and cuisine from Western Europe and North America – have permeated every nook and cranny of our planet. The so-called 'Westernisation' process seems to have accelerated dramatically with the spread of certain types of ideas, values, philosophies, languages, industries, technologies, commercial practices, and lifestyles. One proxy, admittedly an imperfect one, is the McDonald's fast food chain that started with a single store in the late 1930s. Nine decades later, it is a juggernaut: more than 36,000 outlets in over 100 countries, serving roughly 70 million people a day.[2] The impacts of McDonald's are far-reaching – and not just on our diet, health and waistlines. Consider the cultural symbolism of the humble burger and fries. After the Berlin Wall came down in 1989, the opening of McDonald's branches

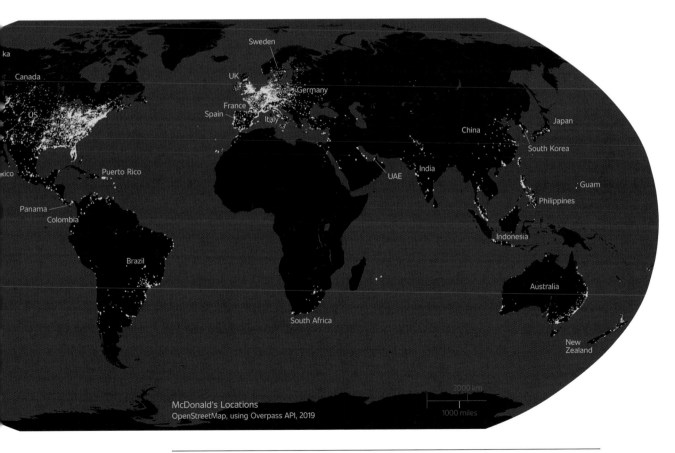

McDonald's Locations
OpenStreetMap, using Overpass API, 2019

assumed global significance – including in East Germany in 1989, Russia in 1990 and China in 1992.

Yet it is also the case that globalisation has mixed effects on culture and is simultaneously reinforcing heterogeneity. What do we mean by this? We mean that the intermingling of cultures that is accelerated by globalisation is spreading diversity. Think about the incredible (and growing) popularity of Bollywood[3] around the world, the massive influence of Chinese video games, the worldwide appeal of K-pop,[4] and the enormous influence of Anime and Manga animation from Japan on North American and European design.[5] Cultural artefacts are emerging with fascinating new blends and permutations. Even the once quintessentially American export – McDonald's – is bending to local preferences. Outlets now serve cherry blossom, radish burgers and seaweed shakes in Japan, raclette in Switzerland and spicy paneer wraps in India.[6] The point is that globalisation is not a one-way street – it is a multi-lane highway running in different directions.

In this chapter we explore some of the complex and contradictory impacts of globalisation on culture. While it is true that the brute force of dominant cultures from China to the US are disruptive, the resilience of local cultural specificities should not be underestimated. The fact is that powerful cultural forms can and do spread; they influence and transform in dramatic fashion, even as their content retains a distinctive native flavour (sometimes literally!). In some instances, cultural collisions can reinforce divides within and between countries and communities. These fault-lines can be exploited and weaponised in fully-fledged culture wars to drive international and domestic political agendas. Culture has always been a battleground, and the twenty-first century is no different from times past. What has changed is the arrival of new technologies – namely the internet and social media – that have accelerated the potential for cultural consensus, and misinformation and manipulation, on a hitherto unimaginable scale.

The globalisation of culture

Culture is the glue that binds our societies and nations together. It is so pervasive that we often forget it even exists. At the most basic level, culture is the means by which we transmit or share objects, ideas and values across time and space. In this way, it facilitates collective action,

Diffusion of the printing press – 1439 and 1500 [7]

The printing press helped disseminate knowledge faster and wider than ever before. Although woodblock printing can be traced back to the ninth century, Johannes Gutenberg's printing press allowed for the mass-production of books and the spread of revolutionary ideas into the hands of literate Europeans. The map highlights the explosion of cities with printing presses – from Mainz in 1439 to hundreds of cities like Cologne, Rome, Paris and London by the end of the fifteenth century.

or co-operation. Along with oversized brains and opposable thumbs, it is what has allowed our species to survive and thrive. But culture doesn't emerge spontaneously. Nor is it fixed in time. New ideas – and especially new technologies – are central to its formation, dissemination and transformation. Stone tools and supercomputers have profoundly shaped the evolution and expansion of cultures worldwide. Consider the water mill, assembly line or internet. Each of these innovations generated leisure time, stimulated creative exchanges, and contributed to an explosion of wealth that in turn triggered booms in cultural formation.

One of the most consequential technologies on the spread of culture was the printing press.[8] Although paper production, ink production and woodblock printing had all been around for hundreds of years, the first genuinely mechanised press was built by Johannes Gutenberg in 1439 in Mainz, Germany. In 1452, Gutenberg printed a version of the Bible in Latin. After the technology went public, presses started popping up in Cologne (1466), Rome (1467), Paris (1470) and London (1477). Within decades more than 270 cities across Europe had printed 20 million books. Following the discovery of sea routes to the Americas in 1492 and expansion of trade links to Asia, printing went viral. By 1500, printing production had increased tenfold. The impacts were

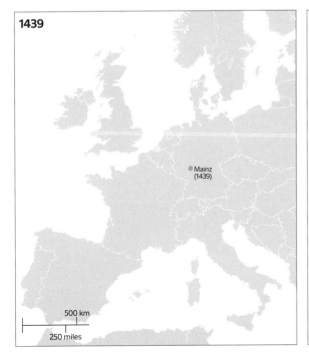

1439

Mainz
(1439)

500 km

250 miles

1500

Proliferation of the early printing press
Based on *The Atlas of Early Printing*,
Greg Prickman, 2008

● Each dot represents the
location of a printing press

London
(1477)

Cologne
(1466)

Paris
(1470)

Rome
(1467)

instantaneous and immensely disruptive. The rapid dissemination of the written word helped stimulate literacy, expand knowledge and triggered the Protestant Reformation and the European Renaissance.[9] Yet because of the way it threatened established power, the printing press also gave rise to mass intolerance, brutal inquisitions, religious wars and extremist violence.

Today, the world is being upended by multiple and simultaneous Gutenberg moments. The speed and scale of technological innovation is profoundly and perceptibly impacting culture just about everywhere you look. One example of this is the World Wide Web, invented in 1989 by Tim Berners-Lee and his colleagues at the CERN laboratories in Switzerland. In its first few years of existence, only a few hundred thousand people were online, most of them living in the US. Today, there are billions of active internet users and another million more come online every day.[10] Social media platforms are spreading more cultural content than ever. Their number of users is truly mind-boggling. As the graphic shows, in 2020, Facebook, Youtube, WhatsApp, Messenger and Instagram collectively drew more than 8 billion users a month. Qzone, TikTok, Weibo and reddit attracted close to 2 billion users, many of them trading ideas, watching movies, listening to music,

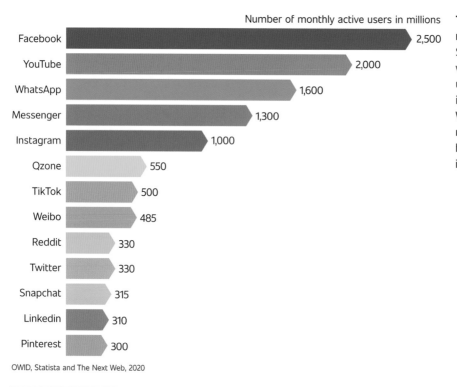

Number of monthly active users in millions

Facebook	2,500
YouTube	2,000
WhatsApp	1,600
Messenger	1,300
Instagram	1,000
Qzone	550
TikTok	500
Weibo	485
Reddit	330
Twitter	330
Snapchat	315
Linkedin	310
Pinterest	300

OWID, Statista and The Next Web, 2020

The distribution of social media use – 2020 [11]
Social media platforms with their monthly active user ratings. While not included in this chart, WeChat, a Chinese multi-media platform, had over 1.2 billion users in 2020.

playing video games, writing blogs and dressing up their avatars for a night on the digital town.[12] The use of these and other platforms has exploded since the COVID-19 pandemic forced even more people to sit in front of their screens.

What are the impacts of these new technologies on cultural identity? Are they bringing cultures closer together or do they distort local identities, driving people, societies and nations apart? The answers depend on how culture is defined. Not surprisingly, the concept is difficult to pin down. There are hundreds of definitions in circulation and virtually no consensus among the experts about what culture means. The only thing most experts seem to agree on is that culture is abstract, complicated and problematic.[14] Some scholars focus narrowly on artistic endeavours and personal edification – what is often referred to as 'high culture'. Others insist it is about something much deeper – our intrinsic systems of knowledge, belief, morality, custom and habit that are passed down generations. While elusive, most interpretations typically home in on three basic characteristics – observable objects, common values and shared assumptions.[15]

Diffusion of internet users around the world – 1990 and 2017 [13]

The internet is profoundly impacting the formation, sharing and consumption of culture around the world. Increased digital connectivity brings culture closer to people and has helped spread new forms of art and expression. It has also generated tensions and anxieties. These maps illustrate the number of internet users by country. Note how the US was the only country with a significant number of internet users in 1990. By 2017, we see that almost every country is online, with billions of users.

Number of internet users by country
World Bank, International Telecommunication Union

All individuals who have used the internet in previous 3 months (in millions)

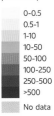

0-0.5
0.5-1
1-10
10-50
50-100
100-250
250-500
>500

No data

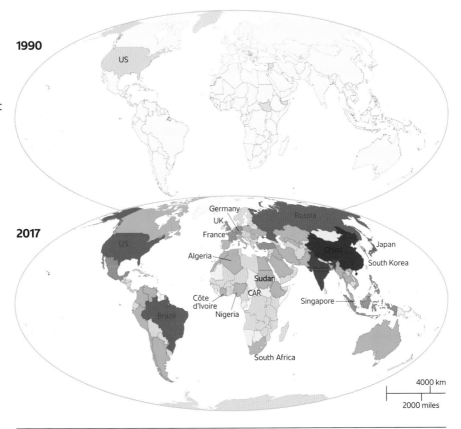

Humans are hardwired for culture. It guides our interactions and understandings about the wider cosmos, our planet, and everything from art, music and food to language, religion, values and moral codes.[16] It also serves an instrumental purpose by facilitating the exchange of ideas and values across multiple generations. From an economic perspective, culture is efficient, helping countries, communities, and households to better divide tasks and collaborate. And from a sociological angle, it is what has allowed our ancestors to develop from small, isolated communities into complex societies, by learning how to grow and store food. In short, culture allows people who might be predisposed to fight to form shared identities instead – it is the software of civilisation.

Culture is not static. Periods of accelerated globalisation can spark cultural revolutions. The rapid spread of people, capital and technologies opens closed societies to entirely new ways of thinking and behaving, as was the case when Gutenberg's printing press started churning out Bibles and political pamphlets. The current phase of digitally-enabled globalisation is no different. It is empowering because it connects the world, creates new communities of belonging and drives economic development. It is also immensely disorientating and disabling, shattering the sense of shared identity and values for some communities. After all, people have a strong attachment to their local artefacts, moral codes and world views, something that evolutionary

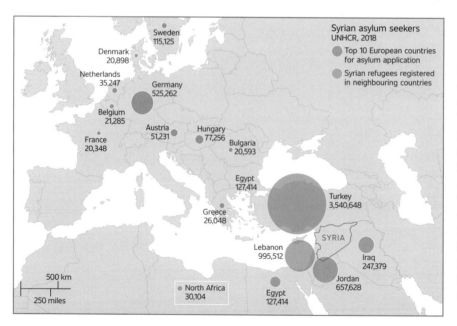

Syrian asylum-seekers in Western Europe [18]
More than 13.5 million Syrians have been displaced since the start of the civil war in 2011. This map shows how many of them sought shelter in refugee camps in Turkey, Lebanon, Jordan and Egypt. Western European countries received less than 10 per cent of them, about 1 million people. This compares to 3.6 million in Turkey alone.

biologist Mark Pagel describes as 'tribal psychology'. Although there are many examples of how some people incorporate 'the other' into their tribe (think of nation states), there are also limits. The sudden mixing of diverse groups of people with distinct cultural currencies is seldom conflict-free.[17]

Consider the case of the Syrians fleeing a brutal civil war that erupted in 2011. As the map shows, millions of terrified Syrian men, women and children sought asylum in European and Middle Eastern countries. The sudden arrival of hundreds of thousands of newcomers to Austria, Germany, Greece and Sweden was initially met with an outpouring of generosity from locals living there. But their welcoming spirit soon faded. Germany, a country of more than 80 million people, received 77,000 refugees of all nationalities in 2012, 126,000 in 2013, 202,000 in 2014, 475,000 in 2015 and 745,000 in 2016. Over time, new asylum-seekers and eventually migrants confronted fear, hostility and xenophobia from some of their local hosts. Far-right and anti-immigrant groups launched street protests, calling for the defence of the homeland from Islamicisation, or anyone looking different. Rumours of 'migrant men' harassing 'local women' inflamed sentiments further still. To be sure, the challenges of assimilating large numbers of new arrivals with different languages, habits and customs are real. Yet it is also worth noting that many times more Syrian refugees were absorbed in neighbouring Jordan, Lebanon and Turkey than in Europe. For the most part, they were integrated without sparking similar levels of ethnocentrism and reactionary nationalism. While the reasons are complex, cultural proximity between Syrians and their neighbours undoubtedly played a role.

The turbocharged spread of telecommunication and broadband services is rapidly rewiring how everyone engages with culture. Consider how film streaming platforms like Netflix have hastened the spread of 'foreign' entertainment around the world. In 1997, the streaming service was only available to US viewers. Today it is operating in more than 190 countries and its content is viewed by hundreds of millions of people. Other streaming services like Amazon, Disney, HBO, and Hulu are making more films and television shows available to more consumers than ever before. In early 2019, *Game of Thrones* – produced by HBO – was the most viewed series in virtually every country on Earth followed closely by *The Walking Dead* (available on Netflix) and *Keeping Up with the Kardashians* (accessible on Hayu or

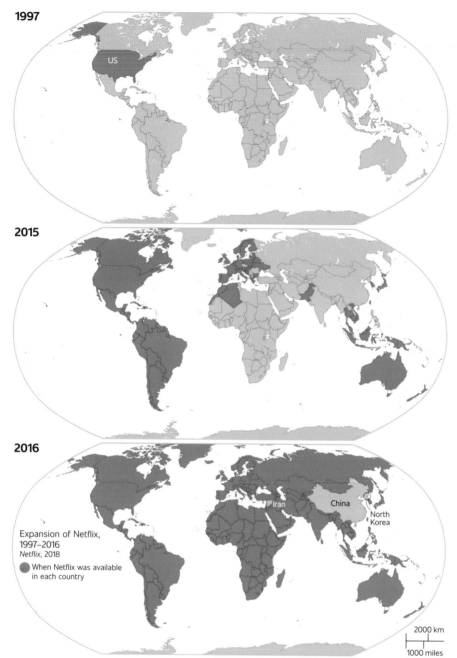

1997

2015

2016

Iran China North Korea

Expansion of Netflix, 1997–2016
Netflix, 2018

⬤ When Netflix was available in each country

2000 km
1000 miles

How Netflix spread from one to 190 countries in ten years[20]

Netflix expanded to more than 190 countries in less than twenty years. More than half of its subscribers are from outside the US. This is an extraordinary achievement for a company that was operating only in the US before 2010. These maps chart the spread of Netflix between 1997 and 2015 and then the explosive growth into 2016. Today, Netflix has more subscribers globally than every other streaming service combined.

Google Play). But midway through the year *The Wandering Earth*, a Chinese blockbuster, closed in on the top spot.[19] Netflix and its rivals are spending tens of billions of dollars to stream content to mobile phones and smart televisions reaching every part of the globe. They are demolishing established entertainment conventions and markets along the way.

The implications of the internet, social media and digitally distributed content for cultural diversity are complex. On the one hand, there are legitimate concerns that the sudden exposure to a waterfall of information could lead to the steamrolling, appropriation and commodification of local cultures. This is happening in unexpected ways. In some of Australia's remote aboriginal communities, for example, social media seems to be contributing to social tensions[21] and negatively affecting mental health.[22] Meanwhile in Mexico, indigenous weavers are taking big-name fashion labels to court for plagiarising their textile patterns. Yet for every example of cultural appropriation, there are instances of new technologies helping minorities to actively preserve their stories, songs, dances, cuisine and rituals.[23] The assumption that globalisation always tramples on local cultures is too simplistic: new technologies can contribute to cultural revival through the sharing of ideas and preservation of traditions.

Digital archives are being developed by governments and indigenous groups around the world. Culture ministries and museums have launched initiatives to digitally map, store and disseminate cultural assets.[24] New open source software has also been deployed to help universities and non-profit organisations to digitally record, preserve and promote archaeology, the arts and languages[25] across Africa, Asia and the Americas.[26] A fascinating example of this is the open source platform, Mukurtu CMS. Established in 2007, it features curated sites run by Native Americans from the Catawba,[27] Spokane[28] and Passamaquoddy peoples allowing them to create, manage and share their cultural heritage online.[29] Similar examples can be found in Canada, where online archives are managed by Inuit groups from Nunatsiavut to Nunavik,[30] and also in New Zealand, where sites like Tāmata Toiere feature Maori songs, chants and dance.[31]

The power of new technologies to spread cultural content is startling. As noted above, Netflix currently has more international subscribers than US ones. The same applies to the other big platforms like Amazon, Apple, and HBO.[32] Precisely because entertainment tastes are strongly determined by local culture and geography, the production and distribution of non-Western content streaming platforms is growing. In 2018, for example, the most searched-for television programme on Google was *Story of Yanxi Palace*, which drew over 15 billion views, most of them in China. In third place was *Love Destiny*, a Thai series,

The spread of the Queen's tongue– 2019 [34]

English has been spreading around the world since the seventeenth century. It was the language of the British Empire and is today the official language of international business and global affairs. Yet off the roughly 1.5 billion people who speak English, less than 400 million use it as their first language. At least 1 billion speak it as a secondary language. This map captures English proficiency around the world. [35]

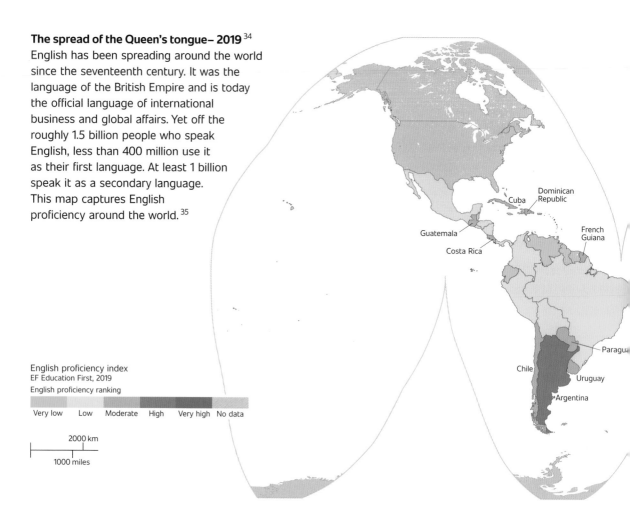

English proficiency index
EF Education First, 2019
English proficiency ranking

Very low | Low | Moderate | High | Very high | No data

2000 km
1000 miles

followed in fourth by *Motu Patlu*, an animated Indian series, and trailing them was the US sitcom *Roseanne*.[33]

Worries about the homogenising effects of globalisation on cultural diversity are hardly new.[36] Nor are these concerns without justification. Consider English, the world's most widely spoken language. It has spread even further with the latest phase of globalisation, or more specifically, in the wake of conquest, colonisation, trade and television. It is currently the official language of almost sixty countries, is spoken in more than forty more, and is regularly used by 1.5 billion native and non-native speakers.[37] Historically it has overtaken Latin, the language of the educated elite during the Middle Ages, and French, which dominated in the nineteenth century in terms of cultural influence.[38] Today it is the language of global business, science, diplomacy, entertainment and the internet.

The continued supremacy of English is far from secure. It was (and continues to be) fiercely resisted in some countries. It has also diversified into distinct dialects. Worth recalling is the fact that English is itself a product of bygone eras of globalisation, being an amalgam of Latin, Greek, French, Hindi and other languages.[39] While English may be top dog at the moment, other language groups are nipping at its heels.[40] Within the next three decades, it is quite possible that Mandarin, Hindi, Spanish, Arabic or even French[41] will be more widely spoken than English. Although more people speak Mandarin than English currently, it is not as widely spread geographically, is hard to read and write, and seldom used in science. On the other hand, French is making a comeback and has been growing quickly in Africa due to rapid population growth. There are projections of as many as 750 million French speakers by 2050.

The most endangered languages around the world [42]

Languages, like plants and animals, can go extinct. This map highlights the most endangered languages around the world. Each dot represents a unique endangered language and the colour represents the level of risk it faces. There are pockets of threatened languages in North America, Mexico and South America, throughout the Sub-Saharan African region, and into South and Southeast Asia and the Pacific. [43]

Endangered languages
Catalogue of Endangered Languages, 2012
● Critically endangered
● Severely endangered
○ Endangered

2000 km

1000 miles

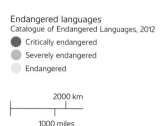

There are roughly 7,100 living languages in the world today.[44] Some of these are incredibly obscure: Njerep recently became extinct in Cameroon and is currently spoken by just four people in Nigeria. Likewise, Kawishana, a Brazilian language, has just one remaining speaker while Paakantyi (Australia), Liki (West Papua) and Chemehuevi (US) are each spoken by a handful of people. About 66 per cent of the world's population speaks just 0.1 per cent of known languages.[45] As noted above, new technologies and preservation societies are helping keep some of them from vanishing entirely. Inspiring examples include Wikitongues and Recovering Voices, projects that revitalise endangered languages.[46] While some endangered American Indian languages and Pacific Island dialects are making a comeback after almost disappearing, most are likely to eventually be permanently extinguished as the map shows.[47]

The speed at which linguistic diversity is declining is beyond words. By some estimates, over a fifth of all the languages spoken in the 1970s have already disappeared. In the US, the most endangered languages

Syria
Iraq

Taiwan

Myanmar
Thailand — Laos
Philippines

Chad   Sudan
Ethiopia
CAR
Malaysia
Indonesia
Papua
New Guinea
Nigeria
Solomon
Cameroon
DRC
Islands

Vanuatu

Australia
New
Caledonia

South Africa

are those spoken by the native peoples on the West Coast and in the Midwest. In Latin America, it is predominantly indigenous languages in Central America and the Amazon basin that are vulnerable.[48] Other threatened languages are found across Sub-Saharan Africa (especially Nigeria, Chad, South Sudan and Ethiopia), South and Southeast Asia (Nepal, Bhutan, Bangladesh, Laos, Malaysia, Indonesia and the Philippines) and Australia and the South Pacific islands.[49] Some linguists fear that between 50 and 90 per cent of all languages could vanish by the end of this century.[50] Already, a third of the world's languages have fewer than 1,000 speakers and every two weeks, a language dies.[51]

Globalisation is often blamed, not just for extinguishing languages, but also for distorting gastronomic preferences, especially because of the way it helps spread Western fast food and processed products at the expense of local fare. Consider the coffee chain, Starbucks, which experienced even more rapid growth than McDonald's. Today there are more than 30,000 Starbucks coffee shops – twice the combined

Top ten countries with Starbucks outlets		
	Company operated stores	Licensed stores
United States	8,575	6,031
China	3,521	—
Japan	1,286	—
Canada	1,109	409
Thailand	352	—
United Kingdom	335	653
South Korea	0	1,231
Indonesia	0	365
Philippines	0	360
Taiwan	0	458
Statista, September 2019		

Growth of Starbucks, 1984–2019
Starbucks, using Overpass API, 2020
● Each dot represents a Starbucks location

2000 km

1000 miles

total of its nearest rivals.[53] The company opened its first store outside the US in 1996.[54] Since then, as the map shows, its international footprint has expanded to cover more than eighty countries.[55] Like McDonald's, it feels as if Starbucks is virtually everywhere.[56] And while the US, China, Canada, Japan and the UK are the countries with the most numbers of stores in absolute terms, Seoul was actually the city with the most outlets in 2019, about 284 of them, more than New York City, with around 241.[57]

Many Western brands have to adjust to local preferences or risk being overtaken by domestic competitors. To help it compete, Starbucks set up eighteen design centres around the world to tailor its offerings to local tastes. In Japan, some shops are designed to look like teahouses featuring low roofs with elements of Shintoism. In China, stores are adapted to handle larger groups of customers (rather than single coffee drinkers, which is more common in Europe and North America). In Saudi Arabia, the logo was changed from a topless mermaid to a crown floating above the waves. While in France, Starbucks launched Viennese coffee and *foie gras* sandwiches. In the UK bacon butties are among the top sellers.[58]

Mapping the spread of Starbucks around the world [52]
There are more than 30,000 Starbucks branches around the world. As this map shows, at least a third of them are in the US, but there are thousands spread out across China, Japan and Canada. Hundreds more are in cities throughout Thailand, the UK, South Korea, Spain, Taiwan, Turkey and Indonesia.

Culture wars

The world is more multicultural than ever. On the next page, the map of cultural diversity – which measures linguistic diversity – shows that North American, African and South and Southeast Asian societies rank especially highly. Meanwhile, countries like Poland, Norway, South Korea and Japan rank lower on the cultural diversity scale. While genuinely welcomed in many societies, multiculturalism has a darker side. In some places, locals embrace difference, while in others it is resented. Cultural fault-lines are often deeply rooted. Well before contemporary debates over political correctness, identity politics, the #MeToo movement and LGBTQ rights, there were bitter disputes about the contraceptive pill, whether or not to sell alcohol, decisions about admitting migrants, the language of education, and the legality of slavery – all of which were heavily influenced by culturally-mediated differences. The sociologist, James Davison Hunter, described them as part and parcel of 'culture wars' after his book of that title in 1991.[59]

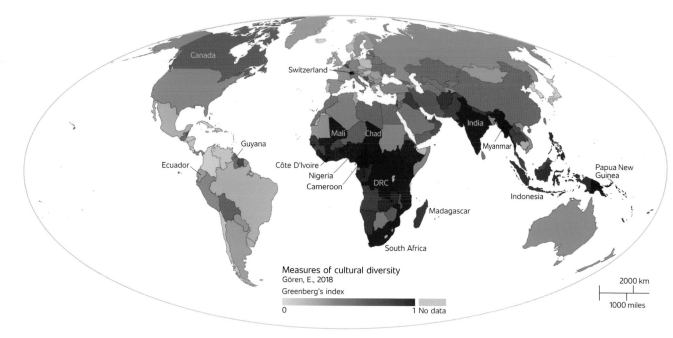

Measures of cultural diversity
Gören, E., 2018
Greenberg's index

0   1 No data

2000 km

1000 miles

The idea of culture wars goes back to the late nineteenth century and the struggle between the German imperial government and the Catholic Church. The German word *kulturkampf* ('culture struggle' in English) refers to the attempts by Austrian, Belgian, German and Swiss secularists to separate church and state. Culture wars then as now were waged along religious and class lines. They were inflamed by political and economic turbulence, sharp increases in immigration and the fiery rhetoric of reactionaries and populists. For its part, the US has been wracked by one culture war or another since its independence. American cultural warriors clamoured for everything from increasing the role of the church in state affairs[60] to allowing state schools the right to restrict teaching to English.[61] All of this can help explain contemporary battles between US Republicans and Democrats, whether over abortion, gay marriage, euthanasia or stem-cell research.[62] Some of the most emotionally charged disagreements in the US legislature or college campuses are not about campaign finance reform, investment in crumbling infrastructure or universal health care but rather same-sex relations, transgender rights and gun control.

Cultures routinely shapeshift and adapt. Take the case of gay marriage, an issue that long divided the US. After decades of protest, struggle and legislative action, the tide has turned. Today at least two-thirds of US citizens believe it should be legally recognised.[63]

Countries reporting the most and least cultural diversity
Greenberg's index estimates the level of 'cultural diversity' based on the extent to which two randomly selected people in a country speak different mother tongues. As such, the map favours countries with high ethnic diversity (including as a result of migration). Countries such as the Democratic Republic of the Congo (DRC), India, South Africa, Switzerland, Myanmar and Papua New Guinea score especially highly.

From bans to legalisation: same-sex marriage in the US, 1995–2015

Most US citizens used to be opposed to same-sex marriage. Yet opinion has changed in recent years. These maps highlight the changes in legislation at the state level. They reveal how, over the course of 20 years, the US shifted from a situation where most states banned same-sex marriage to one where every state had legalised it.

Same-sex marriage in the US
1995–2015
Public Religion Research Institute, 2015

- No statute banning same-sex unions
- Banned by statute
- Banned by constitutional amendment
- Legalized

Compare this to polls back in the 1990s when just over a quarter of them felt that it should be valid under the law.[64] Even more Americans – about 72 per cent – currently support same-sex relationships.[65] It is not only Democrats who approve overwhelmingly of gay marriage and same-sex marriage. Endorsement has tripled among rank-and-file Republicans since the mid-1990s. Support has risen among both Protestants and Catholics to 65 and 55 per cent respectively.[66] This helps explain why most states that once opposed same-sex marriage as recently as 2005 have flipped and now support it.

Social media is helping spread the culture wars in new ways. For decades, right-wing and libertarian radio personalities raged against the so-called liberal elite.[67] Before satellite radio, they had limited geographical reach. But new communications technologies have changed that, amplifying influencers like Alex Jones, Milo Yiannopoulos and Louis Farrakhan who have used Facebook, Twitter and YouTube to amass huge followings.[68] In democratic settings, algorithmic manipulation also helps push out political propaganda, vile extremist content and paedophiliac material. Hyper-partisan networks are peddling alternative facts, conspiracy theories, and digital propaganda about the 'mainstream media'. Meanwhile, in authoritarian countries the screws have been tightened on the cultural sphere. The Chinese government exerts enormous influence due to its control of WeChat (which has more than 1.2 billion accounts and 850 million registered users)[69] and Weibo (with more than 500 million users) through which it suppresses online dissent. Chinese internet providers are required to register the names of bloggers and live-censor conversations, banning words linked to Tiananmen Square, Tibet or Falun Gong.

The availability of the internet is also changing viewer decisions about how and where to access their news. Less than half of adults and a quarter of people aged 18 to 29 in the US say television is their primary source of news.[70] Instead, almost 80 per cent of adults (aged thirty to forty-nine) and close to 90 per cent of people aged eighteen to twenty-nine in the US say social media is their primary source of news. In India and throughout much of Asia, smartphones are the principal (and in some cases only) delivery mechanism to access news.[71] With a few venerable exceptions, print newspapers are in terminal decline. Most traditional media groups have had to dramatically reduce staff and budgets and adopt digital strategies to survive. Local media is being wiped-out in many parts of the world while social media content and digital distribution is exploding.[72] In the US, just 5 per cent of American adults used social media platforms in 2004. By 2018, the proportion had risen to 68 per cent.[73]

A fragmented media landscape can undermine a shared sense of reality and culture. One consequence of this is that communities increasingly speak past each other, grow more intolerant, have their divisions reinforced and retreat to the comfort of their digital echo chambers. Around the world, populist leaders and nationalist politicians are inciting and feeding off cultural divisions. Before and after gaining office in 2016, the US President Donald Trump pointedly reminded voters that he was engaged in a culture war on behalf of his white working-class base against 'politically correct' coastal elites.[74] During a 2018 speech to the United Nations General Assembly, he repeatedly denounced 'globalism' – a vague term to describe political and economic integration – and internationalised his culture war.[75] For several years, the US president and his counterparts from Hungary and Italy to Brazil, India and Turkey ratcheted-up culture wars to entirely new levels, purposefully widening divisions between groups according to their ethnicity, class, sexual preference and religion.

Globalising culture wars

Although culture wars share certain characteristics across countries, they manifest differently from place to place. For example, in India, the current culture war involves concerted efforts by President Modi to rewrite history to reflect an imagined purer Hindu past.[76] In China, culture wars are mediated by the Communist Party, and historically involved disastrous outbreaks of violence (for example, the Cultural Revolution from 1966 to 1976).[77] The latest iteration includes suppression of religious minority groups like Falun Gong to the detainment and re-education of Muslim Uighurs in Xinjiang Province discussed in the chapters on geopolitics and violence. One reason why culture wars are globalising is because ideologues and interest groups are eagerly projecting their views on to the international stage. They are doing this directly through social media, channelling messages to would-be and actual followers, dispatching envoys and organising like-minded networks.

Evangelical Christian churches are exporting devotees, funding missionary zeal across the planet. This map provides a crude estimate of the extent of the Neo-Pentecostal Evangelical presence in proportion to population growth. Developed by the Joshua Project, an evangelical organisation, it purports to show how Christian 'presence' is spreading at a faster rate than the natural population growth in virtually every country outside North America, a smattering of European and African countries, Japan and New Zealand.[78] According to its authors, progress has slowed in North America and some African countries, while it has stalled outright in Japan, Scandinavia and a small number of European countries. Given their provenance, such maps should be interpreted with a large dose of salt.

Evangelical churches have long been active in remote corners of the Amazon Basin, Central Africa and the South Pacific. Faith-based groups such as the Joshua Project have detailed the gradual penetration of Christian missionaries around the world, including in East Papua in Indonesia. Christianity has comparatively deep roots there, having arrived in the middle of the nineteenth century. In recent years, the Muslim population has overtaken the Protestant population in what some locals believe constitutes a deliberate process of 'Islamicisation' and 'de-Papuanisation'. Militant groups are emerging, with periodic

Growth rate of Evangelical Christianity as compared to population growth rate, 2005–2010[79]
Evangelical Christianity has spread rapidly around the world. This map provides information from 2005 to 2010 of countries where there were decreasing numbers of evangelicals, where evangelical growth rates were slower than the national population growth rate, and where evangelical growth rates were faster than country population growth.

Annual growth rate of evangelicals (2005–2010)
Joshua Project, Operation World 2010

- Decreasing evangelical population
- Evangelical growth slower than country population growth
- Evangelical growth faster than country population growth

2000 km
1000 miles

outbursts of violence among the faithful as they fight over their own brand of exclusivist truth. Recent attacks against Christian churches partly explain the expansion of the Evangelical presence in the region.

Culture is often weaponised as part of a wider geopolitical power-play. A prime example comes from Russia's long-serving president, Vladimir Putin. In many of his speeches, Putin calls for a 'return' to traditional Christian family values in contrast to what he describes as the decadence, liberalism and amorality of the West. By supporting conservative values and advocating for 'spiritual security', he is projecting an alternative to what he calls the 'soft liberalism' of North America and Western Europe.[80] This brand of conservatism, traditionalism and illiberalism has attracted followers. Trans-Atlantic far-right coalitions are emerging, some of them committed to eradicating the 'threat' of atheism, socialism and Islam.[81] Meanwhile, governments in France and Germany have urged the European Union to adopt a more assertive multicultural posture on progressive causes ranging from reproductive and gay rights to climate action.

President Putin's call to deepen Christian values and reject homosexuality and moral relativism is welcomed by a growing number of sympathetic governments. For example, Hungary's Prime Minister, Viktor Orbán, describes his political opponents as 'nihilist elites' and rails against the philanthropist and investor, George Soros, and the European Union for supporting migrants, whom he accuses of destroying the ethnic (Christian) foundations of his country. Political and religious leaders from Austria, Czech Republic, Hungary, Poland and the United Kingdom are fighting what some refer to as a 'cultural counter-revolution'. Their primary targets include liberal politicians, foreign and domestic human rights groups and left-leaning pundits. With the aid of sympathetic Western ideologues and Russian intelligence services, there are signs of an incipient transitional nationalist front emerging. The political strategist and former Trump aide, Steve Bannon, even launched a network called 'The Movement' to unite Europe's far-right.[83] The network was supposedly established to rival the Open Society Foundation backed by Soros, which promotes pluralism and liberal values worldwide.

Some far-right parties in European countries are ramping up the culture wars, including with foreign support. For decades, France's National Front has condemned threats to what it calls 'Frenchness', including Islam, globalism and the European Union.[84]

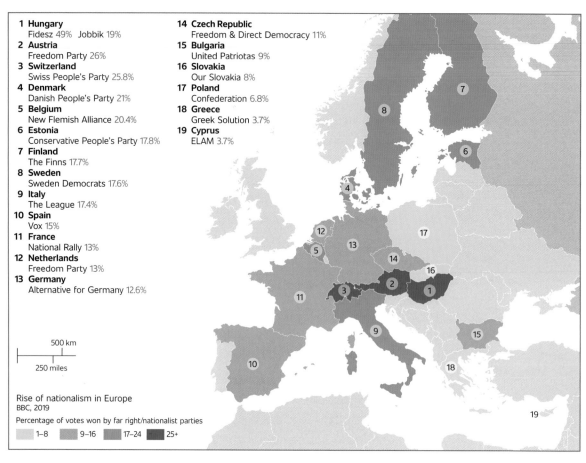

1 Hungary
Fidesz 49%   Jobbik 19%
2 Austria
Freedom Party 26%
3 Switzerland
Swiss People's Party 25.8%
4 Denmark
Danish People's Party 21%
5 Belgium
New Flemish Alliance 20.4%
6 Estonia
Conservative People's Party 17.8%
7 Finland
The Finns 17.7%
8 Sweden
Sweden Democrats 17.6%
9 Italy
The League 17.4%
10 Spain
Vox 15%
11 France
National Rally 13%
12 Netherlands
Freedom Party 13%
13 Germany
Alternative for Germany 12.6%

14 Czech Republic
Freedom & Direct Democracy 11%
15 Bulgaria
United Patriotas 9%
16 Slovakia
Our Slovakia 8%
17 Poland
Confederation 6.8%
18 Greece
Greek Solution 3.7%
19 Cyprus
ELAM 3.7%

500 km
250 miles

Rise of nationalism in Europe
BBC, 2019

Percentage of votes won by far right/nationalist parties
1–8   9–16   17–24   25+

Spreading culture wars in Europe – 2019 [82]

Europe has witnessed a surge in voter support for ultra right-wing nationalist and populist parties. Growing numbers of Europeans are expressing anger at the political establishment, but also concerns about globalisation, migration and the apparent loss of national identity. This map highlights the rise of nationalist parties and the proportion of government seats they occupy in a selection of countries as of 2019. Notice how Austria, Hungary and Switzerland exhibit the highest proportion of ultra-right-wing representatives in government.

In what was memorably described as 'pig whistle politics', the Front's leader, Marine Le Pen, tried banning non-pork substitutes in school cafeterias where her party won seats.[85] Likewise, Germany's anti-immigrant Alternative for Germany (AfD) party is part of a much larger ultra-right movement that has swelled since the refugee crisis of 2015.[86] Its membership is contesting what it describes as an 'international leftist tendency to liquidate Germany and the German cultural nation'.[87] Meanwhile, Geert Wilders, the Netherlands' best-known far-right politician, shares some of the same goals as his French and German counterparts. He rails against Muslims, migrants, homosexuals and routinely singles out the mainstream media for attack.[88] Meanwhile in the UK, values are replacing economics as the key driver of politics, as the case of Brexit amply revealed.

China too has extended its cultural soft power with state support. The country is looking to expand its cultural influence in parallel with the mammoth Belt and Road Initiative described in the chapter on geopolitics. Over the past decade the government has expanded investment in Chinese culture through film and literature, as well as through its network of so-called Confucius Institutes.[89] China has also dramatically scaled up the presence of the Xinhua news agency and China Central Television around the world.[90] In these and other ways, the country's public authorities are crafting what they refer to as 'cultural security'.[91] For years, Chinese leaders have denounced North American and European governments for 'intensifying the strategic plot of Westernising and dividing China'. Communist Party speeches have decried the way 'ideological and cultural fields are the focal areas of [the West's] long-term infiltration'.[92] Unsurprisingly, geopolitical culture wars are also a foil for politicians to shore up domestic support. Whether in China, Russia or the US, few things bring (some) citizens closer together than foreigners speaking a different language.

On the front-line of the global culture wars are not soldiers, but rather scholars, artists and activists as depicted on the map on the following page. Launched in the 1930s, the British Council has 177 offices in more than 100 countries.[93] With its headquarters in Paris, the French Alliance Française has 229 centres in 137 countries. Meanwhile, Germany supports the Goethe-Institut (159 offices), Italy sponsors the Società Dante Alighieri (93 offices), Spain has the Instituto Cervantes (76 offices), and the US has the Voice of America and numerous other organisations designed to project its ideas, values and interests. Most of these countries have recently started reducing the footprint of these institutions. By contrast, the fastest growing cultural institutes are Chinese and Russian. China maintains more than 500 Confucius Institutes in more than 100 countries. These centres promote the Mandarin language and Chinese culture but have also drawn criticism for spreading Communist Party propaganda.[94] Likewise, the Russkiy Mir Foundation has some 171 offices, up from eighty-one in 2013. If trends continue as they are, these new cultural networks could help dramatically reshape soft power projection.[95]

Mapping the spread of soft power – cultural institutes around the world
Global changes are taking place in the soft power landscape. A growing number of countries are investing in this space, including expanding the presence and impact of official bodies for cultural and educational exchange. China has more than 500 Confucius Institutes, an increase from 320 in 2013. Other countries are reducing their footprint. The UK has 177 British Council offices, down from 196 in 2013, and Germany has 159 offices in contrast to 169 in 2013.

De-weaponising culture

The export of culture is often regarded as key to international success in politics and economics. Countries such as France, Germany, the UK and US have long exported the soft power of their culture for wider strategic purposes. Meanwhile, China and Russia have, in different ways, increased their soft power projection. Like their Western counterparts, they are engaging in diverse ways including through

social media, conventional news outlets, cultural exchanges and, of course, via political parties and social movements. Whether pursued by China, Russia, the UK or the US, the weaponisation of culture is a surprisingly cheap and effective way of projecting influence and pursuing interests.[96]

The best way to tame culture wars is by exposing the political and economic imperatives behind them. In the case of Russia, the focus on conservative traditionalist values conceals a regime that resists meaningful political, economic and social change. Despite pushing a 'family-friendly' agenda, Russia's birth rates are comparatively low.[97] As for the US, there are clear political advantages to stirring up identity and racial politics for electoral gain, as the 2016 and 2020 national elections made clear. It is equally important to shed light on the foreign funding and diplomatic support offered by proponents of culture wars. Russia has been actively supporting far-right (and some far-left) parties and political movements in and outside the country, including in the US. Exposing these networks, their digital platforms and their financing is critical to discrediting them.

Greater support for independent media around the world could also help diminish the power and potency of culture wars.[98] The fragmenting media landscape and the rise of populist politics means that both facts and science are under assault, especially with the rise of fake news. Revitalising and supporting independent and impartial news is challenging but vital for democracy and evidence-based policies to thrive. Setting rules and regulations that restrict the hate speech and extremist content that inflames culture wars is no less essential. For example, firewalls are needed that separate independent and free media from elected or unelected politicians and business elites who might manipulate it. Also required is a dramatic increase in the regulation of large technology and social media companies to ensure that hate speech and fake news is dramatically curtailed. Without legal safeguards, the truth can be misrepresented. By contrast, a free press can reveal the ways in which our cultures are being manipulated to sow dissent and division, discredit opponents and undermine governments.

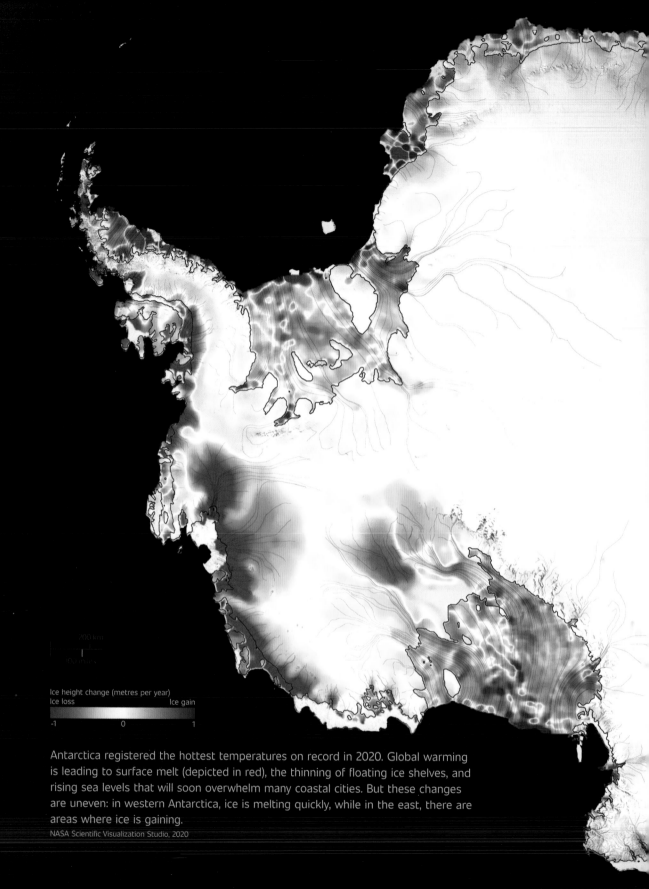

Ice height change (metres per year)
Ice loss   Ice gain

-1   0   1

Antarctica registered the hottest temperatures on record in 2020. Global warming
is leading to surface melt (depicted in red), the thinning of floating ice shelves, and
rising sea levels that will soon overwhelm many coastal cities. But these changes
are uneven: in western Antarctica, ice is melting quickly, while in the east, there are
areas where ice is gaining.
NASA Scientific Visualization Studio, 2020

Conclusion

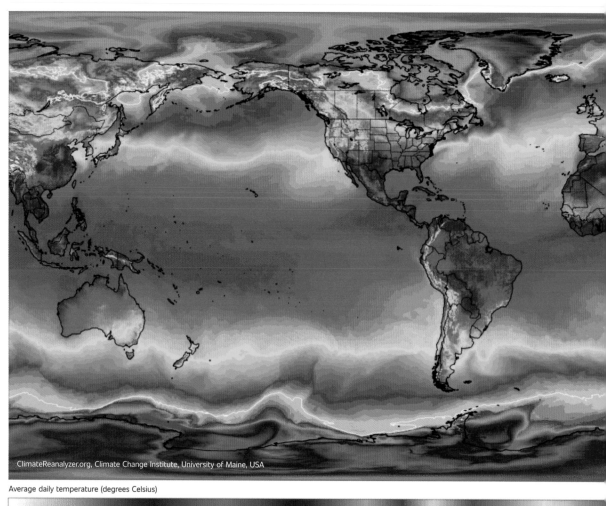

ClimateReanalyzer.org, Climate Change Institute, University of Maine, USA

Average daily temperature (degrees Celsius)

| -60 | -50 | -40 | -30 | -20 | -10 | 0 | 10 | 20 |

Introduction

In spite of achieving monumental progress over the last century, humanity faces threats of biblical proportions in the current one. Many people feel a deepening sense of foreboding. The 2020s certainly began with a bang. The planet recorded unprecedentedly high temperatures and soaring greenhouse gas emissions together with vast forest fires from Australia to Brazil that suffocated cities. Far from the headlines, massive swarms of locusts – hundreds of billions strong – devoured crops and pasture across Ethiopia, Kenya, South Sudan and Tanzania, the worst such crisis in seventy years. These events were stark reminders of just how far off course we

A burning world – 2020[1]
The last ten years have been the hottest ever recorded. The warm temperatures in both the Arctic and Antarctic were unprecedented. The Climate Reanalyzer uses National Oceanic and Atmospheric Administration (NOAA) satellite forecasting systems to record daily temperatures, as this map from Sunday 3 May, 2020 shows.

are in addressing climate change. Meanwhile, tensions between the US and Iran reached a boiling point, threatening to set the Middle East on fire. The US's trade war with China and bitter disputes with Europe also endangered the future of global trade and multilateralism. Global protests erupted around the world over inequality and corruption, contributing to growing unease. And then the coronavirus struck.

The exponential spread of coronavirus, or COVID-19, the worst pandemic since the 1918 Spanish Flu, perfectly illustrates the interconnected challenges we face in a globalised world. Although today a household name, COVID-19 has humble origins. The virus was first reported in Wuhan, a city of 11 million people that most people outside China had never heard of until it dominated the international news cycle. COVID-19 was identified by doctors in early January 2020, but it was several weeks, and only after millions of people had left the city to celebrate Lunar New Year, before a lockdown was imposed. Within months, COVID-19 had spread to at least 188 countries and infected tens of millions of people through community transmission. At first, governments imposed travel bans and repatriated their citizens. Then they started quarantining cruise ships, cancelling events and significantly curtailing air-travel. As the virus circumnavigated the globe, entire cities and then countries were locked-down, schools and universities were temporarily shuttered, and global supply chains were profoundly disrupted. The pandemic cratered global growth and sharpened geopolitical tensions in what became the most consequential financial crisis since the Great Depression.

Although it is a symptom of globalisation, COVID-19, or rather the response to it, is also a reminder of the virtues of global connectivity and international co-operation. Despite fears that the pandemic triggered an 'infodemic' due to the deluge of misinformation, rumour and xenophobia, it was the very abundance of information that allowed data-miners like BlueDot to detect the viral outbreak at least a week ahead of official government sources.[2] Within a month of it being discovered in China, germ-hunting biologists had sequenced the virus and begun sharing synthetic copies of its genetic code, some 30,000 characters long. Biotech companies and research institutes rapidly started designing and testing vaccines, speeding up the process by an order of magnitude compared to the past. Although governments responded slower than they should have, lessons from previous SARS and H1N1 pandemics were shared and implemented. The internet allowed thousands of Chinese, Brazilian, Canadian, French, German, Italian, Japanese and US-based epidemiologists, virologists and geneticists to contribute to the solution in a 24-7 cycle of exchange. Within months, tens of thousands of separate scientific studies on the virus and its transmission were available in over a dozen languages.

At the time this book went to press, it was impossible to determine just how deep and lasting the impacts of COVID-19 would be. What we can say is that deadly pandemics are dramatically illustrative of the many inter-dependencies that will determine whether we survive or thrive in the next decade. We all rely on a healthy planet – clean air, fresh water and healthy ecological systems – to live and breathe. Cities, companies, and communities depend on each other for the transmission of ideas and the basic necessities of life. Because of this, it is in our individual self-interest to cooperate and to strengthen institutions for collective action. We must not just work together, but learn to do so quicker, more efficiently, more effectively and more

The spread of COVID-19: December 2019–June 2020 [3]

COVID-19 spread rapidly around the world in 2020. It reached more than 188 countries within four months of being discovered, killing hundreds of thousands and infecting many millions of people. Maps produced by Johns Hopkins University and ESRI helped raise global awareness and deepen understanding of the speed, scope and scale of the crisis.

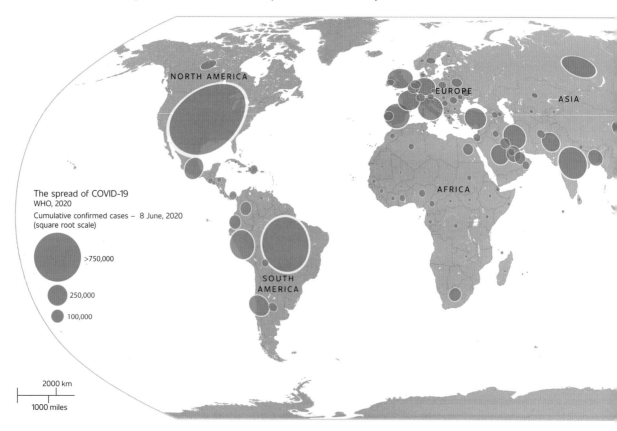

The spread of COVID-19
WHO, 2020
Cumulative confirmed cases – 8 June, 2020
(square root scale)

>750,000

250,000

100,000

2000 km
1000 miles

sustainably than ever before. Yet one of the paradoxes of our era is that many national governments and polarised societies are moving in precisely the opposite direction. The spread of market forces is one of the reasons for the rapid rise in global incomes, but it is also the source of rising individualism, increased inequality and a narrow focus on short-term gains over longer-term priorities.

It is an extraordinary coincidence that the present generation – our generation – holds the future of civilisation – indeed, of life on earth – in its hands. Our planet is roughly 4.5 billion years old. Living organisms appeared around 3.5 billion years ago. *Homo sapiens* emerged roughly 200,000 years ago, living essentially nomadic and primitive lives until approximately 10,000 years back. The industrial revolution only began in the late-1700s and it was not until the latter half of the 1900s that we began to appreciate the full extent of the human impact on climate change. The Intergovernmental Panel on Climate Change (IPCC) has determined that if we do not fundamentally reduce emissions by 2030, then we will experience catastrophic and irreversible climate change. We need to start sustained reversals in emissions now. It took a monumental tragedy to start the process: within months of the COVID-19 pandemic being announced, carbon dioxide emissions plummeted by as much as 17 per cent compared to 2019 averages, the largest single drop in history. The choices we make in the next decade will determine the fate of humanity, and of most living species. If we feel the burden of responsibility it is because we are alive at a crossroads for humanity.

We are not condemned to a race to the bottom. Our fate is not predetermined. This book is replete with examples that demonstrate how it is human actions, not inviolable forces, that shape our destinies. Around the world, more and more individuals and organisations are joining forces to take action, including through civil disobedience and non-violent protest. There is a growing recognition that widening inequalities are a key cause of our predicament. We are learning how some communities are more vulnerable than others due to the brute force of globalisation. The aspiration of the United Nations to 'leave no-one behind' is not an empty slogan – inclusion is fundamental to our survival. But how we move forward will be different from how we worked in the past. Despite paralysis in the United Nations Security Council and faltering co-operation to fight COVID-19, new forms of multilateralism(s) are already emerging that may presage a healthy

AUSTRALIA

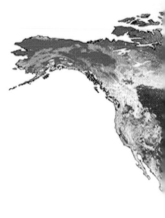

modernisation of global co-operation. Cities are banding together in networks of solidarity while some shareholder activists and investors are pushing for greener business and circular economies, albeit not fast enough. Networks of philanthropic organisations, non-governmental organisations and citizen activists are reimagining networked collaboration in a multipolar age.

We believe that maps can help us navigate an increasingly uncertain world. We are not alone, as the explosion of map-makers from ESRI and Planet Labs to Open Street Map and Our World in Data attest.[4] The sheer complexity and velocity of change requires that we engage with multiple navigational instruments to steer a path forward. We are, after all, living in a ZetaByte era, with more data generated every day than in all of history combined. Quantum computing and AI is increasingly central in shaping how decisions are made – from the functioning of our democracies to our everyday choices. Although maps can help guide us, they are also instruments of political and economic power.[5] Throughout history, cartographers working for empires and colonial powers conjured up sovereign territories by drawing straight lines on maps, often with little regard for existing ethnic or religious boundaries. Today's technology companies also use maps to subtly shape the behaviour of consumers, thus determining whether many businesses succeed or fail. Maps reflect the choices made by the map-makers. In our case, the goal is to help clarify what is happening around us, not just to enlighten, but to inspire action.

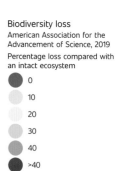

Biodiversity loss
American Association for the Advancement of Science, 2019
Percentage loss compared with an intact ecosystem

- 0
- 10
- 20
- 30
- 40
- >40

The existential threats to progress

We need a baseline from which to move forward. Hard as it is to believe, humans have never been safer or more prosperous than they are right now. As Steven Pinker reminds us, war-related deaths and most forms of violent crime have declined and, over the long term, and in spite of COVID-19, life really is getting better for most people.[6] However, this is of little consolation to the millions of victims and survivors of state repression, extremism and sexual and domestic violence, or for those living in grinding poverty, suffering from stagnating wages or losing their jobs and livelihoods to automation. Personal security and safety are still an aspiration for far too many people. The stubborn persistence of insecurity and the deepening of inequalities within

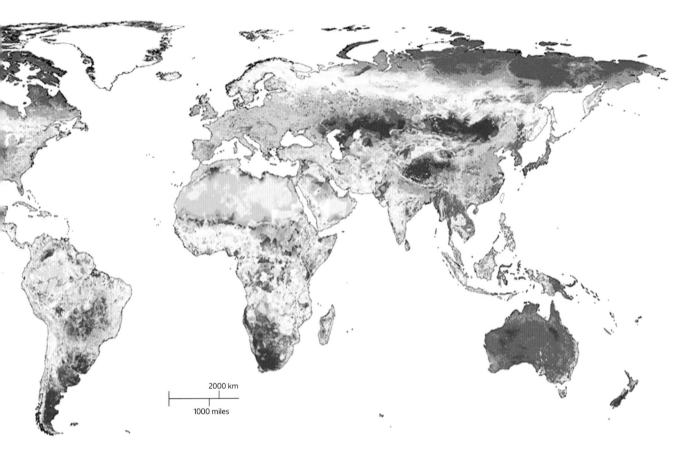

1000 miles | 2000 km

Biodiversity at the edge[8]

Up to one million species are facing extinction owing to human influence. The Intergovernmental Science-Policy Platform on Biodiversity and Ecosystem Services includes scientists from over 130 countries who have warned about 'an imminent and rapid acceleration in the global rate of species extinction'. The speed of loss is already tens to hundreds of times higher than it has been over the previous ten million years.

countries mean that there is absolutely no room for complacency.[7] Progress does not inevitably embrace everyone. It has to be fought for and may well relapse, as is painfully evident when there are pandemic outbreaks, disastrous wars, and economic or ecological collapse. Many risks have not disappeared but are instead compounded by newer risks such as climate change and superbugs that pose catastrophic threats to all of us, rich and poor.

While on average humans have experienced rising incomes, better health, more education and elongated life-spans, the averages mask widening disparities. We are poised at the precipice of multiple and interconnected climate disasters with millions of plant and animal species hovering at the edge of extinction. Many of our fellow humans, especially poorer and vulnerable people living at the margins, are facing a perfect storm of risks that threaten their livelihoods and lives. Although most people have experienced tangible improvements, new rapidly evolving risks, some of which can literally move at the speed of light, are making life more dangerous for us all. There is a real danger that we will succumb to pessimism and paralysis. In his book *Our Final Century*, Martin Rees, the former president of the Royal Society, said he believes our civilisation has only even odds of surviving the present century.[9] He echoes the concerns of philosopher Glenn Albrecht who coined the term 'solastalgia' to describe the anxiety and sorrow arising from environmental calamities.[10] Lord Rees and Glenn Albrecht are hardly alone: eco-anxiety is spreading.[11] We agree with their call for urgent and radical action. Even if the probability that we are heading for catastrophe is low, we should take urgent action to reduce the odds further still. After all, even if you felt there was a slim chance of your house catching fire, surely you would take every step possible to avoid this from happening?

The tremendous achievements of the post-war period have brought most, but certainly not all, of humanity close to conquering some of our most enduring risks from maternal mortality to measles. Greater openness and connectivity have raised incomes and life expectancies, including by increasing the quality and availability of everything from food to phones and vaccines. The spread of new technologies accounts for some of the progress, but the diffusion of ideas is even more important. These cumulative achievements conceal deep and dangerous inequalities. Billions of people still suffer from disease or despair, are at risk of malnutrition and malaria, or are experiencing extreme poverty. Globalisation is part of the problem and the solution. This is because the upsides and downsides of globalisation are transmitted through the world's globally integrated market systems and interconnected digital and physical infrastructure, as described in *The Butterfly Defect*.[12] COVID-19 provides a graphic demonstration of how the hubs and spokes which spread the benefits of globalisation, like airports, can also spread viruses. Before that, it was the 2008 financial crisis that bankrupted pensioners across Europe. Or consider the US–China trade war that has devastated jobs for cobalt miners in the Democratic Republic of the Congo and Zambia[13] and is fuelling Amazonian forest fires and deforestation.[14]

Recalculating risk in an age of hyperconnectivity

We are living in an era of intense turbulence, disillusionment and bewilderment. Deepening geopolitical tensions are transforming international relations, and political tribalism is revealing deep fissures within countries. COVID-19 not only exposed the many inequalities

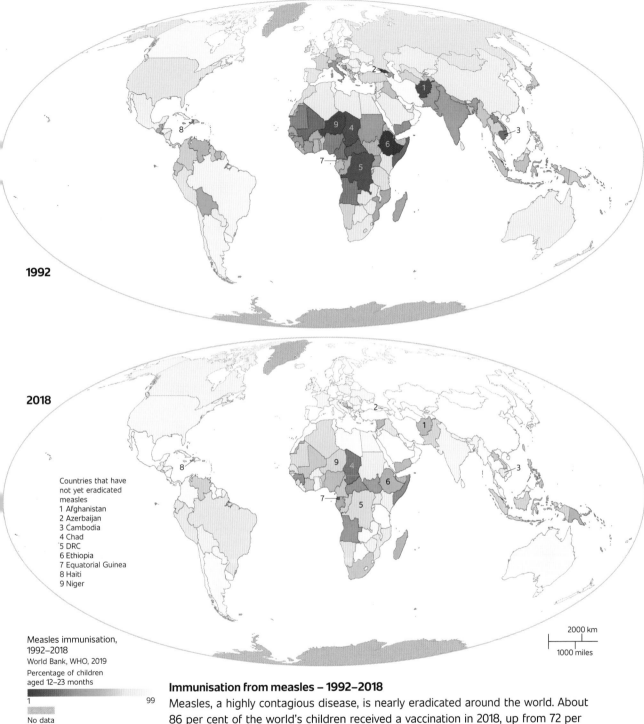

1992

2018

Countries that have
not yet eradicated
measles
1 Afghanistan
2 Azerbaijan
3 Cambodia
4 Chad
5 DRC
6 Ethiopia
7 Equatorial Guinea
8 Haiti
9 Niger

2000 km
1000 miles

Measles immunisation,
1992–2018
World Bank, WHO, 2019
Percentage of children
aged 12–23 months

1   99

No data

Immunisation from measles – 1992–2018

Measles, a highly contagious disease, is nearly eradicated around the world. About
86 per cent of the world's children received a vaccination in 2018, up from 72 per
cent in 2000. Between 2000 and 2018, these vaccinations prevented over 23 million
deaths making it one of the best buys in public health. Even so, about 140,000
people die of measles every year, mostly children under five.[15]

strafing our societies, but it also accelerated a set of structural changes including the digitisation of work. The spread of disruptive new technologies is upending long-held assumptions about security, politics, economics and even what it means to be human. It is hard to know what happens next. Angst has replaced hubris. That all of these huge global challenges are emerging at a time of weakening multilateral and international co-operation is the biggest cause for concern. This is because our future looks infinitely more complex and uncertain than the past. The management of today's and tomorrow's risks requires that we do not look in the rear-view mirror. While learning from the past, we need to focus firmly on the changing nature of both the challenges and the potential solutions. After all, it is the new systemic risks that arise from the growing entanglement of companies, supply chains, markets and patterns of consumption that are escalating most rapidly in front of our eyes and now pose the gravest dangers, as the fallout from the coronavirus pandemic amply demonstrates.[16]

All of this means we need to rethink risk. The principle of subsidiarity should be applied to risk, as it needs to be to other areas of governance. Whatever can be handled effectively at the individual, household, neighbourhood, city or corporate level should be. Wherever possible, designing in resilient and cellular thinking – including investments in self-sufficiency circular economy and adaptive regeneration – will be more important than ever.[17] Where more aggressive levels of coordination and intervention are necessary, there needs to be the political will and financial and other resources available to step in. This can occur at the state, national, regional or even global levels. Indeed, many of the largest risks we are currently facing are planetary in nature and require responses from the hyper-local to the multilateral. When a massive disaster strikes, the ability of the global community to help support local efforts in a timely manner is crucial, as was the case with the Ethiopian famine in 1984–85, the Indonesian tsunami in 2004, and Haiti's earthquake in 2010. The 2008 financial crisis and 2020 coronavirus outbreaks are examples of systemic risks that spiralled out of control owing to a failure to manage the consequences of a local problem.

The lights are certainly flashing red. We considered some of the far-reaching effects of global warming in the climate chapter – from flooding the world's coastal cities to destroying global biodiversity.

We also reviewed the possibly catastrophic global threats to population health posed by pandemics and antimicrobial resistance in the chapter on health. And as we demonstrated in the chapter on inequality, poor people in poorer countries are especially vulnerable to systemic risks. It is precisely because they frequently lack deeds to their homes, land tenure, or savings to bide them through bad times, that they are most exposed to price shocks and market failures. More than a decade after the 2008 financial crisis, many of the poorest families in Europe, the UK and the US are still poorer than before the crisis. Their anger against the elites who repeatedly let them down has fuelled a backlash.[18]

Everyone has a role to play in reducing risk

Risk is too important to be left to experts, whether they are academics, bankers or civil servants. Each and every one of our individual and collective actions shape the evolution of risk and – by extension – the future environment of our children and grandchildren. We should not simply pass the buck upwards to our governments hoping they will 'save us'. We *all* need to take action to reduce the risk of pandemics and climate change, to cut plastic waste, or to curtail our use of antibiotics. This is not just about taking a moral stand – it is about our collective survival. Not all issues come down to the question of global, collective action. Not all big challenges can be solved by governments or international organisations. Citizens and communities have a central role to play. So do companies, since they account for most of what is ultimately produced, traded and consumed in the world. A combination of government rules, corporate leadership and consumer action are all needed to accelerate change towards a more sustainable future.[19]

Almost all the great challenges we face can be dramatically mitigated by a small group of actors. They may not solve the entire problem, but it is generally the case that the *Pareto Principle* applies. That is, around 20 per cent of the actors can resolve 80 per cent of the problems. Climate change is a good example. As the climate chapter made clear, fewer than twenty countries account for eighty per cent of global greenhouse gas emissions,[21] and twenty companies account for one-third of all greenhouse gas emissions (GHGs) since 1965.[22]

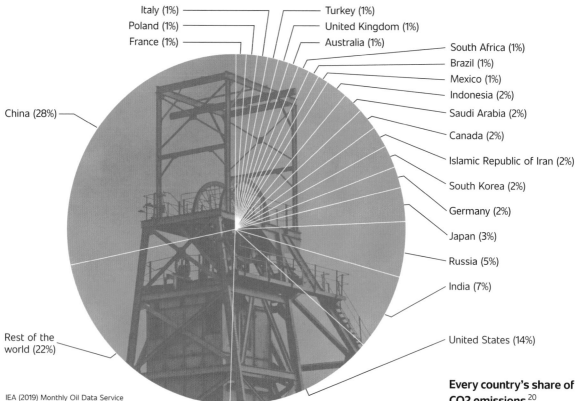

Italy (1%)
Poland (1%)
France (1%)
Turkey (1%)
United Kingdom (1%)
Australia (1%)
South Africa (1%)
Brazil (1%)
Mexico (1%)
Indonesia (2%)
Saudi Arabia (2%)
Canada (2%)
Islamic Republic of Iran (2%)
South Korea (2%)
Germany (2%)
Japan (3%)
Russia (5%)
India (7%)
United States (14%)
China (28%)
Rest of the world (22%)

IEA (2019) Monthly Oil Data Service

While it is not essential that all 193 countries agree to rapidly reduce GHGs, the biggest emitters must immediately move to zero emissions. Agreement under the United Nations umbrella secures legitimacy but is not necessarily a guarantee of success because the signatories of many treaties fail to deliver on their promises.

Because a relatively small group of governments and companies account for most of the problems we face, they are necessarily a bigger part of the solution. However, those that are benefiting from existing rules cannot be expected to change them. Whether it is climate change or ending tax avoidance, the affected need to have their voice heard against powerful vested interests who have benefited from the status quo. In the case of climate change, it is vital that countries who will be particularly adversely affected, such as Bangladesh, the Maldives and the Solomon Islands, are heard. Perverse subsidies for fossil fuels need to be ended. The subsidies, like regulations and bail-outs for the banks, are fuelled by lobbies. These need to be curtailed and the revolving door between Wall Street and the White House bolted shut.[23]

Every country's share of CO2 emissions [20]
Every country emits different amounts of greenhouse gases. This chart, developed by the International Energy Agency, estimates CO2 emissions from oil, coal, natural gas and other waste products. The largest emitters are China, the US and India, followed by Russia, Japan, Germany, South Korea, Iran, Canada and Saudi Arabia.
If measured by per capita emissions, Saudi Arabia, Australia, the US, Canada, South Korea, Russia and Japan are the worst polluters.

Technology companies are now among the biggest lobbyists in Washington DC and Brussels. These companies should not be allowed to subvert legitimate enquiries into their monopoly power, invasions of privacy, access to data or facilitation of extremism, hate speech and fake news. The hacking and manipulation of social media through the creation of false accounts and misrepresentation of information threatens democracy to its very core. The Cambridge Analytica scandal and evidence of the subversion of voting around the world is a reminder of how pervasive these practices have become. The spread of dangerous ideas across cyberspace, from Jihadism to the anti-vax (vaccination) movement, and the amplification of fake news, is greatly complicating the management of global risks. Government regulators and social media firms need to ensure that they are accountable and contribute to the mitigation of risks.

With the spread of complex new viral threats, pharmaceutical companies also need to step up and work with governments and businesses to develop new antibiotics and to be more active in addressing pandemics. In finance, more needs to be done to safeguard financial stability, not least in curtailing the rise of shadow banking, by which a growing range of products are being offered by firms that are not subject to the controls on the big banks. Insurance companies that specialise in managing risk also have a pivotal role to play, not least in helping develop the next generation of tools to anticipate and safeguard against risk. Their ability to increase the coverage of new risks, from climate change to cyber threats, is essential. Rising claims and cascading risks are already testing the reinsurance industry on which all insurers depend, as it is through these reinsurers that risks can be laid off. Governments urgently need to come together to define the terms under which they could provide the necessary support to reinsurers to manage systemic risks.

New technologies are changing the nature of insurance, which is essentially a club in which the misfortune of some is compensated by the better tidings of others. New technologies allow much better predictions of an individual's susceptibility to risk. Some of these are more invasive than others. For example, biometric testing may predict an individual's health needs over the coming years, or sensors in cars could differentiate individual driving habits and in homes identify smokers. The potential to price individual behaviour and give discounts and incentives for good conduct is growing, and for this and other reasons we might well see a narrowing coverage for certain more risky and lower-income people in society, further widening inequalities and subverting norms around privacy.

Adherence to globally agreed and enforced rules and regulations can greatly simplify the complexity facing governments and companies. For example, membership of the World Trade Organization allows countries to establish predictable trade and universal norms. Although the European Union is at times characterised as an overreaching bureaucracy, in practice the single market has aligned regulations in twenty-seven countries, greatly simplifying crucial dimensions of the lives of European citizens, firms and governments. In return, these same nations have handed over significant parts of their sovereignty, in key areas such as defence, foreign affairs, immigration and justice, to the European Commission. The pooling of responsibility has not only provided a significant boost for economies, but has also increased the bargaining power of members when it comes to dealing with highly complex risks, such as cyber or other threats posed by hostile countries or non-state actors. The European Union is not the only example of shared risk management. Take the case of air traffic control, which, despite the rapid evolution of technologies, the creation of 100 new countries over the past 100 years and the 10 per cent annual rise in air traffic, is a case study in effective risk management in a rapidly globalising world.[24] The airline industry faces an entirely different risk since the onset of COVID-19, which has severely curtailed air travel and bankrupted dozens of carriers.

A call for collective action in a risky world

Multiple risks – be they related to infectious diseases, climate change, rapid automation, financial collapse, or geopolitical tension – are rapidly converging. This is overwhelming and unsettling to most of us. When people confront uncertainty and feel vulnerable, they look for familiar handholds and old certainties. Faced with dramatic change, individuals can become more defensive, protectionist and nationalistic. Opportunistic politicians frequently seize on fear, anxiety and doubt, promising simple solutions and appealing to the past. And while their promises may offer temporary consolation, the reality is that there is no wall high enough to keep out pandemics, global warming, ecosystem and ocean collapse, or any of the other dangerous systemic risks reviewed in this volume. What is needed is more co-operation, not less. Withdrawal and fragmentation compound the existential risks we face.

Twelve years before the COVID-19 pandemic, the world came to the brink of financial meltdown. It was saved when an emergency meeting of the US, China, Europe and at least seventeen other leading powers collectively agreed to stave off disaster. The danger now is that the rich countries have not only fired-off all their fiscal ammunition, but they have also lost the political will to cooperate. While China has stepped up to assume a larger role in the management of several global challenges, not least those related to climate change and trade, the US is retreating, and a new Cold War is underway. Rich countries are trapped in a cycle of stagnant growth and loss of trust in the experts who are blamed for sluggish economies. As risks rise, populists and nationalists blame the failure of elites, as well as foreign migrants and refugees. This undermines the will to work together, leading to a further escalation of risks and splintering of social cohesion.

The solutions are to be found not in the retreat from globalisation, but in closer collaboration to overcome our shared trials. More vigilance on the part of regulators and supervisors, greater sophistication in risk management, and closer co-operation of policymakers are all required to meet the complex web of challenges on our horizon. Growing integration brings rising interdependency. Unco-ordinated responses, within countries and between countries, compound the problem and are bound to fail. The greatest threat we face is the short-termism of governments that are not addressing systemic risks and failing to cooperate to resolve them. In every sphere of our lives – individual, community, city, state, corporate, national, regional and global – we need to forge new alliances, commit ourselves to stretch targets and urgently act on our knowledge of the many risks we face. It is only by mitigating the risks that we can harvest the extraordinary opportunities of the twenty-first century. The best way to determine the future is to shape it.

Notes

Preface

1 Coronavirus.jhu.edu, COVID-19 Dashboard, Center for Systems Science and Engineering (CSSE) at Johns Hopkins University (JHU), 2020.

2 Gates, Bill, We're Not Ready for the Next Epidemic, Gates Notes, 2015; Muggah, Robert, Pandemics Are the World's Silent Killers. We Need New Ways to Contain Them, Devex, 2019; Goldin, Ian and Mariathasan, Mike, *The Butterfly Defect*, Princeton University Press, 2014.

3 Altman, Steven A., Will Covid-19 Have a Lasting Impact on Globalization?, *HBR*, 2020.

4 Schipani, Andres, Foy, Henry, Webber, Jude, et al., The 'Ostrich Alliance': The Leaders Denying the Coronavirus Threat, *FT*, 2020.

5 Spreeuwenberg, Peter, Kroneman, Madelon and Paget, John, Reassessing the Global Mortality Burden of the 1918 Influenza Pandemic, *American Journal of Epidemiology*, Vol. 187, 2018; Johnson, Niall P. A. S. and Juergen Mueller, Updating the Accounts: Global Mortality of the 1918–1920 'Spanish' Influenza Pandemic, *Bulletin of the History of Medicine*, Vol. 76, 2002.

6 Based on Nickol, Michaela and Kindrachuk, Jason, *A year of terror and a century of reflection: Perspectives on the great influenza pandemic of 1918–1919*, BMC Infectious Diseases, Vol. 19, 2019; Nicholson, Karl, Webster, Robert G., et al. *Textbook of Influenza*, Blackwell Science, 1998.

7 Brainerd, Elizabeth and Siegler, Mark V., The Economic Effects of the 1918 Influenza Epidemic, CEPR Discussion Papers 3791, 2003; Correia, Sergio, Luck, Stephan and Verner, Emil, Pandemics Depress the Economy, Public Health Interventions Do Not: Evidence from the 1918 Flu, SSRN, 2020.

8 Gyr, Ueli, The History of Tourism: Structures on the Path to Modernity, EGO, 2010.

9 See Fedrico, Giovanni and Tena, Junguito, World Trade Historical Databases.

10 Nielsen.com, Outbound Chinese Tourism and Consumption Trends, Nielsen, 2017.

11 Rabouin, Dion, Coronavirus Has Disrupted Supply Chains for Nearly 75% of U.S. Companies, Axios, 2020.

12 Crow, Alexis, COVID-19 and the Global Economy, Observer Research Foundation, 2020; Goldin, Ian and Muggah, Robert, Viral Inequality, Syndicate Project, 2020.

13 Antonenko, Oksana, COVID-19: The Geopolitical Implications Of A Global Pandemic, Control Risks, 2020.

14 See Binding, Lucia, Coronavirus: Only 9% of Britons Want Life to Return to 'Normal' Once Lockdown is Over, Sky News, 2020.

15 Monks, Paul, Coronavirus: Lockdown's Effect On Air Pollution Provides Rare Glimpse Of Low-carbon Future, The Conversation, 2020.

16 Muggah, Robert, Redesigning The COVID-19 City, NPR, 2020.

17 See Goldin, Ian, Divided Nations: *Why Global Governance Is Failing, and What We Can Do About It*, Oxford University Press, 2013.

18 Foa, Roberto, S., Klassen, Andrew, et al. The Global Satisfaction with Democracy Report 2020, University of Cambridge, 2020.

19 See Pew Surveys at Pewresearch.org, Topics: Political Polarization, Pew Research, 2020.

20 See Goldin, Ian, The Compass: *After the Crash: The Future*, BBC, 2018.

21 Oxfordmartin.ox.ac.uk, Now for the Long Term: The Report of the Oxford Martin Commission for Future Generations, Oxford Martin School, 2013.

Globalisation

1 See Goldin, Ian and Reinert, Kenneth, *Globalization for Development*, Oxford University Press, 2012.

2 Internetworldstats.com, World Internet Usage and World Population Statistics, 2020.

3 Ofcom.org.uk, Achieving Decent Broadband Connectivity for Everyone, 2016.

4 Cable.co.uk, Worldwide Broadband Speed League, 2019.

5 Fukuyama, Francis, *The End of History and the Last Man*, Penguin, 2012.

6 Cairncross, Frances, *The Death of Distance: How the Communications Revolution is Changing Our Lives*, Harvard Business School Press, 2001.

7 Friedman, Thomas, *The World is Flat: The Globalized World in the Twenty-First Century*, Penguin, 2007.

8 Goldin, Ian, Muggah, Robert, How to Survive and Thrive in Our Age of Uncertainty, World Economic Forum, 2019.

9 Inequality.org, Facts: Global Inequality, 2020.

10 Ahmed, Kamal, Workers are £800 a Year Poorer Post-crisis, BBC News, 2018.

11 World Bank Group, World Development Report 2016: Digital Dividends, 2016.

12 Osterhammel, Jurgen and Petersson, Niels, *Globalization: A Short History*, Princeton University Press, 2005; O'Rourke, Kevin H., Williamson, Jeffrey G., When Did Globalization Begin?, NBER No. 7632, 2000.

13 Francis Galton, Isochronic Passage Chart, Proceedings of the Royal Geographical Society, 1881.

14 Vanham, Peter, A Brief History of Globalization, World Economic Forum, 2019.

15 This section draws on Goldin, Reinert, 2012, op. cit., p23.

16 Goldin, Ian and Kutarna, Chris, *Age of Discovery*, Bloomsbury, 2017.

17 Vanham, 2019, op. cit., p23.

18 Beltekian, Diana, Ortiz-Ospina, Esteban, The 'Two Waves of Globalisation', 2018.

19 Keynes, John M., *The Economic Consequences of the Peace*, Harcourt, Brace, and Howe, 1920; See Goldin, Reinert, 2012, op. cit., p23.

20 Eichengren, Barry, *Globalizing Capital*, Princeton University Press, 2008.

21 See Hatton, Timothy and Williamson, Jeffrey, *The Age of Mass Migration*, Oxford University Press, 1998.

22 Smith, Oliver, Switzerland's Out, Albania's In – How the Travel Map Has Changed Since 1990, *Telegraph*, 2020.

23 Data.worldbank.org, Trade (% of GDP), World Bank national accounts data, and OECD National Accounts data files, 2019.

24 Data.worldbank.org, Merchandise Exports, Transparency in Trade initiative, 2019.

25 Data.worldbank.org, Trade, 2019, op. cit., p27.

26 Chatzky, Andrew, McBride, James and Sergie, Mohammed A., NAFTA and the USMCA: Weighing the Impact of North American Trade, Council on Foreign Relations, 2020.

27 Misbahuddin, Sameena, What is the World Wide Web?, BBC Bitesize, 2020.

28 Roser, Max, Democracy, Our World in Data, 2013.

29 See Foa, Roberto, S., Klassen, Andrew, et al. The Global Satisfaction with Democracy Report 2020, University of Cambridge, 2020.

30 Altman, Steven, A., Ghemawat, Pankaj and Bastian Phillip, DHL Global Connected Index 2018 – The State of Globalization in a Fragile World, DHL, 2018.

31 Ibid., p32.

32 Imf.org, World Economic Outlook Database, 2019.

33 Goldin, Reinert, 2012, op. cit., p32.

34 Goldin, Ian and Mariathasan, Mike, *The Butterfly Defect*, Princeton University Press, 2014.

35 Ibid., p32.

36 Murphy, Francois and Wroughton, Lesley, IMF Warns of Financial Meltdown, Reuters, 2008.

37 Wikipedia.org, Stock Market Crash, 2020.

38 Worldbank.org, Record High Remittances Sent Globally in 2018, World Bank Press Release, 2019.

39 Knomad and World Bank Group, Migration and Remittances: Recent Developments and Outlook, Migration and Development Brief 31, 2019.

40 Ibid., p34.

41 Unctad.org, Global Foreign Investment Flows Dip to Lowest Levels in a Decade, UNCTAD News, 2019.

42 Regling, Klaus, 'Cross-border Capital Flows: Theory and Practice' – Speech K. Regling, ESM News, 2017.

43 Ibid., p34.

44 See iif.com, Global Focus: Global Macro Views, Institute of International Finance, 2020.

45 Chappell, Bill, U.S. National Debt Hits Record $22 Trillion, NPR Economy, 2019.

46 Amoros, Raul, The Biggest Foreign Holders of U.S. Debt – In One Chart, howmuch.net article, 2019.

47 Ibid., p35.

48 Ibid., p35.

49 Goldin, Ian, Karlsson, Mats, Stern, Nicholas, Rogers, Halsey and Wolfensohn, James, D., A Case for Aid: Building a Consensus for Development Assistance, World Bank Report, 2002.

50 Goldin, Ian, *Development: A Very Short Introduction*, Oxford University Press, 2018.

51 Ibid., p36.

52 Ibid., p37.

53 Oecd.org, Official Development Assistance 2019: Compare your Country, DAC Statistics, 2020.

54 Ibid., p37.

55 Freund, Caroline and Ruta, Michele, Belt and Road Initiative, World Bank Brief, 2018.

56 See Goldin, 2018, op. cit., p37.

57 Oecd.org, What is ODA?, DAC Report, 2018.

58 Goldin, 2018, op. cit., p38.

59 Roser, Max, Economic Growth, Our World in Data, 2013.

60 Glenny, Misha, *McMafia: A Journey Through the Global Criminal Underworld*, Vintage, 2009.

61 Naim, Moises, *Illicit: How Smugglers, Traffickers, and Copycats are Hijacking the Global Economy*, Penguin Random House, 2006.

Climate

1 Cookson, Clive, Homo Sapiens 100,000 Years Older Than Thought, *Financial Times*, 2017.

2 Washingtonpost.com, Himalayan Death Tolls, *Washington Post*, 2014.

3 Farooq, Mohd, Wagno, Patrick, Berthier, Etienne, et al., Review of the Status and Mass Changes of Himalayan-Karakoram Glaciers, *Journal of Glaciology*, Vol. 64, 2018.

4 Mukherjee, Kriti, Bhattacharya, Atanu, Pieczonka, Tino, et al., Glacier Mass Budget and Climate Reanalysis Data Indicate a Climatic Shift Around 2000 in Lahaul-Spiti, Western Himalaya, *Climate Change*, Vol. 148, 2018.

5 Skymetweather.com, Gangotri Glacier Shrinking, skymetweather, 2014.

6 Dixit, Kunda, In Mount Everest Region, World's Highest Glaciers are Melting, Inside Climate News, 2018.

7 King, Owen, Quincey, Duncan J., Carrivick, Jonathan L., et al., Spatial Variability in Mass Loss of Glaciers in the Everest Region, Central Himalayas, Between 2000 and 2015, *The Cryosphere*, Vol. 11, 2017.

8 Thejournal.ie, Shrinking Himalayan Glaciers Have Been Granted Status of 'Living Entities', theJournal.ie, 2017.

9 Pelto, Mauri, Zemu Glacier, Sikkim Thinning and Retreat, From A Glacier's Perspective: Glacier Change, 2009.

10 Kornei, Katherine, Glacial Outburst Flood Near Mount Everest Caught on Video, Eos, 2017.

11 Ibid., p51.

12 Dixit, 2018, op. cit., p52.

13 UNDP Nepal, Danger In The Himalayas, UNDP Nepal, 2018.

14 Spacedaily.com, Black Carbon Driving Himalayan Melt, Space Daily, 2010.

15 Chao, Julie, Black Carbon a Significant Factor in Melting of Himalayan Glaciers, Berkeley Lab: News Center, 2010.

16 Voiland, Adam, A Unique Geography -- and Soot and Dust -- Conspire Against Himalayan Glaciers, nasa.gov, 2009.

17 Npolar.no, Albedo Effect, Norwegian Polar Institute.

18 Wester, Philipous, Mishra, Aditi, Mukherji, Arun, et al., The Hindu Kush Himalaya Assessment: Mountains, Climate Change, Sustainability and People, Springer, 2019.

19 Ibid., p53; Sahasrabudhe, Sanhita and Mishra, Udayan, Summary of the Hindu Kush Himalaya Assessment Report, ICIMOD, 2019.

20 Wester, Mishra, Mukherji, 2019, op. cit., p53.

21 Cornell University, Rising Seas Could Result in 2 Billion Refugees by 2100, ScienceDaily, 2017.

22 Holmes, Robert M., Natali, Susan, Goetz, Scott, et al., Permafrost and Global Climate Change, Woods Hole Research Center, 2015.

23 Winski, Dominic, Osterberg, Erich, Kreutz, Karl, et al., A 400-Year Ice Core Melt Layer Record of Summertime Warming in the Alaska Range, *JGR Atmospheres*, Vol. 123, 2018.

24 Larsen, Chris F., Burgess, Evan and Arendt, Anthony A., Surface Melt Dominates Alaska Glacier Mass Balance, *Geophysical Research Letters*, Vol. 42, 2015.

25 NASA, World of Change: Columbia Glacier, Alaska, Phys. org, 2018.

26 Trusel, Luke D., Das, Sarah B., Osman, Matthew B., et al., Nonlinear Rise In Greenland Runoff In Response To Post-industrial Arctic Warming, *Nature*, Vol. 564, 2018.

27 Van As, Dirk, Hubbard, Alun L., Hasholt, Bent, et al., Large Surface Meltwater Discharge from the Kangerlussuaq Sector of the Greenland Ice Sheet During the Record-warm Year 2010 Explained by Detailed Energy Balance Observations, *The Cryosphere*, Vol. 6, 2012.

28 Andrews, Lauren C., Methane Beneath Greenland's Ice Sheet is Being Released, *Nature*, 2019.

29 Oltmanns, Marilena, Strane, Fiammetta and Tedesco, Marco, Increased Greenland Melt Triggered by Large-scale, Year-round Cyclonic Moisture Intrusions, *The Cryosphere*, Vol. 13, 2019.

30 Phys.org, Iceberg 4 Miles Wide Breaks off From Greenland Glacier, Phys.org, 2018.

31 Gray, Ellen, Unexpected future boost of methane possible from Arctic permafrost, Global Climate Change, 2018.

32 Portnov, Alexey, Vadakkepuliyambatta, Sunil, Mienert, Jürgen, et al., Ice-sheet-driven Methane Storage and Release in the Arctic, *Nature Communications*, Vol. 7, 2016.

33 Boberg, Frederik, Langen, Peter L., Mottram, Ruth H., 21st-century Climate Change Around Kangerlussuaq, West Greenland: From the Ice Sheet to the Shores of Davis Strait, Arctic, *Antarctic, and Alpine Research*, Vol. 50, 2018.

34 NCDC.noaa.gov, Global Climate Report – Annual 2018, NOAA, 2019.

35 Shen, Lucinda, These 100 Companies Are Responsible for Most of the World's Carbon Emissions, *Fortune*, 2017.

36 Allsopp, Michelle, Page, Richard, Johnston, Paul, et al., State of the World's Oceans, Springer, 2009.

37 NCDC.noaa.gov, 2019, op. cit., p60.

38 European Academies' Science Advisory Council, Leopoldina – Nationale Akademie der Wissenschaften, New Data Confirm Increased Frequency of Extreme Weather Events, Science Daily, 2018.

39 Europarl.europa.eu, Greenhouse Gas Emissions by Country and Sector (Infographic), European Parliament, 2019.

40 Rodgers, Lucy, Climate Change: The Massive CO2 Emitter You May Not Know About, BBC News: Science and Environment, 2018.

41 EPA.gov, Sources of Greenhouse Gas Emissions, EPA, 2018.

42 Griffin, Paul, The Carbon Majors Database: CDP Carbon Majors Report 2017, CDP, 2017.

43 Thompsonreuters.com, Global 250 Greenhouse Gas Emitters, Reuters, 2017.

44 Carr, Mathew, *China's Carbon Emissions May Have Peaked*, Bloomberg, 2018.

45 Guan, Dabo, Meng, Jing, Reiner, David M., et al., Structural Decline In China's CO2 Emissions Through Transitions In Industry And Energy Systems, *Nature Geoscience*, Vol. 11, 2018.

46 Storrow, Benjamin, Global CO2 Emissions Rise after Paris Climate Agreement Signed, Scientific American, 2018.

47 Kurtis, Alexander, Oil Companies Want Sf, Oakland Climate Lawsuits Dismissed, *San Francisco Chronicle*, 2018.

48 Milman, Oliver and Holden, Emily, Lawsuit Alleges Exxonmobil Deceived Shareholders On Climate Change Rules, *Guardian*, 2018.

49 DiChristopher, Tom, Judge Throws Out New York City's Climate Change Lawsuit Against 5 Major Oil Companies, CNBC, 2018.

50 Kurtis, 2018, op. cit., p62.

51 Irfan, Umair, Playing hooky to save the climate: why students are striking on March 15, Vox, 2019.

52 Balch, Jennifer K., Bradley, Bethany A., Abatzoglou, John T., et al., Human-started Wildfires Expand the Fire Niche Across the United States, PNAS, 2017.

53 Wolters, Claire, California Fires Are Raging: Get the Facts on Wildfires, *National Geographic*, 2019.

54 Wall, Mike, Raging California Wildfires Spotted from Space, Space.com, 2018.

55 Miller, Casey and Irfan, Umair, Map: See Where Wildfires Are Causing Record Pollution In California, Vox, 2017.

56 Azad, Arman, Due to Wildfires, California Now Has The Most Polluted Cities In The World, CNN, 2018.

57 Williams, Jeremy, Tracking Coal Power from Space, The Earthbound Report, 2019.

58 Magill, Bobby, U.S. Has More Gas Flares than Any Country, *Scientific American*, 2016.

59 Bismarktribune.com, North Dakota Oil Production Natural Gas Flaring Reach New Highs, Bismarck Tribune, 2018.

60 Kroh, Kiley, Emissions From North Dakota Flaring Equivalent To One Million Cars Per Year, Think Progress, 2013.

61 Finlay, Sarah E., Moffat, Andrew, Gazzard, Rob et al., Health Impacts of Wildfires, *PLoS Currents*, Vol. 4, 2012.

62 WHO.int, Health Topics: Air Pollution, WHO, 2020.

63 Wettstein, Zachary S., Hoshiko, Sumi, Fahimi, Jahan, et al., Cardiovascular and Cerebrovascular Emergency Department Visits Associated With Wildfire Smoke Exposure in California in 2015, Journal of the American Heart Association Vol. 7, 2018.

64 Schultz, Courtney and Moseley, Cassandra, Better Forest Management Won't End Wildfires, But It Can Reduce The Risks – Here's How, The Conversation, 2018.

65 Jenner, Lynn, Agricultural Fires Seem to Engulf Central Africa, NASA, 2018.

66 Tosca, Michael, The Impact Of Savanna Fires On Africa's Rainfall Patterns, The Conversation, 2015.

67 Yang, Yan, Saatchi, Sassan S., Xu, Liang, et al., Post-drought Decline of the Amazon Carbon Sink, *Nature Communications*, Vol. 9, 2018.

68 Butler, Rhett A., Calculating Deforestation Figures for the Amazon, Mongabay, 2018.

69 Lancaster University, Carbon Emissions from Amazonian Forest Fires up to Four Times Worse than Feared, Phys.org, 2018.

70 Sawakuchi, Henrique O., Neu, Vania, Ward, Nicholas D., et al., Carbon Dioxide Emissions along the Lower Amazon River, Frontiers in Marine Science, 2017.

71 Aragao, Luis, Barlow, Jos and Anderson, Liana, Amazon Rainforests that Were Once Fire-proof Have Become Flammable, The Conversation, 2018.

72 The state includes 50 million acres (208,000 square kilometres) of land.

73 WWF.org.co, Brazilian Amazon: Environmental Awareness Higher in Deforested Areas, World Wide Fund for Nature, 2001.

74 Bauters, Marijn, Drake, Travis W., Verbeeck, Hans, et al., High Fire-derived Nitrogen Deposition on Central African Forests, PNAS Vol. 115, 2018; Sinha, Eva, Michalak, Anna M. and Balaji, Venkatramani, Eutrophication Will Increase During The 21st Century As A Result Of Precipitation Changes, *Science*, Vol. 357, 2018.

75 Brazil is losing roughly 1.5 million acres (6,000 square kilometres) of land a year since 2009. McCarthy, Niall, Brazil Sees Worst Deforestation In A Decade, *Forbes*, 2018.

76 Song, Xiao-Peng, Hansen, Matthew C., Stehman, Stephen V., et al., Global land change from 1982 to 2016, *Nature*, Vol. 560, 2018.

77 Pearce, Fred, Rivers in the Sky: How Deforestation Is Affecting Global Water Cycles, Yale Environment 360, 2018.

78 Ibid., p77.

79 Staal, Arie, Tuinenburg, Obbe A., Bosmans, Joyce H. C., et al., Forest-rainfall Cascades Buffer Against Drought Across the Amazon, *Nature Climate Change*, Vol. 8, 2018.

80 Sinimbu, Fabiola and Jade, Liria, Over 850 Brazil Cities Face Major Water Shortage Issues, Agencia Brazil, 2017.

81 Stocker, Thomas F., Qin, Dahe, Plattner, Gian-Kasper, et al., IPCC: Climate Change 2013: The Physical Science Basis (Contribution of Working Group I to the Fifth Assessment Report of the Intergovernmental Panel on Climate Change), Cambridge University Press, 2013.

82 C40.org, Staying Afloat: The Urban Response to Sea Level Rise, C40, 2018.

83 Ibid., p77.

84 Holder, Josh, Kommenda, Niko and Watts, Jonathan, The Three-degree World: The Cities That Will Be Drowned by Global Warming, *Guardian*, 2017.

85 Muggah, Robert, The World's Coastal Cities Are Going Under. Here's How Some Are Fighting Back, World Economic Forum, 2019.

86 IPCC.ch, Global Warming of 1.5 °C: Special Report, IPCC, 2018.

87 Mei Lin, Mayuri and Hidayat, Rafki, Jakarta, The Fastest-sinking City in the World, BBC News, 2018.

88 Ibid., p79.

89 Ibid., p79.

90 Win, Thei L., In Flood-prone Jakarta, Will 'Giant Sea Wall' Plan Sink or Swim?, Reuters, 2017.

91 Ibid., p79.

92 Holder, Kommenda and Watts, 2017, op. cit., p80.

93 Muggah, Robert, The world's coastal cities are going under, 2019, op. cit., p80.

94 Radford, Tim, Coastal Flooding 'May Cost $100,000 bn a Year by 2100', Climate News Network, 2014.

95 Swissre.com, Confronting the Cost of Catastrophe, Swiss Re Group, 2019.

96 Fu, Xinyu, Gomaa, Mohamed, Deng, Yujun, et al., Adaptation Planning for Sea Level Rise: a Study of US Coastal Cities, *Journal of Environmental Planning and Management*, Vol. 60, 2017.

97 Sealevel.climatecentral.org, These U.S. Cities Are Most Vulnerable to Major Coastal Flooding and Sea Level Rise, Surging Seas, 2017.

98 Loria, Kevin, Miami is Racing Against Time to Keep up with Sea-level Rise, Business Insider, 2018.

99 Brasilero, Adriana, In Miami, Battling Sea Level Rise May Mean Surrendering Land, Reuters, 2017.

100 Euronews.com, Rising Sea Levels Threat: a Shrinking European Coastline in 2100?, Euronews, 2018.

101 Brown, Sally, African Countries Aren't Doing Enough to Prepare for Rising Sea Levels, The Conversation, 2018.

102 See globalcovenantofmayors.org, Our Cities, Global Covenant of Mayors, 2020.

103 Ryan, Julie, North Texas Cities Combating Climate Change from Bottom Up, Green Source DFW, 2018.

104 Statesman.com, Austin on Track to Meet Carbon-neutral Goal, Statesman, 2018.

105 Heidrich, Oliver and Reckien, Diana, We Examined 885 European cities' Plans to Tackle Climate Change – Here's What We Found, The Conversation, 2018.

106 Taylor, Lin, Cycling City Copenhagen Sprints to Become First Carbon-neutral Capital, Reuters, 2018.

107 See Carbonneutralcities.org, About, Carbon Neutral Cities Alliance, 2020.

108 Globalcovenantmayors.org, 2020, op. cit., p82.

109 Ibid., p83.

110 CBSnews.com, Sea Change: How the Dutch Confront the Rise of the Oceans, CBS News, 2017.

111 C40.org, C40 Good Practice Guides: Rotterdam – Climate Change Adaptation Strategy, c40 Cities, 2016.

112 Caramel, Laurence, Besieged by the Rising Tides of Climate Change, Kiribati Buys Land in Fiji, Guardian, 2014.

113 Climate.gov.ki, Fiji Supports Kiribati On Sea Level Rise, Climate Change: Republic of Kiribati, 2014.

114 Tooze, Adam, Rising Tides Will Sink Global Order, Foreign Policy article, 2018.

115 Letman, Jon, Rising Seas Give Island Nation a Stark Choice: Relocate Or Elevate, National Geographic, 2018.

116 Dauenhauer, Nenad J., On Front Line Of Climate Change As Maldives Fights Rising Seas, New Scientist, 2018.

117 IUCN.org, Deforestation and Forest Degradation, IUCN: Issues Brief, 2020.

118 Gerretsen, Isabelle, How Climate Change is Fueling Extremism, CNN, 2019.

119 Busby, Joshua and Von Uexkull, Nina, Climate Shocks and Humanitarian Crises, Foreign Affairs article, 2018.

120 Muggah, Robert and Cabrera, José L., The Sahel Is Engulfed by Violence. Climate Change, Food Insecurity and Extremists Are Largely to Blame, World Economic Forum, 2019.

121 IFLscience.com, Engineers Develop Roadmap To Get The US To Run on 100% Renewable Energy By 2050, IFL Science; Jacobson, Mark Z. , Delucchi, Mark A., Bazouin, Guillaume, et al., 100% Clean And Renewable Wind, Water, and Sunlight (Wws) All-sector Energy Roadmaps For the 50 United States, Energy Environ Science, Vol. 8, 2015.

122 Ellenmacarthurfoundation.org, Effective Industrial Symbiosis, Ellen MacArthur Foundation, 2017.

123 Childress, Lillian, Lessons from China's Industrial Symbiosis Leadership, GreenBiz, 2017.

124 Griffin, Paul, The Carbon Majors Database: CDP Carbon Majors Report 2017, CDP, 2017.

125 Byers, Logan, Friedrich, Johannes, Hennig, Roman, et al., A Global Database of Power Plants, World Resources Institute, 2019.

126 Fischetti, Mark, The Top-22 Air Polluters Revealed, Scientific American, 2017.

127 Carrington, Damian, Avoiding Meat and Dairy Is 'Single Biggest Way' To Reduce Your Impact on Earth, Guardian, 2018.

128 Grain.org, Emissions Impossible: How Big Meat and Dairy are Heating Up the Planet, GRAIN and the Institute for Agriculture and Trade Policy (IATP), 2018.

129 IPCC.ch, Summary for Policymakers of IPCC Special Report on Global Warming of 1.5°C Approved by Governments, IPCC, 2018.

130 CAT.org.uk, Zero Carbon Britain, Centre for Alternative Technology, 2020.

131 Stein, Jill and Hawkins, Howie, The Green New Deal, Green Party US, 2018.

Urbanisation

1 Brilliantmaps.com, The 4037 Cities in The World With Over 100,000 People, Brilliant Maps article, 2015.

2 Misra, Tanvi, Half the World Lives on 1% of Its Land, Mapped, CityLab article, 2016.

3 Muggah, Robert, A Manifesto of a Fragile City, Journal of International Affairs, 2015.

4 Galka, Max, Watch as the world's cities appear one-by-one over 6000 years, Metrocosm, 2016; worldpopulationhistory.org, World Population Visualization, 2016; Desjardins, Jeff, These 3 Animated Maps Show the World's Largest Cities Throughout History, Visual Capitalist, 2016.

5 Not all scholars share the view that agriculture was a precondition of city growth. Jane Jacobs (1969), for example, suggests that other forms of subsistence (fishing) may have precipitated the rise of cities.

6 Moore, Andrew, The Neolithic of the Levant, Oxford University Press, 1978; Compton, Nick, What is the Oldest City in the World?, Guardian, 2015.

7 Mumford, Lewis, The City in History, Harcourt, Brace and World, 1961.

8 Evans, Damian, A Cross-Section of Results from the 2015 Lidar Campaign, CALI, 2016.

9 TheGuardian.com, Laser Technology Reveals Lost City Around Angkor Wat, Guardian, 2013; Damian, 2016, op. cit., p102.

10 Wainwright, Oliver, How Nasa Technology Uncovered the 'Megacity' of Angkor, Guardian, 2016.

11 Zimmern, Helen, The Hanseatic League – A History of the Rise and Fall of the Hansa Towns, Didactic Press, 2015.

12 Metrocosm.com, 2016, op. cit., p102.

13 Brilliantmaps.com, 2015, op. cit., p102.

14 Ritchie, Hannah and Roser, Max, Urbanization, Our World in Data, 2018.

15 Hollen Lees, Lynn, World Urbanization – 1750 to Present, Cambridge University Press, 2015.

16 Between 1870 and 1920, 11 million people migrated from rural to urban areas. Most of the 25 million migrants who arrived during this time also settled in cities.

17 Arsht, Adrienne, Urbanization in Latin America, Atlantic Council, 2014.

18 Dikotter, Frank, Mao's Great Famine: *The History of China's Most Devastating Catastrophe*, 1958–1962, Bloomsbury Paperbacks, 2018.

19 Statista.com, Urban and Rural Population of China from 2008 and 2018, Statista Demographics, 2019.

20 Worldbank.org, Urban Development, World Bank article, 2020.

21 Prasad, Vishnu, 'Triumph of the City' – Why Cities Are Our Greatest Invention, Financing Cities IFMR.

22 Ritchie and Roser, 2018, op. cit., p 106.

23 Sassen, Saskia, The Global City: *Introducing a Concept*, *Brown Journal of World Affairs*, Vol. 11, 2005.

24 Much of the debate over how to define cities can be traced to Wirth's classic text – *The City* – in 1938. He described the key characteristics of the 'city' as having four dimensions – (i) population size, (ii) population density, (iii) social heterogeneity, and (iv) an element of permanence. This approach was criticised for being overly geographically and demographically focused, and not accounting for functions of cities and their immediate borders (at the regional, national, global scale).

25 Mark, Joshua J., The Ancient City, *Ancient History Encyclopedia*, 2014.

26 Fang, Chuanglin and Yu, Danlin, Urban Agglomeration: An Evolving Concept of an Emerging Phenomenon, *Landscape and Urban Planning*, Vol. 162, 2017.

27 Ibid., p106.

28 Scott, Allen J. and Storper Michael, The Nature of Cities: The Scope and Limits of Urban Theory, *International Journal of Urban and Regional Research*, Vol. 39, 2015.

29 Atkearny.com, Leaders in a World of Disruptive Innovation, Global Cities 2017 Report, 2017; Sennet Richard, *Classic Essays on the Culture of Cities*, Prentice-Hall, 1969.

30 Bevan, Robert, What makes a city a city – and does it really matter anyway?, 2014; and McClatchey, Caroline, Why do towns want to become cities? *BBC News magazine*, 2011.

31 Ibid., p107

32 The United Nations uses the concept of 'urban agglomeration' defined as a given concentration of population (the 'metropolitan area' rather than strict 'city limits'). In this way, it also can include certain types of agglomerations. For example, the Washington DC metropolitan area is roughly 5.95 million (2014) covering Northern Virginia, Maryland and the District of Columbia. Yet Washington DC city is just 659,000 people. Likewise, the city of Manila is just 1.65 million, metro Manila has 11.8 million and a greater metropolitan area of 25.5 million, which is nested in Mega Manila cluster with 35.6 million. See https://www.un.org/development/desa/en/news/population/2018-revision-of-world-urbanisation-prospects.html

33 Every year this report estimates the global urban population by reviewing demographic data and asking national statistical offices what they think is going on. See https://population.un.org/wup/

34 United Nations Department of Economic and Social Affairs (UNDESA), Population Division, *World Urbanization Prospects: The 2018 Revision*, United Nations Publications, 2019.

35 Ibid., 107.

36 Ritchie and Roser, 2018, op. cit., p107.

37 ghsl.jrc.ec.europa.eu, Testing the degree of urbanization at the global level, CIESIN, 2020.

38 Rather than relying on a wide range of interpretations provided by states, they applied a universal definition of an urban centre (at least 50,000 people and a population density of at least 1,500 people per square kilometres), urban cluster (a minimum of 5,000 inhabitants plus a population density of at least 300 people), and rural areas (fewer than 5,000 people).

39 ghsl.jrc.ec.europa.eu, 2020, op. cit., p108; Dijkstra, Lewis, Florczyk Aneta, et al., Applying the Degree of Urbanization to the Globe, 16th Conference of IAOS, 2018.

40 Some critics believe these figures grossly overestimate city size since they incorporate agriculture employment figures and observed population densities at the peripheries of cities; Angel, Shlomo, Lamson-Hall, Patrick, Guerra, Bibiana et al., Our Not-So-Open World, p42, The Marron Institute of Urban Management, 2018.

41 Cheney, Catherine, Is the world more urban than UN estimates? It depends on the definition, Devex, 2018.

42 Unstats.un.org, Definition of 'Urban', table 6, *United Nations Demographic Yearbook*, 2005.

43 Ibid., p108.

44 Statcan.gc.ca, Population Centre and Rural Area Classification 2016, Statistics Canada, 2017.

45 An urban centre must have a contiguous geographical area with municipal boundary; a population with more than 50,000 people; and a density of at least 1,500 people per square kilometre. A commuting area is one where 15% of the employed population commute to the city.

46 Urbanists and geographers have conceived of cities as a series of concentric land-use zones with an urban core (with central business district), a transitional zone (industrial, residential and commercial bands), a peri-urban zone (including satellite and dormitory settlements) and a rural hinterland on which cities often depend. The city is in some ways the entire system, not just the urban core. The extreme rural zone, while not part of the city, is a zone of influence running in both directions.

47 By contrast, in many developed country contexts, rural life may have many of the same amenities as some of the most well-serviced cities.

48 en.wikipedia.org, Megacity, Wikipedia article, 2020; en.wikipedia.org, List of Cities by GDP, Wikipedia article, 2020.

49 UNDESA, 2019, op. cit., p108.

50 Ibid., p109.

51 Emerging market cities such as Shenzhen, Seoul, and São Paulo are active participants in the post-Ford economy and fomenting dense interlinkages across and within production systems, fostering high-skill employment opportunities and incubating information and capital

flows to stimulate innovation. Trade between cities is allowing more and more of them to specialise and sell with their counterparts. Not surprisingly, city diplomacy is rapidly returning to the significance it had in the past. See Chan, Dan H-K., City diplomacy and 'glocal' governance: revitalizing cosmopolitan democracy, Innovation: *The European Journal of Social Science Research*, Vol. 29, 2016.

52 Metropolitan Manila is over 13.4 million in 2019. See UNDESA, 2019, op. cit., p109.

53 Weller, Chris, Manila is the Most Crowded City in the World — Here's What Life is Like, *Business Insider*, 2016.

54 Metropolitan Paris has some 10.9 million people in 2019. See UNDESA, 2019, op. cit., p109.

55 Hunn Patrick, Australian Cities Among the Largest and Least Densely Settled in the World, ArchitectureAU, 2017.

56 Demographia, World Urban Areas 16th Edition, Demographia Report, 2020.

57 From Earthtime, CREATE Lab, CMU.

58 Murphy Douglas, Where is the World's Most Sprawling City?, *Guardian*, 2017.

59 See Las Vegas in projects.propublica.org; Lasserre, Frederic, Water in Las Vegas: Coping with Scarcity, Financial and Cultural Constraints, City, Territory and Architecture, Vol. 2, 2015.

60 Plummer, Brad, Watch Lake Mead, the Largest Reservoir in the US, Shrink Dramatically over 15 Years, Vox article, 2016.

61 Worldpopulationreview.com, Population of Cities in China (2020), World Population review, 2020.

62 Worldpopulationreview.com, Population of Cities in United Kingdom (2020), World Population review, 2020.

63 Kunshan, China is Trying to Turn Itself Into a Country of 19 Super-regions, *Economist*, 2018.

64 Ward, Jill, Will Future Megacities Be a Marvel or a Mess? Look at New Delhi, Bloomberg article, 2018.

65 Li, He, Zhao, Shichen and Wang Daqiang, Urbanization Patterns of China's Cities in 1990-2010, International Review for Spatial Planning and Sustainable Development Vol. 3, 2015.

66 Kumar, Amit and Navodaya, Ambarish R., Urbanization Process, Trend, Pattern and Its Consequences in India, Neo Geographia Vol. 3, 2015; Prasad, Sangeeta, Why the World Should be Watching India's Fast-growing Cities, World Economic Forum article, 2019.

67 Abraham, Reuben and Hingorani, Pritika, India's a Land of Cities, Not Villages, Bloomberg article, 2019.

68 Ibid., p112.

69 Worldpopulationreview.com, Population of Cities in India (2020), World Population review, 2020.

70 Kumar and Navodaya, 2015, op. cit., p112.

71 Griffiths, James, 22 of the Top 30 Most Polluted Cities in the World are in India, CNN Health article, 2019.

72 McKinsey & Co., India's Urban Awakening: Building Inclusive Cities, Sustaining Economic Growth, McKinsey Global Institute, 2010.

73 Charlton, Emma, India is Building a High-tech Sustainable City from Scratch, World Economic Forum article, 2018.

74 Chandran, Rina, As India adds 100 Smart Cities, One Tells a Cautionary Tale, Reuters article, 2018.

75 Vidal, John, UN report: World's Biggest Cities Merging into 'Mega-regions', *Guardian*, 2010; Mukhopadhyay, Chandrima, Megaregions: Globalization's New Urban Form?, European Planning Studies, Vol. 24, 2016.

76 Rouhana, Salim and Bruce Ivan, Urbanization in Nigeria: Planning for the Unplanned, World Bank Blog, 2016.

77 Worldpopulationreview.com, Population of Cities in Nigeria (2020), World Population review, 2020.

78 Vidal, John, The 100 million city: is 21st century urbanisation out of control?, *Guardian*, 2018.

79 Pope, Kevin and Hoornweg, David, Population predictions for the world's largest cities in the 21st century, *Environment & Urbanization*, Vol. 29, 2017.

80 Leithhead, Alastair, The City that Won't Stop Growing: How can Lagos Cope with its Spiralling Population?, BBC News Report, 2017.

81 Earthtime, CREATE Lab, CMU.

82 Dobbs, Richard, Smit, Sven, Remes, Jaana, et al., Urban world: Mapping the Economic Power of Cities, McKinsey Global Institute, 2011.

83 Florida, Richard, The Economic Power of Cities Compared to Nations, CityLab, 2017.

84 financialexpress.com, With GDP of $370 billion, Delhi-NCR Muscles out Mumbai as Economic Capital of India, Financial Express article, 2016.

85 Florida, Richard, What To Do About the Rise of Mega-Regions, City Lab, 2018.

86 Woetzel, Jonathan, Remes, Jaana, Boland, Brodie, et al., Smart cities: Digital Solutions for a More Livable Future, McKinsey Global Institute Report, 2018.

87 Ibid., p116.

88 Muggah, Robert and Goodman, Marc, Cities are Easy Prey for Cybercriminals. Here's How they can Fight Back, World Economic Forum article, 2019.

89 Florida, Richard, The Real Powerhouses That Drive the World's Economy, City Lab, 2019.

90 xinhuanet.com, 'Jing-jin-ji': China's Regional City Cluster Takes Shape, Xinhua Headlines, 2019.

91 Mongabay.com, Population Estimates for Karachi, Pakistan, 1950-2015, Population Mongabay Data, 2016.

92 Earthtime, CREATE Lab, CMU.

93 McCarthy, Niall, The World's Largest Cities By Area, Statista: Urban Areas, 2018.

94 Arguably the largest urban unit of analysis is the megalopolis. Also described as a megaregion or supercity, the megalopolis is described as a chain of contiguous metropolitan areas. It was cited in Spengler (1918), Mumford (1938) and Gottman (1954) and linked to urban over-development and social decline. They may be linked by ground transportation corridors and can occur both within and across international frontiers. See Florida, Richard, Gulden, Tim and Mellander, Charlotta, The Rise of the Mega-region, *Cambridge Journal of Regions, Economy and Society*, Vol. 1, 2008.

95 Urban conurbations are described as regions comprising a number of adjacent cities, suburbs and their peripheries that have merged to form a continuous urban industrially-developed area. A conurbation is a polycentric urban

agglomeration that helps create a unified labour market. The term was actually coined in 1915 by Geddes in his book – *Cities in Evolution*.

96 Johnson, Ian, As Beijing Becomes a Supercity, the Rapid Growth Brings Pains, *New York Times*, 2015.

97 Roxburgh, Helen, Endless Cities: Will China's New Urbanisation Just Mean More Sprawl?, *Guardian*, 2017.

98 business.hsbc.com, China's Emerging Cities, HSBC Belt and Road article, 2020.

99 Phillips, Tom, 'Forest cities': The Radical Plan to Save China from Air Pollution, *Guardian*, 2017.

100 Other studies have also used mobility data from over 4 million commuter flows to track the parameters of megaregions in the U.S. See Nelson, Garrett D., Rae, Alasdair, An Economic Geography of the United States: From Commutes to Megaregions, *Plos One*, Vol. 11, 2016.

101 Florida, 2019, op. cit., p120.

102 Ibid., p120.

103 Frem, Joe, Rajadhyaksha, Vineet and Woetzel, Jonathan, Thriving Amid Turbulence: Imagining the Cities of the Future, McKinsey article, 2018.

104 Davis, Mike, *Planet of Slums*, Verso, 2007.

105 Wikipedia, 2020, List of Slums, op. cit., p120.

106 UN PSUP Team Nairobi, Slum Almanac 2015–2016, UN Habitat Report, 2016.

107 Ritchie and Roser, 2018, op. cit., p 120; wikipedia.org, List of Slums, Wikipedia article, 2020.

108 Warner, Gregory, In Kenya, Using Tech To Put An 'Invisible' Slum On The Map, NPR article, 2013.

109 Kuffer Monika, Pfeffer, Karen, Sliuzas, Richard V., Slums from Space: 15 Years of Slum Mapping Using Remote Sensing, *Remote Sensing*, Vol. 8, 2016; See impactlab.net, A Satellite Tour of the Biggest Slums, Impact Lab article, 2012.

110 Mberu, Blessing U., Haregu, Tilahun N., Kyobutungi, Catherine and Ezeh, Alex C., Health and Health-related Indicators in Slum, Rural, and Urban Communities: A Comparative Analysis, *Global Health Action*, Vol. 9, 2016.

111 Aggarwalla, Rohit T., Hill, Katie, Muggah, Robert, Smart City Experts Should be Looking to Emerging Markets. Here's Why, World Economic Forum article, 2018.

112 Ritchie and Roser, 2018, op. cit., p 122.

113 Friesen, John, Rausch, Lea, Pelz, Peter F., et al., Determining Factors for Slum Growth with Predictive Data Mining Methods, *Urban Science*, Vol. 2, 2018.

114 Marx, Benjamin, Stoker, Thomas and Suri, Tavneet, The Economics of Slums in the Developing World, *Journal of Economic Perspectives*, Vol. 27, 2013.

115 Florida, Richard, The Amazing Endurance of Slums, City Lab, 2014.

116 Scott, Allen J. and Michael Storper, The Nature of Cities: The Scope and Limits of Urban Theory, *International Journal of Urban and Regional Research*, Vol. 35, 2015.

117 Kopf, Dan, China Dominates the List of Cities with the Fastest Growing Economies, Quartz article, 2018.

118 Muggah, Robert, A Manifesto for the Fragile City, *Journal of International Affairs*, Vol. 68, 2015.

119 Perur, Srinath, What the Collapse of Ancient Capitals Can Teach us About the Cities of Today, *Guardian*, 2015.

120 Muggah, Robert, Where Are the World's Most Fragile Cities?, City Lab, 2017.

121 From EarthTime, CREATE Lab, CMU.

122 From Baltimore to Bogota, unequal cities tend to be more violent than those with a fairer distribution of wealth and basic services. This is because real and relative deprivation of income, property, and social status are associated with lower social capital, social efficacy and social disorganisation. Concentrated disadvantage is strongly associated with underperforming schools, poor housing and health conditions, and higher rates of incarceration and criminality.

123 These findings are similar to a 2018 study by Maplecroft which combined UM projections on annual population growth in over 1800 cities with data from the organisation's Climate Change Vulnerability Index. It found that 84 of the top 100 fastest growing cities are facing extreme risks and another 14 are in the high risk category. Over 95% of the 234 cities in the extreme risk category were in Africa and Asia. Maplecroft, Verisk, 84% of World's Fastest Growing Cities Face 'Extreme' Climate Change Risks, Prevention Web article, 2018.

124 From Earthtime, CREATE Lab, CMU.

125 Muggah, Robert, Kilcullen, David, These are Africa's Fastest-growing Cities – and They'll Make or Break the Continent, World Economic Forum, 2016.

126 Van Leggelo-Padilla, Daniella, Why We Need to Close the Infrastructure Gap in Sub-Saharan Africa, World Bank article, 2017.

127 Statista.com, Annual Average Infrastructure Expenditures as Percent of GDP Worldwide from 2010 to 2015, by Country, Statista Heavy Construction, 2020.

128 Ballard, Barclay, Bridging Africa's Infrastructure Gap, World Finance, 2018.

129 Kenyanwallstreet.com, Africa Infrastructure Index; $108B Financing Gap, The Kenyan Wall Street article, 2018.

130 Gutman, Jeffrey and Patel, Nirav, Foresight Africa viewpoint – Urban Africa: Avoiding the Perfect Storm, Brookings: Africa in Focus, 2018.

131 Williams, Hugo, COP 21: Five Ways Climate Change Could Affect Africa, BBC News article, 2015.

132 Chuttel, Lynsey, How Cape Town Delayed its Water-shortage Disaster—At Least Until 2019, Quartz Africa, 2018.

133 Hill, Tim, Asia's Urban Crunch: What To Do About 900,000 Weekly Arrivals?, Eco-Business, 2018.

134 Oxfordeconomics.org, Which Cities will Lead the Global Economy by 2035, Oxford Economics, 2018.

135 globaldata.com, 60% of the World's Megacities Will be Located in Asia by 2025, Global Data press release, 2018.

136 chinabankingnews.com, China's Second-tier Cities on the Verge of Peak Population Growth, China Banking News: Economy, 2018.

137 Bughin, Jacques, Manyika, James and Woetzel, Jonathan, Urban World: Meeting the Demographic Challenge, McKinsey Global Institute Report, 2016.

138 Hananto, Akhyari, We will See Southeast Asia's Spending Spree in Infrastructure in 2018, Seasia, 2017.

139 From EarthTime, CREATE Lab, CMU.

140 Oxfordeconomics.org, 2018, op. cit., p131.

141 See globalcovenantofmayors.org 'Our Cities', 2020.

142 Knight, Sam, Sadiq Khan Takes On Brexit and Terror, *New Yorker*, 2017.

143 McAuley, James and Rolfe, Pamela, Spain is the Most Welcoming Country in Europe for Migrants. Will it Last?, *Washington Post*, 2018.

144 Horowitz, Jason, Palermo Is Again a Migrant City, Shaped Now by Bangladeshis and Nigerians, *New York Times*, 2018.

Technology

1 Arthur, W. Brian, *The Nature of Technology*, Penguin, 2010.

2 Humanorigins.si.edu, 'What Does it Mean to be Human?': Early Stone Age Tools, Smithsonian, National Museum of Natural History, 2020.

3 Wayman, Erin, 'The Earliest Example of Hominid Fire', *Smithsonian Magazine*, 2012.

4 Roser, Max, Economic growth, Our World in Data, 2013.

5 Arthur, 2010, op. cit., p138.

6 Kendall, Graham, Your Mobile Phone vs. Apollo 11's Guidance Computer, Real Clear Science, 2019.

7 Authors' estimate based on conversations with Oxford computing colleagues.

8 Roser, Max and Ortiz-Ospina, Esteban, Literacy, Our World in Data, 2016.

9 Ibid., p141.

10 Goldin, Ian, Cameron, Geoffrey and Balarajan, Meera, *Exceptional People*, Princeton University Press, 2012.

11 Goldin, Cameron and Balarajan, 2012, op. cit., p141.

12 World Bank Group, World Development Report 2016: Digital Dividends, 2016.

13 Ibid., p141.

14 Ibid., p142.

15 See Release dates and Locations on Apple Store.

16 BBC.co.uk, Using Drones to Deliver Blood in Rwanda, BBC News video, 2019.

17 Aggarwalla, Rohit T., Hill, Katie, Muggah, Robert, Smart City Experts Should be Looking to Emerging Markets. Here's Why, World Economic Forum article, 2018.

18 For example, see Diamandis, Peter and Kotler, Steven, *Abundance*, Free Press, 2012; And more recently pathwayscommission.bsg.ox.ac.uk, The Digital Roadmap: How Developing Countries Can Get Ahead, Pathways for Prosperity Commission, 2019.

19 World Bank Group, World Development Report 2016, op. cit., p142.

20 Geomedici.com, M-Pesa: Mobile Phone-based Money Transfer, Medici article, 2019.

21 Ian Goldin visited a number of these health centres as part of the production of his BBC documentary, *Will AI kill development?*

22 Goldin, Ian, *Will AI kill Development?*, BBC World Service Documentary, 2019.

23 Pasti, Francesco, State of the Industry Report on Mobile Money 2018, GSMA Report, 2019.

24 Data.worldbank.org, GDP per capita (current US$) – United Kingdom, Gambia, World Bank data, 2020.

25 Ibid., p143.

26 Wikipedia.org, Solar Power by Country, Wikipedia article, 2020; Wikipedia.org, Wind Power by Country, Wikipedia article, 2020.

27 Ibid., p148.

28 Ibid., p148.

29 JPmorgan.com, Driving into 2025: The Future of Electric Vehicles, JP Morgan Report, 2018.

30 Bnef.com, Electric Vehicle Outlook 2019, BNEF EVO Report, 2019; Muggah, Robert, Cities Could Be Our Best Weapon in the Fight Against Climate Change, World Economic Forum, 2019.

31 Holland, Maximillian, China 2019 Electric Vehicle Market Share Grows To 4.7% Despite Tighter Incentives, Clean Technica, 2020; Huang, Echo, China Buys One Out of Every Two Electric Vehicles Sold Globally, Quartz article, 2019.

32 Wikipedia.org, Carna Botnet, Wikipedia article, 2019.

33 McKetta, Isla, The World's Internet in 2018: : Faster, Modernizing and Always On, Speedtest: Global Speeds, 2018.

34 National Telecommunications and Information Administration, A Nation Online: How Americans Are Expanding Their Use of the Internet, US Department of Commerce: NTIA, 2002.

35 'McKetta, 2018, op. cit., p152; Speedtest.net, 2019 Mobile Speedtest U.S. Mobile Performance, Speedtest Report, 2019.

36 McKetta, 2018, op. cit., p152.

37 Wikipedia.org, List of countries by Internet Connection Speeds, Wikipedia article, 2020; Speedtest.net, 2019, op. cit., p152.

38 McCann, John, Moore, Mike, Lumb, David, 5G: Everything You Need to Know, Techradar article, 2020; justaskthales.com, What is the Difference Between 4G and 5G?, Just Ask Thales, 2020.

39 Economist.com, Why Does 5G Have Everyone Worried About Huawei?, *Economist*: Business, 2018.

40 Li, Tao, Nearly 60 Percent of Huawei's 50 5G Contracts are from Europe, *South China Morning Post*, Big Tech, 2019.

41 Araya, Daniel, Huawei's 5G Dominance in the Post-American World, *Forbes*, 2019.

42 Vincent, James, Putin Says the Nation that Leads in AI 'will be the ruler of the world', The Verge, 2019.

43 Economist.com, The Technology Industry is Rife with Bottlenecks, *Economist* Pinch Points, 2019.

44 Ibid., p155.

45 Ibid., p156.

46 Levy, Michael, Stewart, Donald E. and Hardy Wise Kent, Christopher, *Encyclopaedia Britannica*, Encyclopaedia Britannica article, 2019.

47 See Graham, Mark, Information Geographies and Geographies of Information, New Geographies Vol. 7, 2015; as well as Stats.wikipedia.org, Wikimedia Traffic Analysis Report, Wikipedia Data Repository, 2015.

48 Ibid., p157.

49 Ibid., p157.

50 Ibid., p157.

51 Ibid., p157.

52 Ibid., p157.

53 Columbus, Louis, 2018 Roundup of Internet of Things Forecasts and Market Estimates, *Forbes*, 2018; Statista.com, Internet of Things (IoT) Active Device Connections Installed Base Worldwide from 2015 to 2025, Statista: Consumer Electronics, 2016.

54 Berger de-Leon, Markus, Reinbacher, Thomas, Wee, Dominik, The IoT as a Growth Driver, McKinsey Digital article, 2018.

55 Muggah, Robert and Goodman, Marc, Cities are Easy Prey for Cybercriminals, World Economic Forum, 2019.

56 Fortney, Luke, Bitcoin Mining, Explained, Investopedia: Bitcoin, 2020.

57 Ibid., p159.

58 Technical University of Munich, Bitcoin Causing Carbon Dioxide Emissions Comparable to Las Vegas or Hamburg', ScienceDaily, 2019; Deign, Jason, Bitcoin Mining Operations Now Use More Energy Than Ireland, Greentechmedia, 2017.

59 See, for example, the work done by deepmind.com.

60 Harris, Ricki, Elon Musk: Humanity is a Kind of 'Biological Boot Loader' for AI, Wired: Business, 2019.

61 Silver, David and Hassabis, Demis, AlphaGo Zero: Starting from Scratch, Deepmind Blog Post, 2017.

62 Suleyman, Mustafa, Using AI to Plan Head and Neck Cancer Treatments, Deepmind Blog Post, 2018.

63 Bostrom, Nick, *Superintelligence*, Oxford University Press, 2014.

64 IFR.org, Executive Summary World Robotics 2019, Industrial Robots, International Federation of Robotics, 2019.

65 BBC.co.uk, Robot Automation Will 'Take 800 Million Jobs by 2030', BBC News, 2017.

66 Frey, Carl B., Berger, Thor, Chen, Chinchih, Political Machinery: Automation Anxiety and the 2016 U.S. Presidential Election, Oxford Martin School paper, 2017.

67 Frey, Carl B., *The Technology Trap*, Princeton University Press, 2019.

68 Ian Goldin interview with former CEO of global mobile phone network, 11 November 2019.

69 Frey, Carl B. and Osborne, Michael A., The Future of Employment, Oxford Martin School paper, 2013.

70 World Bank Group, World Development Report 2016, op. cit., p162.

71 OECD.org, Putting Faces to the Jobs at Risk of Automation, Policy Brief on the Future of Work, 2018.

72 Chui, Michael, Lund, Susan and Gumbel, Peter, How Will Automation Affect Jobs, Skills, and Wages?, McKinsey Global Institute, 2018.

73 World Bank Group, World Development Report 2016: Figure O.18 p23, op. cit., p164.

74 See: Goldin, Ian, Will AI kill development?, BBC World Service Documentary, 2019.

75 AFDB.org, Jobs for Youth in Africa, African Development Bank Group article, 2016.

76 Goldin, *Will AI kill development?*, op. cit., p165.

77 Berg, Andrew, Buffie, Edward F. and Zanna, Luis-Felipe, Should We Fear the Robot Revolution?, IMF Working Paper No. 18/116, 2018.

78 Ibid., p165.

79 Ibid., p165.

80 Obamawhitehouse.archives.gov, Artificial Intelligence, Automation, and the Economy, Executive Office of the President, 2016.

81 Ibid., p165.

82 Berg, Buffie and Zanna, 2018, op. cit., p166.

83 World Bank Group, World Development Report 2016, op. cit., p166.

84 Ibid., p166.

85 Ibid., p166.

86 Ibid., p166.

87 Goldin, Ian, Koutroumpis, Pantelis, Lafond, François, Rochowicz, Nils and Winkler, Julian, The Productivity Paradox, Oxford Martin School Report, 2019; World Bank Group, World Development Report 2016, op. cit., p166.

88 Ibid., p167.

89 Ibid., p167.

90 Ibid., p167.

91 Ibid., p167.

92 Ibid., p168.

93 Schwab, Klaus, The Fourth Industrial Revolution, World Economic Forum, 2016.

94 Ibid., p168.

95 Frey, 2019, op. cit., p168.

96 Keynes, John M., *A Tract on Monetary Reform*, Macmillan, 1923.

97 Goldin, Ian and Kutarna, Chris, *Age of Discovery: Navigating the Storms of Our Second Renaissance*, Bloomsbury, 2017.

Inequality

1 The richest 147 billionaires represent just 0.000002 of the world's population yet control around one per cent of global wealth.

2 Matthews, Dylan, Are 26 Billionaires Worth More Than Half the Planet?, Vox, 2019.

3 Authors' calculations based on Forbes Rich List Profile of Jeff Bezos; And databank.worldbank.org, Gross Domestic Product 2018, World Development Indicators, 2019.

4 Krugman, Paul, The Great Gatsby Curve, The New York Times blog, 201; HDR.undp.org, Income Gini Coefficient, World Development Indicators 2013: Human Development Report, 2013.

5 Worldpopulationreview.com, World City Populations (2020), World Population review, 2020.

6 Energy.gov, State of New York: Energy Sector Risk Profile, U.S. Department of Energy, 2014; Campbell, John, Electricity Distribution Is Holding Nigeria Back, Council on Foreign Affairs, 2018.

7 Ibid., p176.

8 Scotsman.com, Class System Began 7,000 Years Ago, Archaeologists Find, *Scotsman News*, 2012.

9 Roser, Max, Global Economic Inequality, Our World in Data, 2013.

10 Frey, Carl B., *The Technology Trap*, Princeton University Press, 2019.

11 Roser, 2013, op. cit., p177.

12 DeLong, James B., A Brief History of Modern Inequality, World Economic Forum, 2016.

13 DeLong, James B., A Brief History of Modern (In)equality, Project Syndicate article, 2016.

14 Ibid., p178.

15 Delong, World Economic Forum, 2016, op. cit., p178.

16 Ibid., p178.

17 Goldin, Ian, *Development*, Oxford University Press, 2017.

18 Roser, 2013, op. cit., p178.

19 Ibid., p178.

20 Philippon, Thomas, *The Great Reversal: How America Gave Up on Free Markets*, Harvard University Press, 2019.

21 Goldin, 2017, op. cit., p180.

22 Worldbank.org, The World Bank in China: Overview, World Bank, 2020.

23 Sanchez, Carolina, From Local to Global: China's Role in Global Poverty Reduction and the Future of Development, World Bank: Speeches and Transcripts, 2017.

24 Ibid., p180.

25 Roser, 2013, op. cit., p180.

26 Worldbank.org, Poverty: Overview, World Bank, 2020.

27 Ibid., p180.

28 Ibid., p180.

29 Goldin, 2017, op. cit., p181; worldbank.org, Decline of Global Extreme Poverty Continues but Has Slowed, World Bank Press Release, 2018.

30 Worldbank.org, Classification of Fragile and Conflict-Affected Situations, World Bank Brief, 2020.

31 AEAweb.org, The Elephant Curve: Chart of the Week, American Economic Association, 2019.

32 Kharas, Homi and Seidel, Brina, What's Happening to the World Income Distribution? The Elephant Chart Revisited, Brookings Report, 2018.

33 Ibid., p182.

34 Scott, Katy, South Africa is the World's Most Unequal Country. 25 Years of Freedom Have Failed to Bridge the Divide, CNN, 2019.

35 Florida, Richard, The World is Spiky, *Atlantic*, 2005.

36 Brewer, Mike and Robles, Claudia S., Top Incomes in the UK: Analysis of the 2015-16 Survey of Personal Incomes, ISER Working Paper Series, 2019.

37 Neate, Rupert, Bill Gates, Jeff Bezos and Warren Buffett Are Wealthier Than Poorest Half of US, *Guardian*, 2017.

38 Kagan, Julia, How Much Income Puts You in the Top 1%, 5%, 10%?, investopedia, 2019.

39 Inequality.org, Income Inequality in the United States, Inequality: Facts, 2019.

40 Kagan, 2019, op. cit., p186.

41 Piketty, Thomas, Saez, Emmanuel and Zucman, Gabriel, Distributional National Accounts, *Quarterly Journal of Economics*, Vol. 133, 2018.

42 Collinson, Patrick, UK Incomes: How Does Your Salary Compare?, *Guardian*, 2014.

43 Brewer and Robles, 2019, op. cit., p186.

44 Metcalf, Tom and Witzig, Jack, World's Richest Gain $1.2 Trillion in 2019 as Jeff Bezos Retains Crown, Bloomberg News, 2019.

45 Coffey, Clare, Revollo, Patricia E., Rowan, Harvey, et al., Time To Care: Unpaid and Underpaid Care Work and the Global Inequality Crisis, Oxfam Briefing Paper, 2020.

46 Forbes.com, Forbes 33rd Annual World's Billionaires Issue Reveals Number of Billionaires and their Combined Wealth Have Decreased For First Time Since 2016, Forbes Press Release, 2019.

47 Coffey, Revollo, Rowan, et al., 2020, op. cit., p187.

48 Economist.com, A Rare Peep at the Finances of Britain's 0.01%, *Economist*, 2019.

49 Ibid., p188.

50 Desroches, David, Georgetown Study: Wealth, Not Ability, The Biggest Predictor Of Future Success, WNPR article, 2019.

51 Wikipedia.org, Great Gatsby Curve, Wikipedia article, 2020.

52 Atkinson, Anthony, *Inequality: What Can Be Done?*, Harvard University Press, 2015.

53 Ibid., p190.

54 Unicef.org, Annual Results Report: 2017: Education, Unicef Report, 2018.

55 Richardson, Hannah, Oxbridge Uncovered: More Elitist Than We Thought, BBC News, 2017.

56 Suttontrust.com, Elitist Britain 2019, Sutton Trust Report, 2019.

57 See, for example, Ramos, Gabriela, The Productivity and Equality Nexus, *OECD Yearbook*, 2016.

58 Gracia Arenas, Javier, Inequality and Populism: Myths and Truths, Caixa Bank Research, 2017; Funke, Manuel, Schularick, Moritz and Trebesch, Christoph, The Political Aftermath of Financial Crises: Going to Extremes, VOX CEPR Policy Portal, 2015.

59 Rawls, John, *A Theory of Justice*, Harvard University Press, 1979; Sen, Amartya, *The Idea of Justice*, Allen Lane, 2009; Atkinson, Anthony, *Public Economics in an Age of Austerity*, Routledge, 2014.

60 Sen, Amartya, *Development as Freedom*, Oxford University Press, 1999.

61 HDRO Outreach, What Is Human Development?, United Nations Development Programme, 2015.

62 See UN.org, Sustainable Development Goals: Goal 5: Achieve Gender Equality and Empower All Women and Girls, UN SDGs, 2020.

63 Goldin, 2017, op. cit., p192.

64 Wikipedia.org, Gender Pay Gap, Wikipedia article, 2020; weforum.org, The Global Gender Gap Report 2018, World Economic Forum: Insight Report, 2018.

65 Graf, Nikki, Brown, Anna and Patten, Eileen, The Narrowing, But Persistent, Gender Gap in Pay, Pew Research Center, 2019.

66 Weforum.org, The Global Gender Gap Report 2018, op. cit., p193.

67 Roser, Max and Ortiz-Ospina, Esteban, Income Inequality, Our World in Data, 2013.

68 Ibid., p196.

69 Wetzel, Deborah, Bolsa Família: Brazil's Quiet Revolution, World Bank article, 2013.

70 Ceratti, Mariana, How to Reduce Poverty: A New Lesson from Brazil for the World?, World Bank article, 2014.

71 Worldbank.org, Bolsa Família: Changing the Lives of Millions, World Bank article, 2020.

72 Soares, Sergei, Guerreiro Osório, Rafael, Veras Soares, Fabio, et al., Conditional Cash Transfers In Brazil, Chile and Mexico, International Poverty Centre W.P. 35, 2007.

73 Medeiros, Marcelo, World Social Science Report 2016: Income inequality in Brazil: New Evidence from Combined Tax and Survey Data, United Nations: Education, Scientific and Cultural Organisation, 2016.

74 Data.worldbank.org, GINI Index (World Bank Estimate), World Bank Data, 2020.

75 Batha, Emma, These Countries are Doing the Most to Reduce Inequality, World Economic Forum, 2018; oxfamilibrary.openrepository.com, The Commitment to Reducing Inequality Index 2018, Development Finance International and Oxfam Report, 2018; Simson, Rebecca, Mapping Recent Inequality Trends in Developing Countries, LSE International Inequalities Institute W.P. 24, 2018.

76 See, Lagarde, Christine, Tweet 2015: 'Reducing excessive inequality is not just morally and politically correct, but is good economics'.

77 Wolf, Martin, Seven Charts that Show How the Developed World is Losing its Edge, *Financial Times*, 2017.

78 Lagarde, Christine, 'Lifting the Small Boats', Speech by Christine Lagarde (June, 17th 2015), IMF: Speech, 2017.

79 IFO.de, Economic Policy and the Rise of Populism – It's Not so Simple, EEAG Report on the European Economy, 2017.

80 Wolf, Martin, The Economic Origins of the Populist Surge, *Financial Times*, 2017.

81 Desroches, 2019, op. cit., p197.

82 Goldin, Ian and Muggah, Robert, Viral Inequality, Project Syndicate, 2020.

Geopolitics

1 Alevelpolitics.com, Are We Moving Into a New Multipolar World, A-Level Politics, 2020.

2 Law, David, Three Scenarios for the Future of Geopolitics, World Economic Forum article, 2018.

3 Muggah, Robert, The Global Liberal Order is in Trouble – Can it Be Salvaged, or Will it Be Replaced?, World Economic Forum, 2018.

4 Krastev, Ivan and Leonard, Mark, The Spectre of a Multipolar Europe, European Council on Foreign Relations, 2010.

5 weforum.org, The Future of Global Liberal Order, World Economic Forum article, 2018.

6 Mearsheimer, John J., Bound to Fail: The Rise and Fall of the Liberal International Order, *International Security*, Vol. 43, 2019.

7 Auslin, Michael, The Asian Century Is Over, *Foreign Policy*, 2019.

8 Economist.com, Ten Years On, *Economist*, 2007.

9 Romei, Valentina and Reed, John, The Asian Century is Set to Begin, *Financial Times*, 2019.

10 Focus-economics.com, The World's Fastest Growing Economies, Focus Economics, 2020.

11 In China, populist tactics are instead instrumentalised to inflame public sentiment. This compares to more open democratic societies where populism can threaten to run out of control.

12 Araya, Daniel, The Future is Asian: Parag Khanna On the Rise of Asia, *Forbes*, 2019.

13 Khanna, Parag, Why This is the 'Asian Century', Fast Company, 2019.

14 Monbiot, George, The New Political Story That Could Change Everything, TED Summit Talk, 2019.

15 Muggah, Robert, 'Good Enough' Global Cooperation is Key To Our Survival, World Economic Forum, 2019.

16 Partington, Richard, What Became of the G20 Leaders Who Met in 2008 to Avert Financial Crisis?, *Guardian*, 2018.

17 Ibid., p205.

18 Goldstein, Joshua S., *International Relations*, 2013–2014 Update, 10th Edition, American University and University of Massachusetts, Amherst.

19 Credit-suisse.com, 'Getting Over Globalization' – Outlook for 2017, Credit Suisse Press Release, 2017.

20 GPF Team, Is a Multipolar World Emerging?, Geopolitical Futures, 2018.

21 Credit-suisse.com, Which Way to a Multipolar World?, Credit Suisse Research Institute, 2017.

22 Credit-suisse.com, 'Getting Over Globalization', 2017, op. cit., p206.

23 Muggah, Robert, America's Dominance Is Over – By 2030, We'll Have A Handful Of Global Powers, *Forbes*, 2016.

24 Bremmer, Ian, *Every Nation For Itself*, Penguin, 2013.

25 Ibid., p206.

26 Unlike in bipolar systems, deterrence is more straightforward since more states can join together to confront an aggressive state. What is more, there is less hostility in the system because states pay comparatively less attention to one another and spread their attention to all great powers. The many interactions among various states create linkages and cleavages that can mitigate war . Mearsheimer, John J., Structural Realism, International Relations Theories: Discipline and Diversity, Vol. 3, 2013.

27 It is more likely to create a peaceful balance of power if the hegemon distributes rewards appropriately among other states, Wohlforth, William C., The Stability of a Unipolar World, International Security, Vol. 23, 1999.

28 Mowle, Thomas S. and Sacko, David, *The Unipolar World*, Palgrave Macmillan US, 2007.

29 Bipolarity can reduce incentives for conflict and creates opportunities for balancing.

30 Allison, Graham, The Thucydides Trap, Foreign Policy, 2017.

31 Allison, Graham, The Thucydides Trap: Are the U.S. and China Headed for War?, *Atlantic*, 2016.

32 See belfercenter.org, Can America and China Escape Thucydides's Trap, Harvard Kennedy School: Belfer Centre for Science and International Affairs, 2020.

33 Allison, Graham, Is War Between China and the U.S. Inevitable?, TED Talk, 2018.

34 The International Commission and Association on Nobility, Map Congress of Vienna, The International Commission and Association on Nobility, Public Domain, 2015.

35 According to one study by Graham Allison, in 12 of the 16 cases where a dominant power felt threatened by a clear rival over the past 500 years, the result was war. belfercenter.org, 2020, op. cit., p208.

36 Muggah, Robert and Owen, Taylor, The Global Liberal Democratic Order Might be Down, But It's Not Out, World Economic Forum, 2018.

37 Wikipedia.org, Polarity: Spheres of Influence of the Two Cold War Superpowers, Map (1959), Wikipedia article, 2020.

38 With the exception of South Africa, the African continent, India and a few European countries sought to keep their distance and remain unaligned.

39 Conflicts and regime change occurred in Tibet (1950), Iraq (1958), Cuba (1960), Bolivia (1970), Uganda (1971), Argentina (1976), Pakistan (1977), Afghanistan (1978), Iran (1979), the Central African Republic (1979) and Turkey (1980). Both the US and USSR were involved in civil wars in Malaya (1948-60), Laos (1953-1975), Cambodia (1967-75) Ethiopia (1974-91), Lebanon (1975-90) and El Salvador (1980-92). During the Cold War Washington sent American troops into Vietnam (1965-75), the Dominican Republic (1965), Lebanon (1982), Grenada (1983) and Panama (1989).

40 Armscontrol.org, Nuclear Weapons: Who Has What at a Glance, Arms Control Association, 2020.

41 Spielman, Richard, The Emerging Unipolar, *New York Times*, 1990.

42 Leaniuk, Jauhien, The Unipolar World, The Dialogue, 2016; Krauthammer, Charles, The Unipolar Moment, Foreign Affairs article, 1990.

43 Wohlforth, 1999, op. cit., p210.

44 Roser, Max and Nagdy, Mohamed, Nuclear Weapons: Number of Nuclear Warheads in the Inventory of the Nuclear Powers, 1945 to 2014 graph, Our World in Data, 2013.

45 Kristensen, Hans and Norris, Robert, U.S. Nuclear Forces, 2009, *Bulletin of the Atomic Scientists*, 2015.

46 Wohlforth, 1999, op. cit., p210; Hansen, Birthe, The Unipolar World Order and its Dynamics, The New World Order, 2000.

47 David Vine, 'Lists of U.S. Military Bases Abroad, 1776–2020', American University Digital Archive, https://doi.org/10.17606/bbxc-4368.

48 Crawford, Neta C., US Budgetary Costs of Wars through 2016, Watson Institute: Cost of War, 2016.

49 Desjardins, 2017, op. cit., p210.

50 According to one study, the US has been at war for 226 out of its 243 years of existence. globalresearch.ca, America Has Been at War 93% of the Time – 222 out of 239 Years – Since 1776, Global Research, 2019.

51 Clark, David, Like It or Not, the US Will Have to Accept a Multipolar World, *Guardian*, 2007; Monteiro, Numo P., Unrest Assured: Why Unipolarity Is Not Peaceful, International Security, Vol. 36, 2011; Gayle, Damien, Vladimir Putin: US trying to create 'unipolar world', *Guardian*, 2015.

52 Turner, Susan, Russia, China and Multipolar World Order: The Danger in the Undefined. *Asian Perspective*, Vol. 33, 2009; Prashad, Vishay, Trump and the Decline of American Unipolarity, CounterPunch, 2017.

53 Kennedy, Scott, Liang, Wei and Reade, Claire, How is China Shaping the Global Economic Order?, China Power, 2015.

54 Desjardins, 2017, op. cit., p212; McGregor, Sarah and Greifeld, Katherine, China Loses Status as U.S.'s Top Foreign Creditor to Japan, Bloomberg article, 2019.

55 Isidore, Chris, The U.S. is Picking a Fight With its Biggest Creditor, CNN Business, 2018.

56 Chinainvestmentresearch.org, Massive Chinese Lending Directed to Silk Road, China Investment Research, 2015.

57 Ma, Alexandra, This map shows a trillion-dollar reason why China is oppressing more than a million Muslims, Business Insider, 2019.

58 FT.com, China Needs to Act as a Responsible Creditor, *Financial Times*, 2018.

59 Hillman, Jonathan E., How Big Is China's Belt and Road?, Center for Strategic and International Studies, 2018.

60 Kozul-Wright, Richard and Poon, Daniel, China's Belt and Road isn't Like the Marshall Plan, but Beijing Can Still Learn From It, *South China Morning Post*, 2019.

61 According to one observer, 'the equivalent of the mid-twentieth century founding of the United Nations and World Bank, plus the Marshall Plan all rolled into one'.

62 Khanna, Why This is the 'Asian Century', 2019, op. cit., p214.

63 Ma, 2019, op. cit., p214.

64 Reed, John, China Construction Points to Military Foothold in Cambodia, *Financial Times*, 2019.

65 Ng, Teddy, China's Belt and Road Initiative Criticised for Poor Standards and 'Wasteful' Spending, *South China Morning Post*, 2019.

66 Cheong, Danson, Belt and Road Initiative Not a Debt Trap, has Helped Partners Grow Faster, *Straits Times*, 2019; Morris, Scott, China's Belt and Road Initiative Heightens Debt Risks in Eight Countries, Points to Need for Better Lending Practices, Centre for Global Development, 2018.

67 Businesstoday.in, India Rejects China's Invite to Attend Belt and Road Initiative Meet for the Second Time, Business Today, 2019.

68 There are signs of rising tensions between China and its neighbours including Japan, India and Southeast Asian nations, not least with competing claims to the South China Sea. See, Gangley, Declan, Tweet, 17 October 2018.

69 Jiangtao, Shi, Dominance or Development? What's at the End of China's New Silk Road?, *South China Morning Post*, 2019.

70 Fickling, David, China Could Outrun the U.S. Next Year. Or Never, Bloomberg, 2019.

71 About 35 per cent of all global growth from 2017 to 2019 came from China as compared to just 18 per cent from the US, 9 per cent from India and 8 per cent from Europe.

72 Guillemette, Yvan and Turner, David, The Long View: Scenarios for the World Economy to 2060, OECD Economic Policy, 2018.

73 Wearden, Graem, Trump Claims Trade War is Working as China's Growth Hits 27-year low, *Guardian*, 2019.

74 China already spends close to 2.5 per cent of its GDP on public and private research and development (as compared to 2.8 per cent in the US).

75 EY Greater China, China is Poised to Win the 5G Race. Are You Up to Speed?, EY article, 2018.

76 Fayd'Herbe, Nannette H., A Multipolar World Brings Back the National Champions, World Economic Forum article, 2019.

77 Though recall that this brief interlude of peace was punctured by the 1853 Crimean War and collapsed entirely with the outbreak of world wars in the twentieth century.

78 Murray, Donnette and Brown, David, *Power Relations in the Twenty-First Century*, Routledge, 2017.

79 Walt, Stephen M., What Sort of World Are We Headed For?, Foreign Policy article, 2018.

80 Lezard, Nicholas, I Told You So, *Guardian*, 2003.

81 Zhen, Liu, Why 5G, a Battleground for US and China, Is Also a Fight for Military Supremacy, *South China Morning Post*, 2019; Medin, Milo and Louie, Gilman, The 5G Ecosystem: Risks & Opportunities for DoD, 2019.

82 Simons, Hadlee, Trump Signs Order Effectively Banning Huawei Telecom Equipment in US, Android Authority, 2019.

83 Worldview.stratfor.com, The Geopolitics of Rare Earth Elements, Stratfor, 2019.

84 See U.S. Geological Survey, usgs.gov, 2020.

85 Rockwood, Kate, How a Handful of Countries Control the Earth's Most Precious Materials, Fast Company, 2010.

86 Worldview.stratfor.com, 2019, op. cit., p217.

87 Lima, Charlie, Cobalt Mining, China, and the Fight for Congo's Minerals, Lima Charlie News, 2018.

88 Millan Lombrana, Laura, Bolivia's Almost Impossible Lithium Dream, Bloomberg article, 2018.

89 cfr.org, Liberal World Order RIP?, Council on Foreign Relations, 2018.

90 Fraga, Armenio, Bretton Woods at 75, Project Syndicate, 2018.

91 Bremmer, Ian, The End of the American Order: Speech at 2019 GZERO Summit, Eurasia Group, 2019.

92 Desilver, Drew, For most U.S. workers, real wages have barely budged in decades, PEW Research Center, 2018.

93 Freedomhouse.org, Democracy in Retreat, Freedom House, 2019.

94 Ibid., p220.

95 Taylor, Adam, Global Protests Share Themes of Economic Anger and Political Hopelessness, *Washington Post*, 2019.

96 Identity politics in the US, for example, track strongly along partisan lines. In the US House of Representatives, 90% of Republican members are white males. More than 66% of their Democratic counterparts are not.

97 Gallup.com, Global Emotions Report 2018, Gallup, 2018.

98 Ibid., p220.

99 Ibid., p221.

100 Muggah, Robert and Kavanagh, Camino, 6 Ways to Ensure AI and New Tech Works For – Not Against – Humanity, World Economic Forum, 2018.

101 Mounk, Yasha, 'Warning signs are flashing red', *Journal of Democracy*, 2016.

102 Ferguson, Peter Allyn, Undertow in the Third Wave: Understanding the Reversion from Democracy, University of Columbia: Open Collections, 2009.

103 Diamond, Larry, The Democratic Rollback, Foreign Affairs, 2008.

104 Diamond, Larry, Facing Up to the Democratic Recession, *Journal of Democracy*, Vol. 26, 2015.

105 Voytko, Lysette, White House Goes Dark As George Floyd Protests Boil Over, *Forbes*, 2017.

106 Pfeifer, Sylvia, Oil Companies in New Rush to Secure North Sea Drilling Rights, *Financial Times*, 2018; Goldsmith, Arthur A., Making the World Safe for Partial Democracy? Questioning the Premises of Democracy Promotion, *International Security*, Vol. 33, 2008.

107 Arce, Moises and Bellinger, Paul T., Low-Intensity Democracy Revisited: The Effects of Economic Liberalization on Political Activity in Latin America, *World Politics*, Vol. 60, 2007.

108 Leterme, Yves and Eliasson, Jan, Democracy – Is the Glass Half Full or Half Empty?, IDEA, 2017.

109 Zakaria, Fareed, The Rise of Illiberal Democracy, Foreign Affairs, 1997.

110 Eiu.com, Democracy Index, The Economist Intelligence Unit, 2018.

111 Kauffmann, Sylvie, Europe's Illiberal Democracies, *New York Times*, 2016.

112 Project-syndicate.org, Recep Tayyip Erdoğan, Project Syndicate, 2020.

113 CRF-usa.org, Putin's Illiberal Democracy, Bill of Rights in Action Vol. 31, 2016.

114 Muggah, Robert, Can Brazil's Democracy Be Saved?, *New York Times*, 2018.

115 Thompson, Mark R., Bloodied Democracy: Duterte and the Death of Liberal Reformism in the Philippines, *Journal of Current Southeast Affairs*, Vol. 35, 2016.

116 Abramowitz, Michael J., Democracy in Crisis, Freedom House, 2018.

117 Levitsky, Steven and Ziblatt, Daniel, *How Democracies Die*, Penguin Random House, 2019.

118 Mounk, Yascha, Illiberal Democracy or Undemocratic Liberalism?, Project Syndicate, 2016.

119 V-dem.net, Democracy for All?, V-Dem Annual Democracy Report, 2018.

120 Economist.com, Democracy Continues Its Disturbing Retreat, *Economist*, 2018.

121 V-dem.net, 2018, op. cit., p224.

122 Fao, Roberto S., Mounk, Yasha and Inglehart, Ronald F., The Danger of Deconsolidation, *Journal of Democracy*, Vol. 27, 2016.

123 Wike, Richard, Silver, Laura and Castillo, Laura, Many Across the Globe Are Dissatisfied With How Democracy Is Working, 2019.

124 Guilford, Gwinn, Harvard Research Suggests That An Entire Global Generation Has Lost Faith in Democracy, Quartz, 2016.

125 Howe, Neil, Are Millennials Giving Up On Democracy?, *Forbes* article, 2017.

126 Fetterolf, Janell, Negative Views of Democracy More Widespread in Countries with Low Political Affiliation, Pew Research Center, 2018.

127 Wike, Richard, Simmons, Katie, Stokes, Bruce, et al., Democracy Widely Supported, Little Backing for Rule by Strong Leader or Military, Pew Research Center, 2017.

128 Systemicpeace.org, Global Trends in Governance: 1800-2017, Polity IV, 2017.

129 Wike, Richard, Simmons, Katie, Stokes, Bruce, et al., Globally, Broad Support for Representative and Direct Democracy, Pew Research Center, 2017.

130 Desilver, Drew, Despite Global Concerns About Democracy, More Than Half of Countries are Democratic, Pew Research Center, 2019.

131 Wike, Simmons, Stokes, et al., Globally, Broad Support for Representative and Direct Democracy, 2017, op. cit., p225.

132 Desilver, 2019, op. cit., p225.

133 Wike, Silver and Castillo, 2019, p225.

134 Fetterolf, 2018, p226.

135 Washingtonpost.com, The Rise of Authoritarians, *Washington Post*, 2019.

136 Gray, Alex, The Troubling Charts that Show Young People Losing Faith in Democracy, World Economic Forum, 2016.

137 Marshall, Monty G., Gurr, Ted R. and Jaggers, Keith, Political Regime Characteristics and Transitions, 1800–2016, Polity IV Project, 2017.

138 Diamond, Larry, A Fourth Wave or False Start?, Foreign Affairs, 2011.

139 Liddiard, Patrick, Are Political Parties in Trouble?, Wilson Center, 2018.

140 Oecd.org, Trade Unions, Employer Organisations, and Collective Bargaining in OECD Countries, OECD, 2017; McCarthy, Niall, The State Of Global Trade Union Membership, Statista, 2019.

141 Liddiard, op cit., p227.

142 McCarthy op cit., p227.

143 Economist.com, The Arab Spring, Five Years On, *Economist*: Daily Chart, 2016.

144 Yagci, Alper H., The Great Recession, Inequality and Occupy Protests around the World, Government and Opposition, Vol. 52, p2017.

145 Economist.com, 2016, op. cit., p232.

146 Wile, Rob, MAP: The World's Economic Center Of Gravity From AD 1 To AD 2010, Business Insider, 2012.

147 Public.wmo.int, Greenhouse Gas Concentrations Atmosphere Reach Yet Another High, World Meteorological Organisation: Press Release, 2019.

148 President Trump described it as a 'poorly negotiated agreement that sacrificed US growth and employment at the expense of developing countries'.

Violence

1 Hedges, Chris, 'What Every Person Should Know About War', *New York Times*, 2003.

2 BBC.com, Afghanistan: Civilian Deaths at Record High in 2018 – UN, BBC News, 2019.

3 Reuters.com, Syrian Observatory Says War Has Killed More Than Half a Million, Reuters World News, 2018.

4 See Yemendataproject.org data.

5 Palmer, Jason, Call for Debate on Killer Robots, BBC News, 2009.

6 According to psychologist Steven Pinker, we 'may be living in the most peaceful time in our species' existence.'

7 Ourworldindata.org, Global Deaths in Conflict Since 1400, Our World in Data, 2016.

8 CGEH.nl, Conflicts and Wars, Centre for Global Economic History, 2020.

9 Roser, Max, War and Peace, Our World in Data, 2016.

10 WHO.int, Global Health Observatory (GHO) data, WHO, 2020.

11 Ourworldindata.org, Death Rate by Cause, World, 1990 to 2017, Our World in Data, 2017.

12 Keeley, Lawrence H., *War Before Civilization*, Oxford University Press, 1996.

13 A particularly grim example of the human toll of these primitive wars was recently uncovered in a 14,000-year-old cemetery in the town of Jebel Sahaba, what is present-day Sudan. Almost half of all the thousands of skulls and bones recovered from the mass grave there show signs of having been penetrated with spears, arrows and clubs. See Kelly, Raymond C., The Evolution of Lethal Intergroup Violence, PNAS, Vol. 102, 2015.

14 Roser, Max, Archaeological Evidence on Violence, Our World in Data, 2016; Roser, Max, Ethnographic Evidence on Violence, Our World in Data, 2016.

15 ICRC.org, The Roots of Restraint in War, International Committee of the Red Cross, 2018; Farhat-Holzman Laina, Steven A. LeBlanc, Constant Battles: The Myth of the Peaceful, Noble Savage; Matt Ridley, The Red Queen, Sex and the Evolution of Human Nature, *Comparative Civilization Review*, Vol. 52, 2005.

16 Alcantara, Chris, 46 Years of Terrorist Attacks in Europe Visualized, *Washington Post*, 2017; Juengst, Sara L., Hillforts of the Ancient Andes: Colla Warfare, Society, and Landscape, *Journal of Conflict Archaeology*, Vol. 6, 2011.

17 Whipps, Heather, How Gunpowder Changed the World, Live Science, 2008.

18 Roser, Max, Battle-related Deaths in State-based Conflicts Since 1946, by World Region, 1946 to 2016, Our World in Data, 2016; Roser, Max, Average Number of Battle Deaths per Conflict Since 1946, per Type, 2016.

19 PCR.uu.se, Fatal Events in 2018 by Type of Violence, Uppsala Conflict Data Program, 2018.

20 Pettersson, Therese, Hogbladh, Stina and Oberg, Magnus, Organized Violence, 1989–2018 and Peace Agreements, *Journal of Peace Research*, Vol. 56, 2019.

21 Muggah, Robert, The U.N. Can't Bring Peace to the Central African Republic, Foreign Policy article, 2018.

22 Muggah, Robert, Is Kabila Using Ethnic Violence to Delay Elections?, Foreign Policy article, 2018.

23 Muggah, Robert, Mali is Slipping Back Into Chaos, *Globe and Mail*, 2018.

24 ICRC.org, 2018, opo. cit., p242.

25 Muggah, Robert and Sullivan, John P., The Coming Crime Wars, Foreign Policy article, 2018.

26 Rowlatt, Justin, How the US Military's Opium War in Afghanistan Was Lost, BBC News article, 2019.

27 Wellman, Phillip W., US Ends Campaign to Destroy Taliban Drug Labs in Afghanistan, *Stars and Stripes*, 2019.

28 Muggah and Sullivan, 2018, op. cit., p242.

29 Peacekeeping.un.org, Principles and Guidelines, United Nations Peacekeeping Operations, 2008.

30 Autesserre, Severine, *Peaceland: Conflict Resolution and the Everyday Politics of International Intervention*, Cambridge University Press, 2014.

31 See tools provided by Earthtime, CREATE Lab, CMU.

32 Unenvironment.org, The Tale of a Disappearing Lake, UN Environment Programme, 2018; Ross, Will, Lake Chad: Can the Vanishing Lake be Saved?, BBC News article, 2018.

33 Muggah, Robert and Cabrera, Jose L., The Sahel is Engulfed by Violence. Climate Change, Food Insecurity and Extremists are Largely to Blame, World Economic Forum, 2019.

34 Gerretsen, Isabelle, How climate change is fueling extremism, CNN World News article, 2019.

35 May, John F., Guengant, Jean-Pierre and Brooke, Thomas R., Demographic Challenges of the Sahel, *PRB*, 2015.

36 FAO.org, Atlas on Regional Integration in West Africa, OECD, 2008.

37 Giordano, Mark and Bassini, Elisabeth, Climate Change and Africa's Future, Hoover Institution, 2019.

38 Semple, Kirk, Central American Farmers Head to the U.S., Fleeing Climate Change, *New York Times*, 2019.

39 Saha, Sagatom, How Climate Change Could Exacerbate Conflict in the Middle East, Atlantic Council, 2019.

40 Busby, Joshua and Von Uexkull, Nina, Climate Shocks and Humanitarian Crises, Foreign Affairs article, 2018.

41 Ibid., p245.

42 OECD.org, *States of Fragility 2018*, OECD Publication, 2019.

43 Samenow, Jason, Drought and Syria: Manmade Climate Change or Just Climate?, *Washington Post*, 2013.

44 Polk, William R., Understanding Syria: From Pre-Civil War to Post-Assad, *Atlantic*, 2013.

45 Femia, Francesco and Werrell, Caitlin, Syria: Climate Change, Drought and Social Unrest, The Center for Climate & Security, 2012.

46 See data provided by unhcr.org, Situations: Syria, Operational Portal: Refugee Situations, 2020.

47 Bernauer, Thomas and Bohmelt, Tobias, Can We Forecast Where Water Conflicts Are Likely to Occur?, New Security Beat, 2014. See map, Ars.els-cdn.com, Likelihood of Hydro-political Interaction, date not given.

48 Ratner, Paul, Where Will the 'Water Wars' of the Future be Fought?, World Economic Forum, 2018.

49 Peek, Katie, Heat Map: Where is the Highest Risk of Water Conflict?, *Popular Science*, 2014.

50 Worldwater.org, Water Conflict Chronology, Pacific Institute, 2018.

51 Despite many cases of cooperation to manage tensions over water – witness the 1960 Indus Water Treaty between India and Pakistan – these are harder to broker than ever before. See vector-center.com, Water and Conflict, Vector Center, 2018.

52 Farinosi, Fabio, Giupponi, Carlo, Reynaud, Arnaud, et al., An Innovative Approach to the Assessment of Hydro-political Risk: A Spatially Explicit, Data Driven Indicator of Hydro-political Issues, *Global Environmental Change*, Vol. 52, 2018.

53 Bernauer, Thomas and Bohmelt, Tobias, Basins at Risk: Predicting International River Basin Conflict and Cooperation, Global Environmental Risks, *Global Environmental Politics*, Vol. 14, 2018.

54 The regime type (level of democracy), the nature of riparian countries' legal systems, precipitation levels, geographic contiguity, and the number of riparian states are most important to determining conflict risk.

55 BBC.com, The 'Water War' Brewing Over the New River Nile Dam, BBC News article, 2018; EC.europa.eu, Global Hotspots for Potential Water Disputes, EU Science Hub, 2018.

56 See Smallarmssurvey.org, Conflict Armed Violence portal, 2017.

57 Smallarmssurvey.org, Weapons and Markets, Small Arms Survey, 2017.

58 Ibid., p247.

59 See SIPRI Arms Flows visualisation provided by Earthtime, CREATE Lab, CMU; Geary, Will, The United States of Arms, Vimeo video, 2018.

60 Ibid., p248.

61 Trefis.com, What Is Netflix's Fundamental Value Based On Expected FY'19 Results, Trefis: Collaborate on Forecasts, 2019; macrotrends.net, Netflix Net Worth 2006-2020: NFLX, MacroTrends, 2020.

62 Sipri.org, Global Share of Major Arms Exports by 10 Largest Exporters, 2014–2018: Pie Chart, SIPRI Arms Transfers Database, 2019; sipri.org, Trends in World Military Expenditure: 2018, SIPRI Fact Sheet, 2019.

63 Sipri.org, World Military Expenditure Grows to $1.8 Trillion in 2018, SIPRI Press Release, 2019.

64 Ibid., Sipri.org, World Military Expenditure Grows to $1.8 Trillion in 2018, op. cit., p250.

65 Wikipedia.org, Participation in the Nuclear Non-Proliferation Treaty Map, Wikipedia Visual, 2014.

66 Israel also is believed to have between 75 and 400 nuclear weapons but maintains a policy of deliberate ambiguity. disarmament.un.org, Treaty on the Non-Proliferation of Nuclear Weapons, UNODA, 2020.

67 Sanders-Zakre, Alicia, What You Need to Know About Chemical Weapons Use in Syria, Arms Control, 2019.

68 OPCW.org, Fact-Finding Mission, OPCW, 2019.

69 Schneider, Tobias and Lutkefend, Theresa, The Logic of Chemical Weapons Use in Syria, GPPI, 2019.

70 Pandya, Jayshree, The Weaponization Of Artificial Intelligence, *Forbes*, 2019.

71 O'Hanlon, Michael E., The Role of AI in Future Warfare, Brookings Report, 2018.

72 Allen, Gregory C., Understanding China's AI Strategy, Center for New American Study, 2019.

73 McMullan, Thomas, How Swarming Drones will Change Warfare, BBC News article, 2019.

74 Futureoflife.org, An Open Letter to the United Nations Convention On Certain Conventional Weapons, Future of Life open letter, 2017.

75 See Map in Lang, Johannes, Schott, Robin M. and Van Munster, Rens, Four Reasons why Denmark Should Speak

up about Lethal Autonomous Weapons, Danish Institute for International Studies, 2018.

76 Stopkillerrobots.org, Country Views on Killer Robots, Campaign to Stop, 2018.

77 See esri.com, A map of terrorist attacks, according to Wikipedia, ESRI, 2020.

78 Horsley, Richard A., The Sicarii: Ancient Jewish 'Terrorists', *Journal of Religion*, Vol. 59, 1979.

79 Szczepanski, Kallie, Hashshashin: The Assassins of Persia, Thought Co., 2019; Irving, Clive, Islamic Terrorism Was Born on This Mountain 1,000 Years Ago, Daily Beast, 2017.

80 Fraser, Antonia, *The Gunpowder Plot*, Orion Books, 2010.

81 Hoffman, Bruce, Terrorism in History, *Journal of Conflict Studies*, Vol. 27, 2007.

82 Senn, Alfred E., The Russian Revolutionary Movement of the Nineteenth Century as Contemporary History, Wilson Center Report, 1993.

83 Johnston, David, Terror In Oklahoma: The Overview, *New York Times*, 1995.

84 Washingtonpost.com, How Terrorism In The West Compares To Terrorism Everywhere Else, *Washington Post*, 2016.

85 See umd.edu, Global Terrorism Database, National Consortium for the Study of Terrorism and Responses to Terrorism, 2019.

86 Mueller, John and Stewart Mark G., Conflating Terrorism and Insurgency, CATO Institute, 2016.

87 UN.org, International Terrorism Committee Report of 17 December 1996, Ad Hoc Committee Established by General Assembly Resolution, 1996.

88 Muggah, Robert, Europe's Terror Threat Is Real. But Its Cities Are Much Safer Than You Think, World Economic Forum, 2017.

89 Ibid., p256.

90 Wikipedia.org, Organizations Currently Officially Designated as Terrorist by Various Governments, Wikipedia article, 2020.

91 Prnewswire.com, IEP's 2018 Global Terrorism Index: Deaths From Terrorism Down 44 Per Cent in Three Years, but Terrorism Remains Widespread, Institute for Economic Peace, 2018.

92 Muggah, Robert and Aguirre, Katherine, Terrorists Want To Destroy Our Cities. We Can't Let Them, World Economic Forum, 2016.

93 Jipson, Art and Becker, Paul J., White Nationalism, Born in the USA, Is Now a Global Terror Threat, The Conversation article, 2019.

94 Splcenter.org, Hate Map, Southern Poverty Law Center, 2019.

95 Muggah, Robert, Global Terrorism May be Down but is Still a Threat In 2019 – Are We Ready?, *Small Wars Journal*, 2019.

96 Green, Manfred S., LeDuc, James, Cohen, Daniel, et al., Confronting the Threat of Bioterrorism: Realities, Challenges, and Defensive Strategies, Terrorism And Health, *Lancet*, Vol. 19, 2019.

97 Kimball, Daryl and Davenport, Kelsey, Timeline of Syrian Chemical Weapons Activity, 2012-2020, Arms Control Association, 2020.

98 One study of violent extremists in Africa found that state killings and indiscriminate arrests were key factors in converting people with existing grievances into terrorists in over 70 per cent of reported cases. See un.org, Marginalization, Perceived Abuse of Power Pushing Africa's Youth to Extremism, UN News, 2017.

99 Freedomhouse.org, Freedom in the World 2019, Featuring Special Release on United States, Freedom House, 2019.

100 Wikipedia.org, List of Genocides by Death Toll, Wikipedia article, 2020.

101 Evans, Gareth, State Sovereignty Was a Licence to Kill, International Crisis Group, 2008.

102 Rummel, Rudolph J., *Death by Government*, Transaction Publishers, 1994.

103 Barry, Ellen, Putin Criticizes West for Libya Incursion, *New York Times*, 2011.

104 Wee, Sui-Lee, Russia, China Oppose 'Forced Regime Change' in Syria, Reuters, 2012.

105 Economist.com, In Some Countries, Killer Cops are Celebrated, *Economist*, 2018.

106 Muggah, Robert, Brazil's Murder Rate Finally Fell—and by a Lot, Foreign Policy article, 2019.

107 See 38th Session of the Human Rights Council or Number of people shot to death by the police in the United States from 2017 to 2020, by Race: Statista, 2020.

108 Ellis-Petersen, Hannah, Duterte's Philippines Drug War Death Toll Rises Above 5,000, *Guardian*, 2018.

109 Sinyangwe, Samuel, Police killed 1,099 people in 2019: Map, Mapping Police Violence, 2020.

110 Reuters.com, ICC Prosecutor: Examination of Philippines Continues Despite Withdrawal, Reuters, 2019.

111 Wagner, Peter and Sawyer, Wendy, States of Incarceration: The Global Context 2018, Prison Policy Initiative, 2018.

112 Kann, Drew, 5 Facts Behind America's High Incarceration Rate, CNN News, 2019.

113 Walmsley, Roy, World Prison Population List 12th Edition, World Prison Brief, 2018.

114 Walmsley, 2018, p261.

115 Pelaez, Vicky, The Prison Industry in the United States: Big Business or a New Form of Slavery?, Global Research, 2019.

116 Kann, 2019, p261.

117 HRW.org, Monitoring Conditions Around The World, Human Rights Watch Prison Project.

118 OHCHR.org, Convention against Torture and Other Cruel, Inhuman or Degrading Treatment or Punishment, UN Human Rights Office of the High Commisioner, 1984; un.org, United Nations Standard Minimum Rules for the Treatment of Prisoners (the Mandela Rules), UN General Assembly: 70th Session, 2015.

119 See Muggah, Robert, Taboada, Carolina and Tinoco Dandara, Q&A: Why Is Prison Violence So Bad in Brazil?, *Americas Quarterly*, 2019; Muggah, Robert, Opinion: Brazil's Prison Massacres Send A Dire Message, NPR, 2019; Muggah, Robert and Szabó De Carvalho, Ilona, Brazil's Deadly Prison System, *New York Times*, 2017.

120 Wilkinson, Daniel, The High Cost of Torture in Mexico, Human Rights Watch, 2017.

121 Torture and Ill Treatment in Syria's Prisons, The Lancet Editorial, *Lancet*, Vol. 388, 2016.

122 Amnestyusa.org, Senior Members Of Nigerian Military Must Be Investigated For War Crimes, Amnesty International, 2015.

123 Fifield, Anna, North Korea's prisons 'worse' than Nazi camps, judge who survived Auschwitz concludes, *Independent*, 2017.

124 Sudworth, John, China's Hidden Camps, BBC News, 2018; Doman, Mark, Hutcheon, Stephen, Welch, Dylan, et al., China's Frontier of Fear, ABC News, 2018.

125 Sigal, Samuel, Internet Sleuths Are Hunting for China's Secret Internment Camps for Muslims, *Atlantic*, 2018.

126 Foreignpolicy.com, China's War on Uighurs, Foreign Policy article, 2019.

127 Ryan, Fergus, Cave, Danielle and Ruser, Nathan, Mapping Xinjiang's 'Re-education' Camps, Australian Strategic Policy Institute, 2018; Nationalawakening.org, China's Gulag Archipelago in Occupied East Turkistan, Awakening Movement Project.

128 Doman, Hutcheon, Welch, et al., 2018, op. cit., p262.

129 Byler, Darren, China's Hi-tech War on its Muslim Minority, *Guardian*, 2019.

130 Tiezzi, Shannon, Is the Kunming Knife Attack China's 9–11?, *Diplomat*, 2014.

131 Doman, Hutcheon, Welch, et al., 2018, op. cit., p263.

132 Mozur, Paul, Inside China's Dystopian Dreams: A.I., Shame and Lots of Cameras, *New York Times*, 2018.

133 Botsman, Rachel, Big Data Meets Big Brother as China Moves to Rate its Citizens, Wired, 2017.

134 HRW.org, China: Big Data Fuels Crackdown in Minority Region, Human Rights Watch, 2018.

135 Benaim, Daniel and Russon, Gilman, Hollie, China's Aggressive Surveillance Technology Will Spread Beyond Its Borders, Slate, 2018.

136 See redd.it, Map of Illegal Trafficked Goods Around the World, UNODC and Sciences Po, 2014.

137 Gambetta, Diego, *The Sicilian Mafia: The Business of Private Protection*, Harvard University Press, 1996.

138 Chu, Yiu-Kong, *The Triads as Business*, Routledge, 2000.

139 Hill, Peter B. E., The Japanese Mafia: Yakuza, Law, and the State, Oxford Scholarship Online, 2004.

140 Andreas, Peter, Gangster's Paradise, Foreign Affairs article, 2013.

141 Varese, Federico, The Russian Mafia: Private Protection in a New Market, Oxford Scholarship Online, 2003.

142 Clough, Christine, Transnational Crime is a $1.6 trillion to $2.2 trillion Annual 'Business', Global Financial Integrity, 2017.

143 Roos, Dave, How Prohibition Put the 'Organized' in Organized Crime, History article, 2019.

144 Stigall, Dan E., Ungoverned Spaces, Transnational Crime, and the Prohibition on Extraterritorial Enforcement Jurisdiction in International Law, *Journal of International & Comparative Law*, 2013.

145 Winton, Alison, Gangs in Global Perspective, Environment and Urbanization, 2014.

146 Muggah, Robert, Violent crime has undermined democracy in Latin America, *Financial Times*, 2019.

147 IFEX.org, In Mexico, 'Narcopolitics' is a Deadly Mix for Journalists Covering Crime and Politics, IFEX, 2018.

148 Muggah, Robert and Szabo de Carvalho, Ilona, Violent crime in São Paulo has dropped dramatically. Is this why?, World Economic Forum, 2018.

149 Sullivan, John P., De Arimatéia da Cruz, José and Bunker, Robert J., Third Generation Gangs Strategic Note No. 9, *Small Wars Journal*, 2018.

150 Ibid., p266.

151 Drago, Francesco, Galbiati, Roberto and Sobbrio, Francesco, The Political Cost of Being Soft on Crime: Evidence from a Natural Experiment, LSE, 2017.

152 Muggah, Robert, Reviewing the Costs and Benefits of Mano Dura Versus Crime Prevention in the Americas, *The Palgrave Handbook of Contemporary International Political Economy*, 2019.

153 Fraser, Alistair, Global Gangs: Street Violence Across the World, *British Journal of Criminology*, Vol. 56, 2016.

154 Economist.com, Why Prisoners Join Gangs, *Economist*, 2014.

155 Wood, Graeme, How Gangs Took Over Prisons, The Atlantic, 2014.

156 Igarapé Institute (2020) Homicide Monitor, accessed at https://homicide.igarape.org.br/.

157 Ibid., p. 266

158 Kleinfeld, Rachel, Magaloni, Beatriz and Ponce, Alejandro, Reducing Violence and Improving the Rule of Law, Carnegie Endowment for Peace, 2014; Kleinfeld, Rachel, *A Savage Order*, Penguin Random House, 2019.

159 See Earthtime, CREATE Lab, CMU, Map of Blue Helmet Deployment. Data from UN, 2019; Muggah, Robert and Tobon, Katherine A., Citizen security in Latin America: Igarape Institute, 2018.

160 Abt, Thomas, Bleeding Out: The Devastating Consequences of Urban Violence -- and a Bold New Plan for Peace in the Streets, *Basic Books*, 2019; Muggah, Robert and Pinker, Steven, We Can Make the Post-Coronavirus World a Much Less Violent Place, Foreign Policy article, 2020.

161 Muggah, Robert and Abdenur, Adriana, Conflict Prevention is Back in Vogue, and Not a Moment Too Soon, *Hill Times*, 2018; Kleinfield, 2019, op. cit., p268.

162 Collin, Katy, The Year in Failed Conflict Prevention, Brookings, 2017.

163 Rand.org, UN Nation Building Record Compares Favorably with the U.S. in Some Respects, Rand Corporation, 2005.

164 Clarke, Colin P., An Overview of Current Trends in Terrorism and Illicit Finance, Rand Corporation, 2018.

165 Ibid., p270.

166 Alvaredo, Facundo, Chancel, Lucas, Piketty, Thomas, et al., World Inequality Report 2018, World Inequality Lab, 2018.

167 Muggah, Robert and Raleigh, Clionad, Violent Disorder is on the Rise. Is Inequality to Blame?, World Economic Forum, 2019.

168 Muggah, Robert and Velshi, Ali, Religious Violence is on the Rise. What Can Faith-based Communities do About it?, World Economic Forum, 2019.

169 Paasonen, Kari and Urdal, Henrik, Youth Bulges, Exclusion and Instability: The Role of Youth in the Arab Spring, PRIO Policy Brief: Conflict Trends, 2016.

170 Dahl, Marianne, Global Women, Peace and Security Index, PRIO, 2017; Busby, Mattha, First Ever UK Unexplained Wealth Order Issued, OCCRP, 2018.

Demography

1 Ritchie, Hannah, The World Population is Changing: Population by Age Bracket Graph, Our World in Data, 2019.

2 UN.org, World Population Ageing 2017 Report, UN Department of Economic and Social Affairs, 2017; Engel, Pamela, These Staggering Maps Show How Much The World's Population Is Aging, Business Insider, 2014.

3 These arguments are reviewed in Goldin, Ian (ed.), *Is the Planet Full?*, Oxford University Press, 2014

4 Roser, Max, The Global Population Pyramid, Our World in Data, 2019.

5 Goldin, 2014, op. cit., p275.

6 Ibid., p276.

7 UN.org, Total Population by Sex (thousands) dataset, UN Department of Economic and Social Affairs, 2020.

8 Harper, Sarah, Demographic and Environmental Transitions, in Goldin, Ian (ed.), *Is the Planet Full?*, Oxford University Press, 2014.

9 Ibid., p276.

10 Ibid., p277.

11 Roser, Max, Ritchies, Hannah and Ortiz-Ospina, Esteban, World Population Growth, Our World in Data, 2019.

12 Lutz, Wolfgang, Goujon, Anne, KC, Samir, et al., Demographic and Human Capital Scenarios for the 21st Century, European Commission, 2018.

13 Roser, Max, *Future Population Growth*, Our World in Data, 2019.

14 Gallagher, James, 'Remarkable' Decline in Fertility Rates, BBC News, 2018.

15 Craig, J., Replacement Level Fertility and Future Population Growth, Population Trends, 1994.

16 Worldpopulationreview.com, Total Fertility Rate (2020), World Population Review, 2020. For a discussion on different explanations of low fertility see Harper, 2014, op. cit., p278.

17 Ibid., p278.

18 Ibid., p278.

19 Goldin, Ian, *Development: A Very Short Introduction*, Oxford University Press, 2017

20 Worldbank.org, Fertility rate, total (births per woman) data, World Bank, 2018.

21 Ibid., p279.

22 Data.worldbank.org, Fertility rate, total (births per woman), World Bank with UN Population Division, 2019.

23 Ibid., p279.

24 Worldbank.org, Fertility rate, 2018, op. cit., p279.

25 Ibid., p279.

26 Worldpopulationreview.com, 2020, op. cit., p279; Romei, Valentina, Italy Registers Lowest Number Of Births Since At Least 1861, *Financial Times*, 2019.

27 Worldbank.org, Fertility rate, 2018, op. cit., p279; Livingston, Gretchen, Is U.S. Fertility At An All-time Low? Two Of Three Measures Point To Yes, Pew Research Center, 2019.

28 Worldpopulationreview.com, 2020, op. cit., p279.

29 Ibid., p279.

30 Ibid., p279.

31 UN.org, Growing At A Slower Pace, World Population Is Expected To Reach 9.7 Billion In 2050 And Could Peak At Nearly 11 Billion Around 2100, UN Department of Economic and Social Affairs, 2019.

32 Worldpopulationreview.com, Africa Population (2020), World Population Review, 2020.

33 UN.org, Growing At A Slower Pace, 2019, op. cit., p281.

34 Worldbank.org, Life expectancy at birth, total (years) – Nigeria data, World Bank, 2019.

35 Ibid., p281.

36 Chappel, Bill, U.S. Births Dip To 30-Year Low; Fertility Rate Sinks Further Below Replacement Level, NPR, 2018.

37 Census.gov, Historical National Population Estimates: July 1, 1900 to July 1, 1999, U.S. Census Bureau, 2000.

38 UN.org, Growing At A Slower Pace, 2019, op. cit., p282.

39 Unicef.org, MENA Generation 2030, UNICEF, 2019.

40 Ourworldindata.org, Increase of life expectancy in hours per day: 2015, Our World in Data, 2017.

41 Wyss-Coray, Tony, Ageing, Neurodegeneration And Brain Rejuvenation, *Nature*, Vol. 539, 2016; royalsociety.org, The Challenge of Neurodegenerative Diseases in an Aging Population, G7 Academies' Joint Statements, 2017.

42 Goldin, Ian, Pitt, Andrew, Nabarro, Benjamin, et al., Migration And The Economy: Citi GPS, Oxford Martin School Report, 2018.

43 Bankofengland.co.uk, Procyclicality And Structural Trends In Investment Allocation By Insurance Companies And Pension Funds, Bank of England Working Paper, 2014.

44 Worldbank.org, Age Dependency Ratio (% of Working-age Population), World Bank, 2018.

45 https://data.worldbank.org/indicator/SP.POP. DPND?view=chart

46 Ibid., p284; oecd.org, Pensions at a Glance 2017, OECD and G20 Indicators at a Glance, OECD, 2017.

47 Harding, Robin, Japan's Population Decline Accelerates Despite Record Immigration, *Financial Times*, 2019.

48 Ibid., p284.

49 Ibid., p284; Worldpopulationreview.com, Total Population by Country, 2020, op. cit., p284.

50 Harding, 2019, op. cit., p284.

51 Ibid., p284.

52 Tanase, Alexandru M., Slowing Down Romania's Demographic Exodus Would be a Historic Achievement, Emerging Europe, 2019. Goldin, Pitt, Nabarro, et. al., 2018, p284.

53 Ibid., p285.

54 Roser, Max, Ortiz-Ospina, Esteban and Ritchie, Hannah, Life Expectancy, Our World in Data, 2019.

55 Roser, Max, Fertility Rate, Our World in Data, 2017.

56 Roxby, Philippa, Why Are More Boys than Girls Born Every Single Year?, BBC News, 2018.

57 Worldbank.org, The World Bank in China: Overview, World Bank, 2020; oecd.org, China: Science And Innovation: Country Notes, OECD Science, 2010.

58 See ceicdata.com, Country Profile: India, CEIC, 2020; Worldbank.org, Fertility rate, 2018, op. cit., p287.

59 Roser, Ortiz-Ospina, Ritchie, 2019, p287.

60 Clark, Peter K., Investment in the 1970s: Theory, Performance, and Prediction, Brookings Institution, 1979.

61 Roser, Ortiz-Ospina, Ritchie, 2019, p288.

62 Dimson, Elroy, Marsh, Paul and Staunton, Mike, *Credit Suisse Global Investment Returns Yearbook 2018*, Credit Suisse: Research Institute, 2018.

63 Golin, Ian and Mariathasan, Mike, *The Butterfly Defect*, Princeton University Press, 2014.

64 Norton, Robyn, Safe, Effective and Affordable Health Care for a Bulging Population, in Goldin, Ian (ed.), *Is the Planet Full?*, Oxford University Press, 2014.

Migration

1 IOM.int, World Migration Report 2020, UN International Organization for Migration, 2019.

2 UNrefugees.org, What is a Refugee?: Refugee Facts, UN High Commissioner for Refugees.

3 Muggah, Robert, A Critical Review of Displacement Regimes, in Hampson, Fen O., Ozerdem, Alpaslan, and Kent, Jonathan, *Routledge Handbook of Peace, Security and Development*, Routledge, 2020.

4 IOM.int, Migration In An Interconnected World: New Directions For Action, Report Of The Global Commission On International Migration, 2005.

5 Toronto.ca, Toronto at a Glance: Social Indicators, City of Toronto, 2020; toronto.ca, World Rankings for Toronto, City of Toronto, 2020.

6 Goldin, Ian, *Exceptional People: How Migration Shaped Our World and Will Define Our Future*, Princeton University Press, 2012.

7 Ibid., p298.

8 Hatton, Timothy, and Williamson, Jeffrey, *The Age of Mass Migration*, Oxford University Press, 1998

9 Ibid., p298.

10 Ibid., p299.

11 Ibid., p299.

12 Ibid., p299.

13 Ibid., p299.

14 Ibid., p299.

15 History.com, Chinese Exclusion Act, History article, 2018.

16 Routley, Nick, Map: All of the World's Borders by Age, Visual Capitalist, 2018.

17 Wikipedia.org, List of Sovereign States by Date of Formation, Wikipedia article, 2020.

18 OM.int, World Migration Report 2020: Chapter 2, 2019, op. cit., p300.

19 It is worth noting that the proportion of immigrants has increased only slightly over the past 100 years – up from 2.9 per cent in 1990 and 2.3 per cent in 1965. The big difference is in the direction of migration, with most migrants now moving from the south to the north in contrast to the past century.

20 Goldin, 2012, op. cit., p300.

21 OM.int, World Migration Report 2020, 2019, op. cit., p301.

22 Ibid., p302.

23 Pison, Gilles, Which Countries Have the Most Immigrants?, World Economic Forum, 2019.

24 OM.int, World Migration Report 2020, 2019, op. cit., p303.

25 Ibid., p303.

26 Wikipedia.org, Schengen Area, Wikipedia article, 2020.

27 EC.europa.eu, Migration and Migrant Population Statistics: Statistics Explained, eurostat, 2020.

28 Wilson, Francis, International Migration in Southern Africa, *The International Migration Review*, Vol. 10, 1976.

29 Goldin, Ian, Pitt, Andrew, Nabarro, Benjamin, et al., Migration And The Economy: Citi GPS, Oxford Martin School Report, 2018.

30 OECD.org, Is Migration Good for the Economy?, Migration Policy Debates: OECD, 2014.

31 Goldin, Pitt, Nabarro, et al., 2018, op. cit., p 306.

32 Ibid., p306.

33 Ibid., p306.

34 Ibid., p306.

35 Ibid., p306.

36 Ibid., p307; oecd-ilibrary.org, International Migration Outlook 2013: The fiscal impact of immigration in OECD countries, OECD, 2013.

37 Ibid., p307.

38 Ibid., p307.

39 Vargas-Silva, Carlos, The Fiscal Impact of Immigration in the UK, The Migration Observatory, 2020.

40 Storesletten, Kjetil, Fiscal Implications of Immigration – a Net Present Value Calculation, IIES Stockholm University and CEPR, 2013.

41 OECD.org, Is Migration Good for the Economy?, 2014, op. cit., p307.

42 Ibid., p307.

43 Goldin, Pitt, Nabarro, et al., 2018, op. cit., p 307.

44 See, for example, Warwick-Ching, Lucy, A Fifth of Over-45s Expect to Leave Work to Become Carers, *Financial Times*, 2019; Romei, Valentina and Staton, Bethan, How UK Social Care Crisis Is Hitting Employment Among Older Workers, *Financial Times*, 2019.

45 Ibid., p308.

46 Ibid., p308.

47 Goldin, Pitt, Nabarro, et al., 2018, op. cit., p 308.

48 Ibid., p308.

49 Ibid., p308.

50 Callahan, Logan D., Are Immigrants The Next Great Appliance?, Bard Digital Commons, 2017.

51 Ibid., p308.

52 Giuntella, Osea, Nicodemo, Catia and Vargas Silva, Carlos, The Effects of Immigration on NHS Waiting Times, Blavatnik School of Governance W.P. 5, 2015.

53 Ibid., p308.

54 Goldin, Pitt, Nabarro, et al., 2018, op. cit., p309.

55 Ibid., p309.

56 Ibid., p309.

57 Ibid., p309.

58 Ibid., p309.

59 Goldin, 2012, op. cit., p310.

60 Woetzel, Jonathan, Madgavkar, Anu, Rifai, Khaled, et al., Global Migration's Impact and Opportunity, McKinsey Global Institute Report, 2016.

61 Goldin, Pitt, Nabarro, et al., 2018, op. cit., p 310.

62 Kerr, William R. and Lincoln, William F., The Supply Side of Innovation: H-1B Visa Reforms and US Ethnic Invention, NBER W.P. 15768, 2010.

63 Ibid., p310.

64 Ibid., p310.

65 Goldin, Pitt, Nabarro, et al., 2018, op. cit., p 310.

66 Winder, Robert, Bloody Foreigners: The Story of Immigration into Britain, Little, Brown, 2004.

67 Guest, Robert, Borderless Economics, Palgrave, 2011.

68 Goldin, Pitt, Nabarro, et al., 2018, op. cit., p 310.

69 Ibid., p310.

70 Ibid., p310.

71 Ibid., p310.

72 Ibid., p310.

73 Ibid., p310.

74 Ibid., p310.

75 Ibid., p311.

76 Alesina, Alberto, Miano, Armando and Stantcheva, Stefanie, Immigration And Redistribution, NBER W.P. 24733, 2018.

77 Ibid., p312.

78 Roser, Max, Fertility Rate, Our World in Data, 2017.

79 Goldin, Pitt, Nabarro, et al., 2018, op. cit., p 313.

80 Ibid., p313.

81 Ibid., p313.

82 UNHCR.org, Figures at a Glance, UN High Commissioner for Refugees, 2020.

83 Ibid., p314; Koser, Khalid and Martin, Susan, The Migration-displacement Nexus, Studies in Forced Migration, Vol. 32, 2011.

84 UNHCR.org, Global Trends: Forced Displacement in 2017, UN High Commissioner for Refugees, 2017.

85 ESA, Zaatari Refugee Camp, Jordan, European Space Agency, 2014. ©KARI/ESA

86 UNHCR.org, Zaatari General Infrastructure Map, UN High Commissioner for Refugees, 2014.

87 Guttridge, Nick, Cologne Rapists WERE refugees: Prosecutor slams reports exonerating migrants as 'nonsense', Express, 2016.

88 Adelman, Robert, Williams, Reid, Markle, Gail, et al., Urban Crime Rates and the Changing Face of Immigration: Evidence Across Four Decades, Journal of Ethnicity in Criminal Justice, Vol. 15, 2017; Kubrin, Charis, Exploring the Connection Between Immigration and Violent Crime Rates in U.S. Cities, 1980–2000, Social Problems, Vol. 56, 2009.

89 Newamericaneconomy.org, Is there a Link Between Refugees and U.S. Crime Rates?, New American Economy: Research Fund, 2017.

90 Ibid., p318.

91 DW.com, Are refugees more criminal than the average German citizen?, Deutsche Welle, 2017.

92 Dutchnews.nl, Refugee Centres Don't Lead to Rising Crime, Dutch research shows, Dutch News, 2018.

93 Ellingsen, Nora, It's Not Foreigners Who are Plotting Here: What the Data Really Show, Lawfare Blog, 2017.

94 Carrion, Doris, Are Syrian Refugees a Security Threat to the Middle East?, Reuters, 2017; Idean, Salehyan and Skrede Gleditsch, Kristian, Refugee Flows and the Spread of Civil War, Oslo, Norway: Peace Research Institute, 2000; Muggah, Robert, No Refuge: The Crisis of Refugee Militarisation in Africa, Zed Books, 2006.

95 Mayda, Anna Maria, The Labor Market Impact of Refugees: Evidence from the U.S. Resettlement Program, US Department of State, Office of the Chief Economist W.P. 4, 2017.

96 Fakih, Ali and Ibrahim, May, The Impact of Syrian Refugees on the Labor Market in Neighboring Countries: Empirical Evidence from Jordan, IZA D.P. 9667, 2016.

97 UNESCO.org, Cities Welcoming Refugees and Migrants: Enhancing effective urban governance in an age of migration, UNESCO, 2016.

98 Of these, 4,000 slots would be reserved for Iraqis who worked with the U.S. military, 1,500 for people from Central America, 5,000 for people persecuted by religion and 7,500 slots for those seeking family unification. This is a fraction of the 110,000 the previous Obama administration said should be allowed in 2016. See, Shear, Michael D. and Kanno-Youngs, Zolan, Trump Slashes Refugee Cap to 18,000, Curtailing U.S. Role as Haven, New York Times, 2019.

99 See welcomingamerica.org, Map: Our Network: Municipalities, Welcoming America, 2020.

100 See Dinan, Stephen, Number of Sanctuary Cities Nears 500, Washington Times, 2017.

101 See eurocities.eu home webpage.

102 Muggah, Robert and Barber, Benjamin, Why Cities Rule the World, TED Ideas, 2016.

103 See uclg.org, Migration, United Cities and Local Governments article, 2020; urbancrises.org, What is different about crises in cities?, Urban Crises, 2020; Solidaritycities.eu/, website, Solidarity Cities, 2020; urban-refugees.org, The NGO network: Map, Urban Refugees, 2017.

104 This and the subsequent paragraphs on policy lessons draw heavily on pp144-148 of Goldin, Pitt, Nabarro, et al., 2018, op. cit., p324.

Food

1 Worldwildlife.org, Which Everyday Products Contain Palm Oil?, World Wide Fund for Nature, 2020.

2 Willett, Walter, Rockstrom, Johan, Loken, Brent, et al., Food in the Anthropocene: the EAT–Lancet Commission on Healthy Diets from Sustainable Food Systems, Lancet Commissions, Vol. 393, 2019.

3 See healthdata.org, Diet, IHME, 2020 and healthdata.org, Diet, Global Burden of Disease, 2020.

4 Gilbert, Natasha, One-third of our Greenhouse Gas Emissions Come From Agriculture, Nature, 2012.

5 Ibid., p330.

6 Ibid., p330.

7 Mateo-Sagasta, Javier, Zadeh, Sara M. and Turral, Hugh, Water Pollution from Agriculture: a Global Review, Food and Agriculture Organisation of UN, 2017.

8 GBD 2017 Collaborators, Health Effects of Dietary Risks in 195 countries, 1990–2017: a Systematic Analysis for the Global Burden of Disease Study 2017, *Lancet Commissions*, Vol. 393, 2019; wfp.org, Zero Hunger, World Food Programme, 2020.

9 Willett, Rockstrom, Loken, et al., 2019, op. cit., p332.

10 Ibid., p332.

11 Hamzelou, Jessica, Overeating Now Bigger Global Problem Than Lack of Food, *New Scientist*, 2012.

12 Willett, Rockstrom, Loken, et al., 2019, op. cit., p332.

13 Sen, Amartya, *Poverty and Famines*, Oxford University Press, 1981. This issue is discussed in Goldin, Ian, *Development: A Very Short Introduction*, Oxford University Press, 2018, pp42-43.

14 Tufts.edu, Famine Trends Dataset, Tables and Graphs, World Peace Foundation, 2020.

15 NYtimes.com, Famine: The Man-Made Disaster, *New York Times*, 1998.

16 Who.int, Overweight and Obesity, World Health Organization, 2020.

17 CDC.gov, The Health Effects of Overweight and Obesity, Centers for Disease Control and Prevention, 2020.

18 Wikipedia.org, Epidemiology of Obesity, Wikipedia article, 2020.

19 Renee, Janet, The Average Calorie Intake by a Human Per Day Versus the Recommendation, SF Gate, 2018.

20 Health.gov, Dietary Guidelines for Americans: 2015–2020: 8th Edition: Appendix 2. Estimated Calorie Needs per Day, by Age, Sex, and Physical Activity Level, USDA, 2015.

21 Donnelly, Laura and Scott, Patrick, Fat Britain: Average Person Eats 50pc More Calories Than They Realise, *Telegraph*, 2018.

22 Renee, 2018, op. cit., p334.

23 Donnelly and Scott, 2018, op. cit., p334.

24 Royalsociety.org, Reaping the Benefits: Science and the Sustainable Intensification of Global Agriculture, Royal Society, 2009.

25 Meadows, Donella H., Meadows, Dennis L., Randers, Jorgen, et al., *The Limits to Growth: A Report for The Club Of Rome's Project On The Predicament of Mankind*, Universe Books, 1972.

26 FAO.org, Crop Production and Natural Resource Use, Food and Agriculture Organisation of the UN, 2015.

27 FAO.org, The Future of Food and Agriculture – Trends and Challenges, Food and Agriculture Organisation of the UN, 2017.

28 Ibid., p338; Royalsociety.org,, 2009, op. cit., p338.

29 Ibid., p338.

30 IPCC.ch, Climate Change and Land: Special Report, IPCC, 2019.

31 Worldwildlife.org, Soil Erosion and Degradation, World Wide Fund for Nature: Overview, 2020.

32 IPCC.ch, Climate Change and Land, 2019, op. cit., p338.

33 Ibid., p338.

34 Piore, Adam, The American Midwest Will Feed a Warming World. But For How Long?, *MIT Technology Review*, 2019.

35 Economist.com, Warmer Temperatures Could Play Havoc with Crops, *Economist*, 2019.

36 Ibid., p339.

37 Ibid., p339.

38 UNfccc.int, Why Methane Matters, UN Climate Change, 2014.

39 IPCC.ch, Land is a Critical Resource, IPCC report says, IPPC News, 2019.

40 Royalsociety.org, 2009, op. cit., p339.

41 Notaras, Mark, Does Climate Change Cause Conflict?, Our World, 2009.

42 Fao.org, The State of Agricultural Commodity Markets 2018: Agricultural Trade, Climate Change and Food Security, Food and Agriculture Organisation of the UN, 2018.

43 Ibid., p339.

44 IPCC.ch, Climate Change and Land, 2019, op. cit., p340.

45 Springmann, Marco, Clark, Michael, Mason-D'Croz, Daniel, et al., Options for Keeping the Food System Within Environmental Limits, *Nature*, Vol. 562, 2018.

46 Globalagriculture.org, Meat and Animal Feed, Global Agriculture, 2020.

47 Economist.com, Global Meat-eating is on the Rise, Bringing Surprising Benefits, *Economist*, 2019.

48 OECD.org, Meat Consumption data, OECD, 2020.

49 Ibid., p341.

50 Ibid., p341.

51 Ritchie, Hannah and Roser, Max, Crop Yields, Our World in Data, 2017.

52 Bloomberg.com, China Ramps Up Brazil Soybean Imports, Rebuffing U.S. Crops, Bloomberg News, 2019.

53 Worldatlas.com, Top Palm Oil Producing Countries In The World, World Atlas, 2018.

54 Rainforest-rescue.org, Questions and Answers About Palm Oil, Rainforest Rescue, 2011.

55 Ibid., p344; Moss, Catriona, Peatland Loss Could Emit 2,800 Years' Worth of Carbon in an Evolutionary Eyeblink: Study, Forest News, 2015; UCSUSA.org, Palm Oil and Global Warming, Union of Concerned Scientists, 2013.

56 Rainforest-rescue.org, 2011, op. cit., p344.

57 FAO.org, The State of the World Fisheries and Aquaculture: Report 2018, Food and Agriculture Organisation of the UN, 2018.

58 This paragraph is based on Zeller, Dirk and Pauly, Daniel, Viewpoint: Back to the Future for Fisheries, Where Will We Choose to Go?, *Global Sustainability*, Vol. 2, 2019.

59 Worldview.stratfor.com, China Sets a Course for the U.S.'s Pacific Domain, Stratfor, 2019.

60 Royalsociety.org, What is ocean acidification and why does it matter?, The Royal Society, 2020; EDF.org, Overfishing: The Most Serious Threat to Our Oceans, EDF, 2020.

61 Zeller, Pauly, 2019, op. cit., p345.

62 FAO.org, The State of the World Fisheries and Aquaculture: Report 2018, Food and Agriculture Organisation of the UN, 2018.

63 Ibid., p347.

64 Tacon, Albert G.J., Hasan, Mohammad R. and Metian, Marc, Demand and Supply of Feed Ingredients for Farmed Fish and Crustaceans, Food and Agriculture Organisation of the UN, 2011.

65 FAO.org, The State of the World Fisheries and Aquaculture, 2018, op. cit., p347.

66 Ibid., p347.

67 Ceballos, Gerardo, Ehrlich, Paul R. and Dirzo, Rodolfo, Biological Annihilation via the Ongoing Sixth Mass Extinction Signaled by Vertebrate Population Losses and Declines, PNAS, Vol. 114, 2017.

68 Willett, Rockstrom, Loken, et al., 2019, op. cit., p347.

69 Willett, Rockstrom, Loken, et al., 2019, op. cit., p347.

70 WWF.org.uk, A Warning Sign From Our Planet: Nature Needs Life Support, World Wild Fund of Nature, 2018.

71 IUCN.org, IUCN Red List of Threatened Species, IUCN, 2018.

72 Ibid., p348.

73 Black, Richard, Bee Decline Linked To Falling Biodiversity, BBC News, 2010.

74 Regan, Shawn, What Happened to the 'Bee-pocalypse'?, PERC, 2019.

75 FAO.org, Declining Bee Populations Pose Threat to Global Food Security and Nutrition.

76 Including from the Highly Respects, Food and Agriculture Organization of the UN, 2019.

77 Willett, Rockstrom, Loken, et al., 2019, op. cit., p349.

78 Willett, Rockstrom, Loken, et al., 2019, op. cit., p349.

79 Willett, Rockstrom, Loken, et al., 2019, op. cit., p349.

80 Willett, Rockstrom, Loken, et al., 2019, op. cit., p349.

81 IPCC.ch, Land is a Critical Resource, 2019, op. cit., p 349.

82 Economist.com, Gloom From the Climate-change Front Line, Economist, 2019.

83 This paragraph draws on Willett, Rockstrom, Loken, et al., 2019, op. cit., p349.

Health

1 See, davos2019.earthtime.org, Life Expectancy by Country, CREATE Lab: CMU, 2020. Data sourced from UN Population Division, Life Expectancy projections, UN Department of Economic and Social Affairs: Population Division: World Population Prospects, 2019; wikipedia.org, Countries by Average Life Expectancy According to the World Health Organization, Wikipedia map, 2015.

2 UNFPA.org, Ageing in the Twenty-First Century: Chapter 1: Setting the Scene, UNFPA, 2011.

3 Dong, Xiao, Milholland, Brandon and Vijg, Jan, Evidence for a Limit to Human Lifespan, Nature, Vol. 538, 2016; Sample, Ian, Geneticists Claim Ageing Breakthrough but Immortality Will Have to Wait, Guardian, 2005.

4 Federation of American Societies for Experimental Biology, Longevity Breakthrough: Scientists 'activate' Life Extension in Worm, Discover Mitochondria's Metabolic State Controls Life Span, ScienceDaily, 2010; Chen, Alice L. , Lum, Kenneth M., Lara-Gonzalez, Pablo, et al., Pharmacological Convergence Reveals a Lipid Pathway that Regulates C. Elegans Lifespan, Nature Chemical Biology, 2019; Faloon, William, Fahy, Gregory M. and Church, George, Age-Reversal Research at Harvard Medical School, Life Extension Magazine, 2016.

5 Hughes, Bryan G. and Hekimi, Siegfried, Many Possible Maximum Lifespan Trajectories, Nature, Vol. 546, 2017.

Slagboom, Eline P., Beekman, Marian, Passtoors, Willemijn, et al., Genomics of Human Longevity, Philos Trans R Soc Lond B Biol Sci, Vol. 366, 2011; Diamandis, Peter H., Extending Human Longevity With Regenerative Medicine, Singularity Hub, 2019.

6 Slagboom, Beekman, Passtoors, et al., 2011, op. cit., p353.

7 Yet one more challenge was one's sex. Males tended to have unhealthier lifestyles and shorter lifespans. They smoked and drank, got into more accidents and tended to die in battle and crime in greater abundance.

8 Global life expectancy increased by more than 5 years between 2000 and 2016, the fastest increase since the 1960s. Those gains slowed in the 1990s due to the HIV-AIDS epidemic in Africa and Eastern Europe. The largest increase since 2000 was in Africa where life expectancy increased by more than 10 years (to 61) due to improved access to anti-retrovirals for treating HIV and also improvements in child survival, who.int, Global Health Observatory (GHO) data: Life Expectancy: Situation, WHO, 2020.

9 Worldpopulationreview.com, Life Expectancy (2020), World Population Review, 2020.

10 CDC.gov, United States Life Tables: Life Expectancy, CDC, 2017; Woolf, Steven H. and Aron, Laudan, Failing Health of the United States, BMJ, Vol. 360, 2018.

11 And the odds improve as one ages: the average 65 year old woman can expect to live another 27.5 years. Kontis, Vasilis, Bennett, James E., Mathers, Colin D., et al., Future Life Expectancy in 35 Industrialised Countries: Projections with a Bayesian Model Ensemble, Lancet, Vol. 389, 2017.

12 Preston, Samuel H., The Human Population: Chapter: Human Mortality Throughout History and Prehistory, Freeman, 1995.

13 Roser, Max, Ortiz-Ospina, Esteban and Ritchie, Hannah, Life Expectancy: Life Expectancy, 1543 to 2015 graph, Our World in Data, 2019.

14 Roser, Max, Ritchie, Hannah and Dadonaite, Bernadeta, Child and Infant Mortality: Child Mortality, 1800 to 2015 Graph, Our World in Data, 2019. Roser, Max, Ritchie, Hannah and Dadonaite, Bernadeta, Child and Infant Mortality: Child Mortality by Country: Kenya: Graph, Our World in Data, 2019.

15 Davos2019.earthtime.org, 2020, op. cit., p355; Gurven, Michael and Kaplan, Hillard, Longevity among Hunter-Gatherers: A Cross-Cultural Examination, Population and Development Review, Vol. 33, 2007.

16 Roser, Max, Ritchie, Hannah and Dadonaite, Bernadeta, Child and Infant Mortality: Child Mortality Graph, Our World in Data, 2019.

17 Alemu, Aye M., To What Extent Does Access to Improved Sanitation Explain the Observed Differences in Infant Mortality in Africa?, Afr J Prim Health Care Fam Med, Vol. 9, 2017.

18 Liu, Li, Oza, Shefali, Hogan, Daniel, et al., Global, Regional, and National Causes of Child Mortality in 2000–13, With Projections to Inform Post-2015 Priorities: an Updated Systematic Analysis, Lancet, Vol. 383, 2015.

19 See, Hill, Kenneth and Amouzou, Agbessi, Disease and Mortality in Sub-Saharan Africa. 2nd ed.: Chapter 3: Trends in Child Mortality, 1960 to 2000, The World Bank, 2006. Roser, Ritchie and Dadonaite, Child Mortality Rate: 1800–2015, 2019, op. cit., p356; Ahmad,Omar B., Lopez, Alan D., and Inoue, Mie, The Decline In Child Mortality: A Reappraisal, *Bulletin of the World Health Organization*, Vol. 78, 2000.

20 WHO.int, Health in the post-2015 Development Agenda: Need For A Social Determinants Of Health Approach, WHO, 2015.

21 Infoplease.com, Health and Social Statistics, Infant Mortality Rates of Countries, 2016.

22 Roser, Max and Ritchie, Hannah, Maternal Mortality, Our World in Data, 2013.

23 WHO.int, Maternal Mortality, WHO, 2019.

24 Roser, Max and Ritchie, Hannah, Maternal Mortality: Maternal Mortality Ratio, 2015.

25 The maternal mortality rate is 12 per 100,000 as compared to 239 per 100,000.

26 Ehling, Holger, No Condition Is Permanent: An Interview With Chinua Achebe, *Publishing Research Quarterly*, Vol. 19, 2003.

27 Who.int, Maternal Mortality, 2019, op. cit., p359.

28 McFadden, Clare and Oxenham, Marc F., The Paleodemographic Measure of Maternal Mortality and a Multifaceted Approach to Maternal Health, *Current Anthropology*, Vol. 60, 2019.

29 Aminov, Rustam I., A Brief History of the Antibiotic Era: Lessons Learned and Challenges for the Future, Frontiers in Microbiology, Vol. 1, 2010.

30 Wikipedia.org, Penicillin Was Being Mass-produced in 1944: Public Domain image, Wikipedia, 2020.

31 He described the process as 'chemotherapy', though the term only caught on decades later in the treatment of cancer.

32 Kardos, Nelson and Demain, Arnold L., Penicillin: The Medicine With the Greatest Impact on Therapeutic Outcomes, *Applied Microbiology and Biotechnolgy*, Vol. 92, 2011.

33 Aminov, 2010, op. cit., p361.

34 AMR-review.org, Antimicrobial Resistance: Tackling a Crisis for the Health and Wealth of Nations, AMR, 2014.

35 Lobanovska, Mariya and Pilla, Giulia, Penicillin's Discovery and Antibiotic Resistance: Lessons for the Future?, *Yale Journal of Biology and Medicine*, Vol. 90, 2017.

36 McKenna, Maryn, What Do We Do When Antibiotics Don't Work Anymore?, TED Talk, 2015.

37 Pearson, Carole, Antibiotic Resistance Fast-Growing Problem Worldwide, Voice of America, 2009.

38 Larson, Elaine, Community Factors in the Development of Antibiotic Resistance, Annual Review of Public Health Vol. 28, 2007.

39 Tete, Annie, Sir Alexander Fleming's Ominous Prediction, The National WWII Museum Blog, 2013.

40 Bowler, Jacinta, The CDC Is Warning About Resistant 'Nightmare Bacteria' Spreading Through The US, Science Alert, 2018; emro.who.int, What is the Difference Between Antibiotic and Antimicrobial Resistance?, WHO EMRO, 2020.

41 WHO.int, What is the Difference Between Antibiotic And Antimicrobial Resistance?, WHO, 2020.

42 BSAC.org.uk, Antimicrobial Resistance Poses 'Catastrophic Threat', says Chief Medical Officer, BSAC, 2013.

43 Telegraph.co.uk, Antibiotic Resistance 'Could Kill Humanity Before Climate Change Does', Warns England's Chief Medical Officer, *Telegraph*, 2019.

44 Hampton, Tracy, Novel Programs and Discoveries Aim to Combat Antibiotic Resistance, *JAMA*, Vol. 313, 2015.

45 Cassini, Alessandro, Högberg, Liselotte D., Plachouras, Diamantis, et al., Attributable Deaths And Disability-adjusted Life-years Caused By Infections With Antibiotic-resistant Bacteria in the EU and the European Economic Area in 2015, *Lancet*, Vol. 19, 2018.

46 AMR-review.org, 2014, op. cit., 362.

47 De Kraker, Marlieke E. A., Stewardson, Andrew J. and Harbart, Stephan, Will 10 Million People Die a Year due to Antimicrobial Resistance by 2050?, *PLoS Med*, Vol. 13, 2016.

48 Wikipedia.org, Map of Cholera Cases, Wikipedia article, 2016. Originally published in 'On the Mode of Communication of Cholera' by John Snow, in 1854 by C.F. Cheffins, London, now in public domain.

49 Koch, Tom, Visualizing Disease: Understanding epidemics through maps, ESRI, 2011.

50 Brown, Lisa, 2009 H1N1 Influenza Pandemic 10 Times More Deadly Than Previously Estimated, NACCHO, 2013.

51 NHS.uk, The History of Swine Flu, NHS, 2009.

52 Simonsen, Lone, Spreeuwenberg, Peter, Lustig, Roger, et al., Global Mortality Estimates for the 2009 Influenza Pandemic from the GLaMOR Project: Figure 5, *PLoS Med*, Vol. 10, 2013.

53 Zepeda-Lopez, Hector M., Perea-Araujo, Lizbeth, Miliar-Garcia, Angel, et al., Inside the Outbreak of the 2009 Influenza A (H1N1)v Virus in Mexico, *PLoS One*, Vol. 5, 2010; Hsieh, Ying-Hen, Ma, Stefan, Velasco Hernandez, Jorge X., Early Outbreak of 2009 Influenza A (H1N1) in Mexico Prior to Identification of pH1N1 Virus, *PLoS One*, Vol. 6, 2011.

54 Fraser, Christophe, Donnelly, Christ A., Cauchemez, Simon, et al., Pandemic Potential of a Strain of Influenza A (H1N1): Early Findings, *Science Magazine*, Vol. 324, 2009.

55 Euro.who.int, WHO Director-General declares H1N1 Pandemic Over, WHO, 2010.

56 Viboud, Cecile and Simonsen, Lone, Global Mortality of 2009 Pandemic Influenza A H1N1, *Lancet*, Vol. 12, 2012; cdc.gov, First Global Estimates of 2009 H1N1 Pandemic Mortality Released by CDC-Led Collaboration, CDC, 2012; Dawood, Fatimah S., Iuliano, A Danielle, Reed, Carrie, Estimated Global Mortality Associated with the First 12 Months of 2009 Pandemic Influenza a H1N1 Virus Circulation: A Modelling Study, *Lancet*, Vol. 12, 2012.

57 Fang, Li-Qun, Li, Xin-Lou, Liu, Kun, et al., Mapping Spread and Risk of Avian Influenza A (H7N9) in China, *Scientific Reports*, Vol. 3, 2013.

58 Ibid., p365.

59 CDC.gov, Asian Lineage Avian Influenza A(H7N9) Virus, CDC, 2018.

60 Czosnek, Hali, Predicting the Next Global Pandemic, Global Risks Insights, 2018.

61 Butler, Declan, Mapping the H7N9 avian flu outbreaks, *Nature*, 2013.

62 Ibid., p366.

63 Smith, Katherine F., Goldberg, Michael, Rosenthal, Samantha, et al., Global Rise In Human Infectious Disease Outbreaks, *Journal of the Royal Society Interface*, Vol. 11, 2014.

64 Declan, 2013, op. cit., p365.

65 Butler, 2013, op. cit., p365; Tatem, Andrew J., Huang, Zhuojie and Hay, Simon I., Spread of H7N9, Unpublished data, 2013.

66 Pathologyinpractice.com, Measles: The Importance of Vaccination, Disease Monitoring and Surveillance, Pathology in Practice, 2018.

67 Muggah, Robert, Pandemics Are the World's Silent Killers. We Need New Ways to Contain Them, Devex, 2019.

68 Ourworldindata.org, Share of population with mental health and substance use disorders Map, Our World in Data, 2017.

69 Ourworldindata.org, Number of people with mental and substance use disorders, World, 1990 to 2017, Our World in Data, 2017.

70 The latest estimate from the Institute for Health Metrics Evaluation suggests a much lower figure – about 971 million people – around 13 per cent of the global population – that suffer from some kind of mental disorder. This includes 300 million people experiencing anxiety, 160 million affected by depressive disorders and another 100 million from milder forms of depression known as dysthymia. See Rice-Oxley, Mark, Mental illness: is there really a global epidemic?, *Guardian*, 2019.

71 Who.int, Mental Disorders, WHO, 2019.

72 Ourworldindata.org, Share of the Population with Depression by Average Country Income, Our World in Data, 2017.

73 Who.int, Age-standardised Suicide Rates Both Sexes, WHO, 2018.

74 Trautmann, Sebastian, Rehm, Jürgen and Wittchen, Hans-Ulrich, The economic Costs of Mental Disorders, *EMBO Reports*, Vol. 17, 2016.

75 McManus, Sally, Meltzer, Howard, Brugha, Traolach, et al. Adult Psychiatric Morbidity in England, 2007: Results of a Household Survey, NHS, 2009.

76 Who.int, Breaking the Vicious Cycle Between Mental Ill-health and Poverty, WHO, 2007.

77 Who.imt, Mental Disorders, op. cit., p368.

78 Rice-Oxley, 2019, op. cit., p368.

79 Thelancet.com, The Lancet Commission on Global Mental Health and Sustainable Development, *Lancet*, 2018.

80 Ferrari, Alize J., Norman, Rosana E., Freedman, Greg, et al., The Burden Attributable to Mental and Substance Use Disorders as Risk Factors for Suicide: Findings from the Global Burden of Disease Study 2010. *PLoS One*, Vol. 9, 2014.

81 The WHO's International Classification of Diseases (ICD-10) defines this set of disorders ranging from mild to moderate to severe. The IHME adopts such definitions by disaggregating to mild, persistent depression (dysthymia) and major depressive disorder (severe).

82 Ferrari, Norman, Freedman, et al., 2014, op. cit., p368.

83 Khazaei, Salman, Armanmehr, Vajihe, Nematollahi, Shahrzad, et al., Suicide Rate in Relation to the Human Development Index and Other Health Related Factors: A Global Ecological Study From 91 Countries, *Journal of Epidemiology and Global Health*, Vol. 7, 2017.

84 Ferrari, Norman, Freedman, et al., 2014, op. cit., p369.

85 Worldpopulationreview.com, Suicide Rate by Country (2020), World Population Review, 2020.

86 Who.int, Mental Health: Suicide data, WHO, 2020.

87 Bantjes, Jason, Iemmi, Valentina, Coast, Ernestina, et al., Poverty and Suicide Research in Low- and Middle-income Countries: Systematic Mapping of Literature published in English and a Proposed Research Agenda, *Global Mental Health*, Vol. 3, 2016.

88 Prasad, Ritu, Why US Suicide Rate is On the Rise, BBC News, 2018.

89 Case, Anne and Deaton, Angus, Rising Morbidity and Mortality In Midlife Among White Non-Hispanic Americans in The 21st Century, PNAS Vol. 112, 2015; Case, Anne and Deaton, Angus, Mortality and Morbidity in the 21st Century, Brookings Papers on Economic Activity, 2017.

90 Oi, Mariko, Tackling the Deadliest Day for Japanese Teenagers, BBC News, 2015.

91 Helpage.org, Global Age Watch Index 2015: Population Ageing Maps, Global Age Watch, 2015.

92 UN.org, World Population Prospects: Key Findings and Advance Tables W.P. 248, UN DESA, 2017.

93 who.int, Fact Sheet: Ageing and Health, WHO, 2018.

94 In upper-income settings, women have a low fertility rate of just 1.6 while in parts of Africa it is still upward of 5. Today, more than 50 per cent of the global population live in a country with fertility rates below the replacement rate. This has created a phenomenon known as 'sub replacement fertility' which means that wealthier low fertility countries are not just ageing, but they are shrinking. UN.org, Global Issues: Ageing, UN, 2019.

95 Haseltine, William A., Why Our World Is Aging, *Forbes*, 2018.

96 Magnus, George, *The Age of Aging*, Wiley, 2008.

97 Population.un.org, Average Annual Rate of Population, Population Change 2020 – 2025: Maps, UN DESA, 2019.

98 Data on total fertility is obtained from three sources: civil registration systems, sample surveys and censuses.

99 Theguardian.com, Russian Men Losing Years to Vodka, *Guardian*, 2014.

100 Bernstein, Lenny, U.S. Life Expectancy Declines Again, a Dismal Trend Not Seen Since World War 1, *Washington Post*, 2018.

101 CGDEV.org, Zimbabweans Have Shortest Life Expectancy, Center for Global Development, 2006.

102 Bor, Jacob, Herbst, Abraham J., Newell, Marie-Louise, et al., Increases in Adult Life Expectancy in Rural South Africa: Valuing the Scale-up of HIV Treatment, *Science*, Vol. 339, 2013.

103 Ibid., p375.

104 Siddique, Haroon, Life Expectancy in Syria Fell by Six Years at Start of Civil War, *Guardian*, 2016.

105 Muggah, Robert, How to Protect Fast Growing Cities from Failing, TED Global, 2014.

106 Haseltine, William A., 2018, op. cit., p375.

107 The projections for diabetes are based on the estimated prevalence of obesity and overweightness, and for road traffic injuries, on the estimated number of cars on the road assuming current economic trends. Who.int, Projections of mortality and causes of death: 2016 to 2060, WHO, 2016.

108 Who.int, Fact Sheets: Malaria, WHO, 2020.

109 There was also a laboratory accident in 1978 in the UK where two people were infected and one person died.

110 PBS.org, Stamping Out Smallpox is Just One Chapter of His Brilliant Life Story, PBS News, 2017.

111 Ourworldindata.org, Number of Reported Smallpox Cases: 1943, Our World in Data, 2018.

112 Ourworldindata.org, Decade in Which Smallpox Ceased to be Endemic by Country, Our World in Data, 2018.

113 Loria, Kevin, Bill Gates Revealed a Scary Simulation that Shows how a Deadly Flu Could Kill More Than 30 Million People Within 6 Months, Business Insider, 2018.

114 WHO.int, Blueprint: Prioritizing Diseases for Research and Development in Emergency Contexts, WHO, 2018.

115 Branswell, Helen, The Data are Clear: Ebola Vaccine Shows 'Very Impressive' Performance in Outbreak, STAT, 2019.

116 Scott, Clare, New 3D Printing Method Combines Multiple Vaccines into One Shot, 2017.

117 Muggah, 2019, op. cit., p378.

118 Tekin, Elif, White, Cynthia, Manzhu Kang, Tina, et al., Prevalence and Patterns of Higher-order Drug Interactions in Escherichia coli, *npj Systems Biology and Applications*, Vol. 4, 2018.

119 Cassella, Carly, Experimental Antibiotic Gives New Hope Against Superbugs in Clinical Trials, Science Alert, 2018; Dockrill, Peter, We Just Found a Game-Changing Weapon Against Drug-Resistant Superbugs, Science Alert, 2018.

120 Gazis, Olivia, Author Jamie Metzl Says the 'Genetic Revolution' Could Threaten National Security, CBS, 2019; Beyret, Ergin , Liao, Hsin-Kai, Yamamoto, Mako, et al., Single-dose CRISPR–Cas9 Therapy Extends Lifespan of Mice with Hutchinson–Gilford Progeria Syndrome. *Nature Medicine*, Vol. 25, 2019; University of Rochester, 'Longevity Gene' Responsible for More Efficient DNA Repair.' ScienceDaily, 2019; Wray, Britt, The Ambitious Quest to Cure Ageing Like a Disease, BBC: Ageing, 2018.

Education

1 Thygesen, Tine, The One Thing You Need To Teach Your Children To Future-Proof Their Success, *Forbes*, 2016.

2 WEForum.org, White Paper: Realizing Human Potential in the Fourth Industrial Revolution, World Economic Forum, 2017.

3 Hanushek, Eric A. and Woessmann, Ludger, Education and Economic Growth, *Economics of Education*, 2010.

4 UNESCO Institute for Statistics, A Growing Number Of Children And Adolescents Are Out of School As Aid Fails to Meet the Mark, UNESDOC Policy Paper 32, 2015.

5 UNESCO.org, Themes: Literacy, UNESCO, 2020.

6 Lee, Jong-Wha and Lee, Hanoi, Human Capital in the Long Run, *Journal of Development Economics*, Vol. 122, 2016.

7 Hao, Karen, China Has Started a Grand Experiment in AI Education. It Could Reshape How the World Learns, *MIT Technology Review*, 2019.

8 Roser, Max and Ortez-Ospina, Esteban, Literacy: Literacy rate, 1475 to 2015: Graph, Our World in Data, 2018.

9 Roser, Max and Ortez-Ospina, Esteban, Literacy: Literate and Illiterate World Population: Graph, Our World in Data, 2018.https://ourworldindata.org/grapher/literate-and-illiterate-world-population

10 Among the first records of dye being applied to papyrus are from around 3,500 BC. The Phoenician writing system emerged soon thereafter. And from there, the script was adapted by the Greeks, which later resulted in the Latin and Cyrillic alphabets, among others. In China, written characters were etched in bone around 1,400 BC during the Shang Dynasty and soon spread.

11 Buringh, Eltjo and Van Zanden, Jan Luiten, Charting the 'Rise of the West': Manuscripts and Printed Books in Europe, A Long-term Perspective From the Sixth Through Eighteenth Centuries, *Journal of Economic History*, Vol. 69, 2009.

12 Van Zanden, Jan Luiten, Baten, Joerg, Mira d'Ercole, Marco, et al., How Was Life?: Global Well-being Since 1820, OECD Publishing, 2014.

13 Roser, Max and Ortez-Ospina, Esteban, Literacy: Population Having Attained at Least Basic Education by Region, 1870-2010, Our World in Data, 2013. Based on Van Zanden, Baten, Mira, et al., 2014, op. cit., p389.

14 OHCHR.org, Adopted and Opened for Signature, Ratification and Accession by General Assembly Resolution 2200A (XXI) of 16 December 1966, International Covenant on Economic, Social and Cultural Rights, 1966.

15 Roser, Max and Ortez-Ospina, Esteban, Global Education, Our World in Data, 2016.

16 Roser, Max and Ortez-Ospina, Esteban, Global Education: Mean years of schooling, 2017, Our World in Data, 2018.

17 Byun, Soo-yong and Park, Hyunjoon, When Different Types of Education Matter: Effectively Maintained Inequality of Educational Opportunity in South Korea, *American Behavioral Scientist*, Vol. 61, 2017.

18 Sistek, Hanna, South Korean Students Wracked with Stress, Aljazeera, 2013.

19 CWUR.org, World University Rankings 2018-19, CWUR, 2019.

20 ICEF.com, South Korea: Record Growth in International Student Enrolment, ICEF Monitor, 2018.

21 Roser, Max and Ortez-Ospina, Esteban, Primary and Secondary Education: Share of the Population with No Formal Education, Projections by IIASA: 1970, 2050.

22 Roser, Max and Ortez-Ospina, Esteban, Global Education, 2016, op. cit., p392.

23 Roser, Max and Ortez-Ospina, Esteban, Government Spending: Share of Government Expenditure Spent on Education: 2016, Our World in Data, 2016.

24 The International Commission on Financing Global Education Opportunity, The Learning Generation, 2018.

25 Worldbank, Learning to Realize Education's Promise, World Development Report, 2018

26 Evans, David, Education Spending and Student Learning Outcomes, Development Impact, 2019.

27 Morais de Sa e Silva, Michelle, Conditional Cash Transfers and Improved Education Quality: A Political Search for the Policy Link, *International Journal for Educational Development*, Vol. 45, 2018; Bertrand, Marianne, Barrera-Osario, Felipe, Linden, Leigh, et al., Improving the Design of Conditional Transfer Programs: Evidence from a Randomized Education Experiment in Colombia, *American Economic Journal: Applied Economics*, Vol. 3, 2011.

28 Patrinos, Harry A. And Psacharopoulos, George, Strong Link Between Education and Earnings, World Bank Blogs, 2018.

29 Patrinos, Harry A. and Psacharopoulos, George, Returns to Investment in Education: A Decennial Review of the Global Literature, World Bank: Education Global Practice W.P. 8402, 2018; and associated data file for Annex 2.

30 Blundell, Richard, Costa Dias, Monica, Meghir, Costas, et al., Female Labor Supply, Human Capital, and Welfare Reform, *Econometrica*, Vol. 84, 2016.

31 Patrinos, and Psacharopoulos, 2018, op. cit., p395.

32 Lutz, Wolfgang, Crespo Cuaresma, Jesus and Sanderson, Warren, et al., The Demography of Educational Attainment and Economic Growth, *Science*, Vol. 319, 2008. Krueger, Alan B. and Lindahl, Mikael, Education for Growth: Why and For Whom?, NBER W.P. 7591, 2000.

33 Hanushek, Eric A. and Woessmann, Ludger, *Universal Basic Skills: What Countries Stand to Gain*, OECD Publishing, 2015.

34 Valero, Anna and Van Reenen, John, The Economic Impact of Universities: Evidence from Across the Globe, NBER W.P. 22501, 2016.

35 Easterly, William, *The Elusive Quest for Growth: Economists' Adventures and Misadventures in the Tropic*, MIT Press, 2001.

36 Hanushek, Eric A. and Woessmann, Ludger, Do Better Schools Lead to More Growth? Cognitive Skills, Economic Outcomes, and Causation, NBER W.P. 14633, 2009.

37 Hanushek, Eric A. and Woessmann, Ludger, Schooling, Cognitive Skills, and the Latin American Growth Puzzle, NBER W.P. 15066, 2009.

38 OECD-ilibrary.org, Education at a Glance 2015: OECD Indicators, 2015.

39 Browne, Angela W., and Barrett, Hazel R., Female Education in Sub-Saharan Africa: The Key to Development?, *Comparative Education*, Vol. 27, 1991; Fishetti, Mark, Female Education Reduces Infant and Childhood Deaths, *Scientific American: Health*, 2011.

40 Glaeser, Edward L., Ponzetto, Giacomo A. M. and Shleifer, Andrei, Why Does Democracy Need Education?, *Journal of Ecomomic Growth*, Vol. 12, 2007; Lutz, Wolfgang, Crespo Cuaresma, Jesus and Abbasi-Shavazi, Mohammad Jalal, Demography, Education, and Democracy: Global Trends and the Case of Iran, *Population and Development Review*, Vol. 36, 2010.

41 Glaeser, Edward A., Want a Stronger Democracy? Invest in Education, *New York Times*, 2009.

42 Lutz, Crespo Cuaresma and Abbasi-Shavazi, 2010, op. cit., p396.

43 Norrlof, Carla, Educate to Liberate: Open Societies Need Open Minds, Foreign Affairs article, 2019.

44 Drutman, Lee, Diamond, Larry and Goldman, Joe, Follow the Leader: Exploring American Support for Democracy and Authoritarianism, Democracy Fund: Voter Study Group, 2018.

45 Roser, Max and Ortiz-Ospina, Esteban, Population Breakdown by Highest Level of Education Achieved for Those Aged 15+, Brazil, 1970 to 2050 Chart, Our World in Data, 2016.

46 Patrinos, Harry A., The Skills that Matter in the Race Between Education and Technology, World Bank Blogs, 2017.

47 OECD.org, *The Future of Education of Skills: Education 2030*, OECD Publishing, 2018.

48 Russell, Stuart, *Human Compatible: Artificial Intelligence and the Problem of Control*, Penguin, 2019.

49 Weforum.org, System Initiative: Shaping the Future of Education, Gender and Work, World Economic Forum, 2018.

50 Saavedra, Jaime, Alasuutari, Hanna and Gutierrez, Marcela, Finland's Education System: The Journey to Success, World Bank Blog, 2018.

51 OPH.fi, Finnish Education in a Nutshell, Finnish National Agency for Education, 2017.

52 NCEE.org, Finland: Teacher and Principal Quality, Center on International Benchmarking, 2018.

53 Hancock, LynNell, Why Are Finland's Schools Successful?, *Smithsonian Magazine*, 2011.

54 Factsmaps.com, PISA Worldwide Rankings, Factmaps.com sourced from OECD, 2016.

55 LynNell, 2011, op. cit., p399.

56 Madden, Duncan, Ranked: The 10 Happiest Countries In The World In 2019, *Forbes*, 2019.

57 THL.fi, Alcohol Consumption in Finland has Decreased, But Over Half a Million are Still at Risk From Excessive Drinking, Finnish Institute for Health and Welfare, 2018.

58 Kingsley, Sam, Finland: From Suicide Hotspot to World's Happiest Country, *Jakarta Post*, 2019.

59 Constituteproject.org, Finland's Constitution of 1999 (with amendments through 2011), Constitution Project, 2011.

60 OPH.fi, Education System: Finnish Education System, Finnish National Agency for Education, 2020.

61 Economist.com, It Has the World's Best schools, But Singapore Wants Better, *Economist*, 2018.

62 Hogan, David, Why is Singapore's School System so Successful, and Is It a Model for the West?, The Conversation, 2014.

63 Including the Primary School Leaving Examination which shapes their educational pathway.

64 Liew, Maegan, The Singaporean Education System's Greatest Asset is Becoming its Biggest Weakness, Asean Today, 2019.

65 Smartnation.sg, Smart Nation: The Way Forward, Smart Nation and Digital, 2018.

66 Keating, Sarah, The Most Ambitious Country in the World, BBC News, 2018.

67 Ourworldindata.org, Average Learning Outcomes vs GDP per Capita, 2015, Our World in Data, 2015.

68 Gatesfoundation.org, Education Research and Development: Learning From the Field, Gates Foundation, 2019.

69 Houser, Kristen, China's AI Teachers Could Revolutionize Education Worldwide, Futurism, 2019.

70 Holoniq.com, Global Education Technology Market to Reach $341B by 2025, Holon IQ, 2018.

71 Ritchie, Hannah and Roser, Max, Share of the population using the Internet, 2015, Our World in Data, 2017.

72 Esposito, Mark, This is How New Technologies Could Improve Education Forever, World Economic Forum, 2018.

73 Lynch, Matthew, What Is the Future of STEM Education?, Education Week, 2018.

74 Siliconrepublic.com, What's Driving STEM Education? Emerging Trends on the Road Ahead, Silicon Republic, 2017.

75 Reynard, Ruth, Technology and the Future of Online Learning, Campus Technology, 2017; Kak, Subbash, Universities Must Prepare for a Technology-enabled Future, The Conversation, 2018.

76 Kaplan, Andreas M. and Haenlein, Michael, Higher Education and the Digital Revolution: About MOOCs, SPOCs, Social Media, and the Cookie Monster, *Business Horizons*, Vol. 59, 2016.

77 Reich, Justin and Ruiperez-Valiente, Jose A., The MOOC Pivot, *Science*, Vol. 363, 2019.

78 McCubbin, James, 4 Predictions for the Future of Technology in Education, Campus Blog, 2018.

79 Lederman, Doug, 'Clay Christensen, Doubling Down', Inside Hired, 2017.

80 Tanzi, Alexandre, U.S. Student Loan Debt Sets Record, Doubling Since Recession, Bloomberg article, 2018.

81 Economist.com, A Booming Population is Putting Strain on Africa's Universities, *Economist*, 2019.

82 Johnson, Theodore R., Did I Really Go to Harvard If I Got My Degree Taking Online Classes?, *Atlantic*, 2013.

83 See, Earthtime, CREATE Lab, CMU, Nobel Prize Winners sourced from worldmapper.org.

84 Economist.com, A Booming Population is Putting Strain on Africa's Universities, 2019, op. cit., p408.

85 McLeod, Scott and Fisch, Karl, Chapter 1: The Future of Jobs and Skills, World Economic Forum, 2018.

86 Pinder, Reuben, Fancy life As A Human Body Designer Or Rewilding Strategist? These 10 Creepy-sounding Job Titles Will Exist by 2025, Cityam, 2016.

87 Bernard, Zoe, Here's How Technology is Shaping the Future of Education, Business Insider, 2017.

Culture

1 Mendelssohn, Scott, 'Avengers: Endgame' Tops 'Avatar' At Worldwide Box Office, *Forbes*, 2019.

2 Rosenberg, Matt, Number of McDonald's Restaurants Worldwide, Thought Co., 2020.

3 BBC.com, Bollywood's Expanding Reach, BBC News, 2012.

4 Zhou, Zier, The global influence of K-pop, *Queen's Journal*, 2019.

5 Jpninfo.com, How Has Japanese Anime Influenced the World?, Japan Info, 2015.

6 Tsunagujapan.com, 50 McDonald's Menu Items Only in Japan, Tsunagu Japan, 2014.

7 Prickman, Greg, 'The Atlas of Early Printing', University of Iowa Libraries, 2008. Retrieved from atlas.lib.uiowa.edu; Wikipedia.org, 15th Century Printing Towns of Incunabula Map, Wikipedia article, 2011. Based on Incunabula Short Title Catalogue of the British Library.

8 And while the Chinese had invented a printing technique almost one thousand years earlier, it was cumbersome and didn't catch on. Dewar, James A., The Information Age and the Printing Press, CA: RAND Corporation, 1998.

9 Dittmar, Jeremiah, Information Technology and Economic Change, 2011, op. cit., p416.

10 Lua, Alfred, 21 Top Social Media Sites to Consider for Your Brand, Buffer, 2019.

11 Statista.com, Global Digital Population as of April 2020, Statista, 2020; Wearesocial.com, Global Digital Report 2019, WeAreSocial Report, 2019.

12 Lua Alfred, 21 Top Social Media Sites to Consider for Your Brand, Buffer, 2019.

13 Ourworldindata.org, Number of Internet Users by Country: 1990, Our World in Data, 2017.

14 Barber, William L. and Badre, Albert N., Culturability: The merging of culture and usability. Presented at the Conference on Human Factors and the Web, Basking Ridge, New Jersey: AT&T Labs, 1998.

15 Spencer-Oatey, Helen, What is Culture? A Compilation of Quotations, GlobalPAD Open House, 2012.

16 CARLA.umn.edu, What is Culture?, CARLA University of Minnesota, 2019.

17 Pagel, Mark, Does Globalization Mean We Will Become One Culture?, BBC: Future, 2014.

18 BBC.com, Islamic State and the Crisis in Iraq and Syria in Maps, BBC News, 2018.

19 Travis, Clark, Netflix Quietly Debuted Sci-fi Movie 'the Wandering Earth', The Second-biggest Chinese Blockbuster Of All Time, Business Insider, 2019; Sharf, Zack, What If Netflix Released a $700 Million Blockbuster and No one Noticed? Oh Wait, It Just Did, Indie Wire, 2019.

20 See Geoawesomeness.com, Netflix Expansion gif, Geo Awesomeness, 2018.

21 Rice, Emma S., Haynes, Emma and Royce, Paul, Social Media and Digital Technology use Among Indigenous Young People in Australia: A Literature Review, *International Journal for Equity in Health*, 2016.

22 Carlson, Bronwyn L., Farrelly, Terri, Frazer, Ryan, et al., Mediating Tragedy: Facebook, Aboriginal Peoples and

Suicide, *Australian Journal of Information Systems*, Vol. 19, 2015.

23 Dunklin, A. L., Globalization: A Portrait of Exploitation, Inequality and Limit, 2005.

24 Gov.uk, Policy Paper: Culture is Digital, Department for Digital, Culture, Media & Sport, 2019.

25 Ward, Peter, Using Today's Technology to Preserve the Past, Culture Trip, 2018.

26 Adzaho, Gameli, Can Technology Help Preserve Elements of Culture in the Digital Age?, Diplo, 2013.

27 See catawbaarchives.libraries.wsu.edu resource.

28 See plateauportal.libraries.wsu.edu resource.

29 See passamaquoddypeople.com resource.

30 See guides.library.ubc.ca, Indigenous Librarianship, University of British Columbia, 2020.

31 See waiata.maori.nz songs resource.

32 Manish, Singh, Global Video Streaming Market is Largely Controlled by the Usual Suspects, Venture Beat, 2019.

33 Google, See what was trending in 2018 – Global, Google, 2018.

34 Kovalchick, Shae, The Spread of the English Language, Geo 106 Human Geography, 2013; Beauchamp, Zack, The Amazing Diversity of Languages Around the World, in One Map, Vox 2015.

35 Eberhard, David M., Simons, Gary F. and Fennig, Charles D., Ethnologue: Languages of the World. Twenty-second edition. Dallas, Texas: SIL International, 2019.

36 Kull, Steven, Culture Wars? How Americans and Europeans View Globalization, Brookings, 2001.

37 Mikanowski, Jacob, Behemoth, Bully, Thief: How the English Language is Taking Over the Planet, *Guardian*, 2018.

38 Noack, Rick, The Future of Language, *Washington Post*, 2015.

39 Mikanowski, 2018, op. cit., p423.

40 Noack, 2015, op. cit., p423.

41 Gobry, Pascal-Emmanuel, Want To Know The Language Of The Future? The Data Suggests It Could Be...French, *Forbes*, 2014.

42 Noack, Rick and Gamio, Lazaro, The World's Languages in 7 Maps and Charts, *Washington Post*, 2015. See map by endangeredlanguages.com, Endangered Languages Project (ELP), 2020.

43 The Catalogue of Endangered Languages. The Catalogue of Endangered Languages is under the direction of Lyle Campbell (University of Hawai'I Mānoa) and Anthony Aristar and Helen Aristar-Dry (LINGUIST List/Eastern Michigan University).

44 Eberhard, Simons, and Fennig, 2019, op. cit., p423.

45 Noack and Gamio, 2015, op. cit. p424.

46 See wikitongues.org; Smithsonian: National Museum of Natural History, Recovering Voices, Smithsonian, 2020.

47 Arnold, Carrie, Can an App Save an Ancient Language?, *Scientific American*, 2016.

48 Endangeredlanguages.com, 2020, op. cit., p425.

49 UNESCO.org, UNESCO Atlas of the World's Languages in Danger, UNESCO, 2010.

50 Mikanowski, 2018, op. cit., p425.

51 Strochlic, Nina, The Race to Save the World's Disappearing Languages, *National Geographic*, 2018.

52 Ben, Reynolds, Starbucks: Aggressive Global Expansion Means Growth Percolating, NewsMax, 2018.

53 Ibid., op. cit., p425.

54 Knoema.com, Number of Starbucks Stores Globally: 1992 – 2019, KNOEMA, 2020.

55 Statista.com, Number of Starbucks Stores Worldwide From 2003 to 2019, Statista, 2020.

56 Yanofsky, David, A Cartographic Guide to Starbucks' Global Domination, Quartz, 2014.

57 Statista.com, Cities with the Largest Number of Starbucks in the United States as of April 2019, Statista, 2019.

58 Flanagan, Jack, How Starbucks Adapts to Local Tastes When Going Abroad, Real Business, 2014.

59 Willick, Jason, The Man Who Discovered 'Culture Wars', *Wall Street Journal*, 2018.

60 Heyrman, Christine L., The Separation of Church and State from the American Revolution to the Early Republic, National Humanities Center, 2008.

61 For example, a bitterly contested state law passed in Wisconsin in 1889 required the use of English to teach key subjects in public and private schools.

62 Castle, Jeremiah, New Fronts in the Culture Wars? Religion, Partisanship, and Polarization on Religious Liberty and Transgender Rights in the United States, *American Politics Research*, Vol. 47, 2019.

63 McCarthy, Justin, U.S. Support for Gay Marriage Edges to New High, Gallup, 2017.

64 Even in 2004, weeks before Massachusetts became the first state to legalise same-sex marriage, just 42 per cent were supportive.

65 Ibid., p430.

66 Ibid., p430.

67 Lewis, Helen, Culture Wars Cross The Atlantic To Coarsen British Politics, *Financial Times*, 2018.

68 Newton, Casey, Far-right Extremists Keep Evading Social Media Bans, The Verge, 2019.

69 McDonell, Stephen, China Social Media: WeChat and the Surveillance State, BBC News, 2019.

70 Matsa, Katerina E., Fewer Americans Rely on TV News; What Type They Watch Varies by Who They Are, Pew Research Center, 2018.

71 Thehindu.com, For Indians, Smartphone is Primary Source of News, *Hindu*, 2019.

72 Wakefield, Jake, Social Media 'Outstrips TV' as News Source for Young People, BBC News, 2016.

73 Rainie, Lee, Americans' Complicated Feelings About Social Media in an Era of Privacy Concerns, Pew Research Center, 2018.

74 Scher, Bill, The Culture War President, Polito, 2017.

75 Borger, Julian, Trump Urges World to Reject Globalism in UN Speech that Draws Mocking Laughter, *Guardian*, 2018.

76 NPR.org, A Lavish Bollywood Musical Is Fueling A Culture War In India, NPR, 2018.

77 Ma Damien, Beijing's Culture War Isn't About the US It's About China's Future, *Atlantic*, 2012.

78 Lewis, Martin W., Mapping Evangelical Christian Missionary Efforts, GeoCurrents, 2013.

79 Legacy.joshuaproject.net, Evangelical Growth Rate Map, Joshua Project, 2010.

80 Van Herpen, Marcel H., *Putin's Propaganda Machine*, London: Rowman and Littlefield, 2016.

81 Haaretz.com, How Steven Bannon's 'The Movement' Is Uniting the Far Right in Europe, Haaretz, 2018.

82 BBC.com, Europe and Right-wing Nationalism: A Country-by-country Guide, BBC News, 2019.

83 BBC.com, The Movement: Steve Bannon Role in 2019 EU Elections, BBC Video Report, 2019.

84 Coman, Julian, Marine Le Pen and Emmanuel Macron Face Off for the Soul of France, *Guardian*, 2017.

85 Titley, Gavan, Pork is the Latest Front in Europe's Culture Wars, *Guardian*, 2014.

86 Chazan, Guy, Germany's Increasingly Bold Nationalists Spark A New Culture War, *Financial Times*, 2018.

87 Angelos, James, The Prophet of Germany's New Right, *New York Times*, 2017.

88 Meeus, Tom-Jan, The Wilders Effect, Politico, 2017.

89 Reuters.com, China Aims to 'optimize' Spread of Controversial Confucius Institutes, Reuters, 2019; Flew, Terry, Entertainment Media, Cultural Power, And Post-globalization: The Case of China's International Media Expansion and the Discourse of Soft Power, Pew Research Center, 2016.

90 Rosen, Stanley, Berry, Michael, Cai, Jindong, et al., Xi Jinping's Culture Wars, China File, 2014.

91 Wong, Edward, China's President Lashes Out at Western Culture, *New York Times*, 2012; Wong, Edward, Pushing China's Limits on Web, if Not on Paper, *New York Times*, 2011.

92 Osnos, Evan, China's Culture Wars, *New Yorker*, 2012.

93 Britishcouncil.org, International Development: Regions, British Council, 2020.

94 Washingtonpost.com, Why US Universities Have Shut Down China-funded Confucius Institutes, *Washington Post*, 2019.

95 McDonald, Alistair, *Soft Power Superpowers: Global Trends in Cultural Engagement and Influence*, London: British Council, 2018.

96 Kreko, Peter, Gyori, Lorant and Dunajeva, Katya, Russia is Weaponizing Culture in CEE by Creating a Traditionalist 'Counter-culture', StopFake.org, 2016.

97 Ibid., p437.

98 Ibid., p437.

Conclusion

1 See ClimateReanalyzer.org, Climate Change Institute, University of Maine, 2020.

2 Niler, Eric, An AI Epidemiologist Sent the First Warnings of the Wuhan Virus, Wired, 2020.

3 See Coronavirus COVID-18 Global Cases by Johns Hopkins CSSE, Data Visualization, 2020.

4 See ESRI, esri.com, The Power of the Map and Planet Lab at planet.com.

5 Nagaraj, Abhishek and Stern, Scott, The economics of Maps, *Journal of Economic Perspectives*, Vol. 34, 2020.

6 Pinker, Steven, *Enlightenment Now*, Penguin, 2019.

7 Muggah, Robert and Wabha, Sameh, How Reducing Inequalities Will Make Our Cities Safer, World Bank, 2020.

8 IPBES.net, Global Assessment, IPBES Media Release, 2019.

9 Rees, Martin, *Our Final Century*, Heinemann, 2004.

10 Grose, Anouchka, How the Climate Emergency Could Lead to a Mental Health Crisis, *Guardian*, 2019.

11 Castelloe, Molly, Coming to Terms With Ecoanxiety, *Psychology Today*, 2019.

12 Goldin, Ian and Mariathasan, Mike, *The Butterfly Defect*, Princeton University Press, 2014.

13 Nyabiage, Jevans, US-China Trade War Hits Africa's Cobalt and Copper Mines with 4,400 Jobs Expected to Vanish, *South China Morning Post*, 2019.

14 Muggah, Robert, It Isn't Too Late to Save the Brazilian Rainforest, Foreign Policy article, 2019.

15 See WHO website at who.int, Fact Sheets: Measles.

16 Goldin, Ian, *Divided Nations: Why Global Governance is Failing, and What We Can Do About It*, Oxford University Press, 2013.

17 Muggah, Robert and Kosslyn, Justin, Why the Cities of the Future are 'Cellular', World Economic Forum Agenda, 2019.

18 Goldin, Ian and Kutarna, Chris, *Age of Discovery: Navigating the Storms of Our Second Renaissance*, Bloomsbury, 2017.

19 Oxfordmartin.ox.ac.uk, Now for the Long Term: The Report of the Oxford Martin Commission for Future Generations, Oxford Martin School, 2013.

20 Energyatlas.iea.org, CO2 Emissions from Fuel Combustion: Map, IEA Atlas of Energy, 2020.

21 Ucsusa.org, Each Country's Share of CO2 Emissions, Union of Concerned Scientists, 2019.

22 Taylor, Matthew and Watts, Jonathan, Revealed: The 20 Firms Behind a Third of All Carbon Emissions, *Guardian*, 2019.

23 Philippon, Thomas, *The Great Reversal: How America Gave Up on Free Markets*, Harvard University Press, 2019.

24 USCA.es, Profession: History of Air Traffic Control, USCA, 2020.

Index

Mayors Climate Protection Agreement (USA), 82
Mayors Migration Council, 323
McDonald's, *412–13*, 413–14, 425
McKinsey & Co., 115, 162, 310
#MeToo movement, 166
measles, 362, 446, *447*
Mechelen Declaration (2017), 322
megacities, *see* urbanisation
Mellander, Charlotta, 119
Mendenhall glacier, 56
mental health, *see* health
Mercator, Gerardus, 5, 9, 15
Mercator projection, *5–6, 7*
Merck, 309
methicillin, 361
Mexico, 28
    and climate change, 62
    and culture, 421
    and demography, 285
    and food, 335
    and health, 364
    and incarceration, 261
    and inequality, 184, 196
    and organised crime, 265
    and urbanisation, 108
    and violence, 237, 242, 267
Mexico City, Mexico, 109, 364
Miami, Florida, 78, *80–81*, 81–2
Middle East, 100, 192, 220, 242, 244, 246–7, 255, 333, 339, 387, 389, 395, 441
MiG, 251
migration, 141, 177, 194, 223, 290, 294–325, *294–5, 300, 301, 302–3, 304–5, 309*
    Age of Mass Migration, 23, 298
    anti-immigration backlash, 311–13, *313*, 322
    asylum seekers, 10–11, *10*, 133, 314, 318
    and cities, 320–3
    and crime, 318
    and culture, 418, 433–4
    early migration, 297–302, *297*
    and economic growth, 306–8, *307*
    and innovation, 308–10
    and terrorism, 318, 320
    refugees, 10–11, *10*, 37–8, 132–3, 194, 296, 314–23, *314–16, 317, 319, 418, 419*
Milanovic, Branko, 182
Mishra, Pankaj, 221
mobile banking, 143–4
mobile phones, 139, 142, 143–4, *144–5*, 166–7, 430
Modi, Narendra, 223, 431
Mogadishu, Somalia, 124, 126
Monaco, 353
Monbiot, George, 203
Monte Verde, Chile, 298
MOOCs (Massive open online courses), 405–6
Moore, Gordon, 139
Moore's Law, 139
Morocco, 192, 196, 318, 365

Moscow, Russia, 115
Mosul, Iraq, 124, 127
Motu Patlu (TV series), 422
Mounk, Yascha, 223
Mount Hunter glacier, 56
Movement, The, 433
MS-13 gang, 266
Mukurtu CMS, 421
multilateralism, 20, 105–6, 133, 203–6, 221, 268, 441, 443, 448
Mumbai, India, 78, 121, *123*, 132
Münster, Germany, 100
Musk, Elon, 159, 251
Myanmar, 181, 279, 322, 365

Nairobi, Kenya, 121, 322
Namibia, 184
narcotics, 242, 261, 264–5
Narodnaya Volya, 253
NASA, 65
nation states, 100–104
National Electric Power Authority, Nigeria (NEPA PLC), 176
National Front party (France), 311, 433–4
National Health Service (NHS), 308
National Oceanic and Atmospheric Administration, 58
nationalism, 10, 24, 42, 197, 223, 256, 300, 311, 431
Native Americans, 421, 424
Nature Climate Change journal, 60
Nature magazine, 58
Negative Experience Index, 220, *220*
Nepal, 51–3
Netflix, 159, 419–20, *420*
Netherlands, 37, 63, 82–4, 225, 312, 369, 434–5
Nevada, USA, 148
New Development Bank, 204
New Guinea, 298
New York, USA, 62, 81, 101, 109–10, 115, 196, 296
New York State, USA, 174, *175*, 194
New York Times, 207
New Zealand, 189, 192, 233, 386, 421, 432
newspapers 430
Neza, Mexico, 121, *123*
Nicaragua, 192, 261
Niger, 63, 244, 333, 391, 392
Nigeria, 63, 244, 423
    and demography, 281
    and food, 333
    and incarceration, 261
    and inequality, 174–6
    and terrorism, 255
    and urbanisation, 109, 112, 114, *114–15*, 122
    and warfare, 236, 242
Nile, 247
Nixon, Richard, 219

Njerep language, 423
Nobel Prizes, 309, *406–7, 408–9*
non-proliferation treaty (NPT), 250–51, *250–51*
North America, 10, 19–20, 26, 28, 37, 40, 77, 220–21
    and culture, 427, 432
    and education, 386, 408
    and food, 330, 336, *337*, 338, 341
    and health, 356
    and inequality, 172, 177
    and migration, 294, 299, *304*
    and urbanisation, 104, 107–10
    see also Canada; United States
North American Free Trade Agreement (NAFTA), 28
North Atlantic Treaty Organisation (NATO), 204, 218
North Dakota, USA, 66–71, *67*
North Korea, 210, 261
Northern Ireland, 254
Norway, 37, 149, 150, 192, 427
nuclear weapons, *see* weaponry

obesity, *see* food
Occupy Wall Street movement, 230
Ofcom, 19
'official development assistance' (ODA), 36
oil production, xii–xiii, 67–8
Oklahoma bombing (1995), 253
Oman, 249
Open Society Foundation, 433
Open Street Map, 444
OpenAI, 159
'Opportunity NYC', 196
Orangi Town (Karachi), Pakistan, 121, *122*
Orbán, Viktor, 223, 433
Orbis Terrae Compendiosa Descriptio (Mercator), *4–5*
Organisation for Economic Cooperation and Development (OECD), 36–7, 162, 399
organised crime, *see* violence
Osborne, Michael, 162
Oslo, Norway, 82, 127
Osnabrück, Germany, 100
Our Final Century (Rees), 445
Our World in Data, 444
Oxfam, 335

Paakantyi language, 424
Pacific Islands, 334, 424–5, 432
Pagel, Mark, 419
Pakistan
    and climate change, 51, 53
    and geopolitics, 210
    and health, 365
    and inequality, 192
    and migration, 303, 314
    and terrorism, 255
    and urbanisation, 118

and weaponry, 249
palm oil, *see* food
Panama, 261
pandemics, *see* health
Papua New Guinea, 37, 424
Paraguay, 343
Pareto Principle, 449
Paris, France, 109–10, 115, 415
Paris Climate Agreement (2016), 62–3, 82, 88–9, 132, 144, 232
Paris-Amsterdam-Brussels-Munich region (ParAmMun), 120
Passamaquoddy peoples, 421
Pasteur, Louis and Marie, 360
PayPal, 310
peace, 240–41
peacekeeping, 268–70, *268*
Peloponnesian War (431–404 BC), 206
penicillin, 361
pensions crisis, 288–9
permafrost, 55, 58
Peru, 73, 264, 268
Petare, Venezuela, 121
Philippines, 109, 126, 130, 202, 223, 237, 257, 261
Philippon, Thomas, 180
Physiologie du Goût (Brillat-Savarin), 328
Piketty, Thomas, 186
Pinker, Steven, 40, 444
Planet Lab, 444
Planet of Slums (Davis), 120
plastics, *40–41*, 42–3
Poland, 152, 223, 249
    and culture, 427, 433
    and demography, 279, 285
    and migration, 312, 313, 322
    and urbanisation, 107
political parties, 227–30
Pompeii, Italy, 124
Population Bomb, The (Ehrlich), 275
population, *see* demography
populism, 9, 11, 20, 42–3, 197, 202, 220, 223, 227–8, 431
Portugal, 108
poverty, 38–41, 120–24, 177, 180–81
Poverty and Famines (Sen), 332
Primeiro Comando da Capital gang, 266
printing press, 415–16, *415, 418*
prison, *see* incarceration
Programme for International Student Assessment (PISA), 397–9, *398*
Prohibition, 264
Protestantism, *see* religion
Prussia, 208, 388–9
Ptolemy, Claudius, 2, 9
Putin, Vladimir, 155, 223, 233, 433

Qatar, 37
QQ, 416
Qualcomm, 155, 309

1984

Dubai is a key hub of globalisation and one of the fastest growing cities in the world. The city's population expanded tenfold from roughly 325,000 people in 1984 to 3.4 million today. Spanning 4,100km², the city's two massive artificial archipelagos – the Palm Jebel Ali and the Palm Jumeirah – are visible from space.